KU-544-984

Moon

A BRIEF HISTORY

Bernd Brunner

Withdrawn From Stock
Dublin Public Libraries

ale UNIVERSITY PRESS

new haven and london

Published with assistance from the foundation established in
memory of Philip Hamilton McMillan of the Class of 1894,
Yale College.

Copyright © 2010 by Bernd Brunner.
All rights reserved.
This book may not be reproduced, in whole or in part, including
illustrations, in any form (beyond that copying permitted by
Sections 107 and 108 of the U.S. Copyright Law and except by
reviewers for the public press), without written permission from
the publishers.

Yale University Press books may be purchased in quantity for
educational, business, or promotional use. For information, please
e-mail sales.press@yale.edu (U.S. office) or sales@yaleup.co.uk
(U.K. office).

Designed by Sonia L. Shannon
Set in Stempel Schneidler type by Tseng Information Systems, Inc.
Printed in the United States of America.

The Library of Congress has cataloged the hardcover edition
as follows:
Brunner, Bernd, 1964–
Moon : a brief history / Bernd Brunner.
    p.   cm.
Includes bibliographical references and index.
ISBN 978-0-300-15212-8 (cloth : alk. paper) 1. Moon. I. Title.
QB581.B78 2010
523.3—dc22

                                           2010015126

ISBN: 978-0-300-17769-5 (pbk.)

A catalogue record for this book is available from the British
Library.

10 9 8 7 6 5 4 3 2 1

Contents

Brainse Fhionnglaise
Finglas Library
Tel: (01) 834 4906

Introduction

The light of the sun is our essential source of energy. Without it, not only would Earth's temperature drop to an unimaginable level, but a thick crust of ice would soon cover the planet's surface. Few microorganisms, if any, would survive. If the sun disappeared altogether, our world would even lose its gravitational anchor. The world is simply inconceivable without the sun. But what would the planet be like without the moon? We might assume that the absence of our satellite would affect us less dramatically than the loss of the sun. But the more we recognize how intimately life on Earth is connected with the moon, the more unsettling the idea of life without it becomes. By stabilizing our planet's inclination, the moon has prevented it from going off on seasonal excursions that might have led to completely other consequences for the evolution of life on Earth.

Without our moon, Earth would be a vastly different place. Just *how* different it is difficult to know, but if we imagine Earth with a much weaker ebb and flow of the tides, we get an inkling of the importance of the moon's role. Life itself may not have been possible without the moon. The strong ocean tides have created pools considered essential for complex biological systems to arise. According to the theory now generally accepted, a body with the dimensions of Mars collided with our planet, flinging off the disk of material that created the moon and gave Earth its axial tilt. What would our planet be like if such an incident had never taken place?

Although there are limits to the credibility of speculative history, some scientists find such thought experiments irresistible. The American astronomer and physics professor Neil F. Comins, for example, draws an elaborate comparison between an Earth on which no impact has occurred, which he calls Solon, and Earth as we know it. According to Comins's hypotheses, Solon rotated much faster — possibly at three times Earth's current speed — which would cause correspondingly stronger movements in the planet's atmosphere. Tall trees or delicate, large leaves would be a rarity in such a world, as would fragile animals with long legs or wings. Humanoid creatures could exist, but their appearance would be quite different. If Comins is right, the moon has therefore played a pivotal role in our development as a species. Although the precise extent of its influence can probably never be determined, Earth's lifeless satellite plays a central part in who we are — along with the sun, the atmosphere, our oceans, and the animals and plants living beside us. The moon's history is intimately related to the Earth's.

Yet the moon's central role in life on Earth is difficult to square with the simple facts we know about it. Our moon is a bleak, gloomy, lifeless celestial body just a quarter of Earth's size and one-eighty-first of its weight, with a gravitational pull of only one-sixth. It rotates on its own axis once approximately every twenty-eight days, which is very slow compared with the current twenty-four-hour rotation period of the Earth. Its surface is a bit larger than Africa's and Australia's combined. Its very thin atmosphere means that there is no sound and, without a means to retain the warmth of the sun, the surface temperature fluctuates considerably.

In some ways, viewing and scrutinizing the moon gives us a look back to the beginning of the solar system. The position of the moon relative to Earth has changed, though. According to computer models, two billion years ago the moon would have been 24,000 miles away from Earth, orbiting it 3.7 times per day, and causing tides up to a thousand times higher than those observed today. Now at an average distance of 238,900 miles from the Earth—or the equivalent of thirty times the Earth's diameter—the moon is losing energy, slowing down, and receding from us as its orbit expands by 1.5 inches each year, or 250 feet in two thousand years. The tides act as a brake on the rotational speed of both the moon and the Earth. At some distant point in the future, the sun will engulf the Earth and the moon. But for now, the sun will continue to shine, the wind to blow, the seas to surge.

These physical facts and figures alone do not explain what the moon means to us. The impor-

fig. 52. Wie der Mond um die Erde kreist und ihr immer dieselbe Seite zukehrt.

The moon circles the Earth, while always exposing the same side. The tension of the cord stands for the attraction of the Earth, locking the spin of the moon.

tance of the moon has less to do with its proximity to Earth than with its centrality in the human imagination. The moon's face has been known to inspire admiration, sadness, joy, even longing, and, under certain conditions, fear. Even though we may think we can lasso the moon, it often eludes our grasp. Close, yet really far away— the moon is a paradox. And when we study it, we are also studying an aspect of ourselves.

If Earth had been shrouded in clouds, objects in the sky would never have evolved into symbols, but because we can see the moon wax and wane each month, for example, we take meaning from its periodic journey between complete darkness and a bright, full disk. The moon's nearness also encourages us to ponder what might be "out there." Are there other worlds among the distant stars? Is there another sphere similar or not so similar to ours? The moon's proximity also made it natural that we should make it the destination of our first venture beyond our home planet.

The moon is the most examined object in the sky. For millennia it has been an enigma, a *luna incognita*. The Greek writer Aeschylus (525–456 B.C.) saw in it "the eye of the night." Lucian of Samosata (ca. A.D. 120–185) wrote, "I found the stars dotted quite casually about the sky, and I wanted to know what the Sun was. Especially the phenomena of the Moon struck me as extraordinary, and quite passed my comprehension; there must be some mystery to account for those many phases, I conjectured." For a long time, the moon had counted among the seven planets moving about a fixed Earth, along with Mercury, Venus, the sun, Mars, Jupiter, and Saturn. In the seventeenth century, as the heliocentric idea of the universe gain wider acceptance, the moon was relegated to a less important position. Now it was just a satellite, not even unique, since most of the planets circling the sun have one or more moons.

This book is a brief history of the imprint the moon has left on the human imagination and of the enduring fascination it has provoked. It is an appreciation of the contributions of various cultures—and of both popular and scientific traditions—in shaping our sense of the moon. An appreciation, too, of the moon's unique power to inspire our capabilities for invention and to nourish our drive to self-understanding. Over the millennia the moon has been the focus of a vast array of human practices. This book is devoted to many questions regarding the moon's role in our lives and the lives of those who went before us. How, for example, has the moon been used to structure time? What kinds of life have both scientists and writers imagined on the moon? How has the moon's origin been accounted for? Why do some people—all the evidence notwithstanding—still claim that the moon landings never happened? My hope is that these investigations, taken together, will be more than a motley collection, that they will provide a sense of the amazing continuity, from the first imagined lunar journeys to the Apollo program, of the moon's eternal place in the life of humankind.

Moon

CHAPTER ONE

Gazing at the Moon

While the sun is too bright for us to look at it directly, the moon lends itself to gazing and contemplating. Over the course of about one month, the moon makes a perceptible trip through the sky. Its phases are more easily distinguishable than its motion. On the third day after the new moon, its visible surface starts to take the form of a thin semicircle, easily lending to a comparison with a pair of horns or a boomerang. The next night it will be higher above the western horizon than the night before, and not as thin. It also sets later as its month moves on. It develops into a half-moon. During the next seven to eight days, its light increases until its image becomes circular. At the time of the full moon, it is directly opposite the sun, so that the latter illuminates the full surface of the moon visible from Earth throughout the whole night. Next the moon goes again through the same shapes as before: now from oval to last quarter, the phases of the waning moon mirror those of the waxing moon.

The last quarter diminishes, further taking the shape of a crescent, the horns of which are raised on the side farthest from the sun. At some point, by the twenty-seventh day, the moon is visible only for a short period of time before sunrise. During the last hours of darkness it can still be

Moon gazing was all the rage in nineteenth-century Europe.

seen, but it's clearly fading. It approaches the sun and finally gets lost in its rays. The moon is actually part of the daytime sky as often as the nighttime sky, even if it is hardly noticed or sometimes mistaken for a soft cloud. Finally, for the duration of three days, the moon is no longer visible in the sky, neither at day nor at night—except during a solar eclipse, when a sliver of the moon is visible. These regularly recurring phases are a consequence of the moon's movement around the Earth, and the phase of the moon as seen from the Earth is always complementary to that of the Earth from the moon. The moon moves along its path about thirteen times faster than the sun, covering the distance in four weeks that the sun travels in a year.

During the full moon, details of the lunar surface are indistinguishable; even the mountains of the moon barely cast a shadow. Tycho, for example, the brightest and most conspicuous crater on the moon—named after the Danish astronomer Tycho Brahe (1546–1601)—is eighty-five kilometers wide and, at perhaps 108 million years old, the youngest of the craters on the near side of the moon. It was formed a long time before humans walked the Earth, but dinosaurs could have observed the moment of impact. As the moon waxes,

The development of the lunar phases as illustrated in a print by John of Sacrobosco (also known as John of Holywood) from *Tractatus de sphaera,* an important astronomy book of the Middle Ages

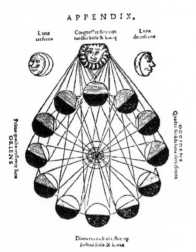

the appearance of Tycho experiences a remarkable transformation. Initially it is a great gaping crater—reminiscent of the Greek root of the word, meaning "cup" or "bowl"—then it slowly grows to become the epicenter of a complex system of rays extending hundreds or thousands of kilometers into the area around it. It is said that Tycho makes the moon resemble a peeled orange. The lunar scientist Thomas Gwyn Elger (1836–1897) called it "the Metropolitan crater of the Moon."

Long before the discovery of the telescope, humans pondered the particular pattern of lighter and darker areas on the lunar surface, much as they have always studied the shapes of clouds in the sky. The human face imagined on the lunar surface was probably the most ancient, and certainly the most anthropomorphic, perception. In this case, the major dark spots on the lunar surface—the maria—would be associated with certain facial features like eyes, eyebrows, nose, cheek, and lips. When we look at the moon with the naked eye, the maria vaguely suggest a human face. Sometimes the pattern of dark and bright spots is identified with the contours of a woman's face, with her hair bound up on the top of the head. Armed with a little imagination, observers could project a vast repertoire of images on the pattern of dark and light areas: from a broad grinning face or a rabbit with long ears to a crab or even a man with a dog. Often they would discern the features of the proverbial Man in the Moon.

The Tycho crater and the surrounding area

Among the most obvious phenomena related to our neighbor in space are moonrise and moonset. Both can be impressive, though the amount of light involved cannot produce a colorful optical effect as spectacular as the interplay of intense red and orange hues at sunrise or sunset. The moon appears much larger—double or even triple in size—when it rises or sets, dwarfing the houses and trees surrounding it. After its ascension to the sky, this impression disappears. Sev-

A hare in the moon?

eral possible explanations have been advanced to account for the "moon illusion." One is that the close juxtaposition of objects such as houses or trees with the brightly illuminated moon fools us into exaggerating its size compared with the foreground objects. But the illusion may also plausibly be related to the perceived distance of the moon: when it is on the horizon, the brain interprets it to be farther away than when it is above us.

As incredible as it may seem, the French astronomer Frédéric Petit, then the director at the Toulouse Observatory, was convinced he had discovered a second moon with an elliptical orbit while looking through his telescope early one evening in 1846. Jules Verne took up this idea in his book *From the Earth to the Moon*. We cannot reconstruct how Petit arrived at such a conclusion, but a small line of astrologers after him claimed to have seen second or even multiple moons. One was the German Georg Waltemath, who, shortly before the close of the nineteenth century, reported seeing an entire group of midget moons. Two decades later, Walter Gornold gave the name Lilith to what he claimed was a dark moon visible only when it crossed the sun.

How about other phenomena associated with the moon? Both myth and history bear witness to the overwhelming sensations inspired by the beauty of the full moon. It provided the backdrop for holy marriages between gods and goddesses, as well as for coronations and dancing rituals. Gautama Buddha is believed to have attained enlightenment on the day of the full moon while sitting under the Bodhi Tree. Many took the full moon as an occasion for unusual behavior. The Chuckchee shamans from northeast Siberia reportedly undressed and exposed themselves to its light, thereby obtaining magical powers. It is doubtful whether we should credit Martin P. Nilsson's assertion that "half Africa dances in the light in the nights of full moon." On the other hand, we know that the boys and girls of the Shona, an ethnolinguistic group centered in Zimbabwe, still like to dance by the light of the full moon to the sound of drums and rattles.

In contrast, the absence of the moon from the sky was frequently accompanied by the fear—contrary to empirical evidence—that it would die, never to return. The Aztecs of central Mexico, for instance, believed that they recognized death in the dark of the moon. Sometimes the phase of the new moon was perceived as a liminal period; it was a time of prayer for the moon to return. And when the silvery disk finally appeared in the sky, it was greeted joyfully, with exclamations of salvation.

Other spectacular lunar phenomena include eclipses that are caused by or that affect the moon. The word *eclipse* goes back to the Greek *ekleipsis,* meaning "omission" or "abandonment." A total solar eclipse counts among the grandest and most dramatic sights in nature. Solar eclipses occur when the moon happens to pass between the Earth and the sun, with the moon fitting more or less over the sun's face. The phenome-

In this illustration by the French satirical illustrator Grandville (1803–1847) the solar eclipse is represented as a conjugal embrace of the sun and moon (1844).

non, which can last for up to seven minutes, is possible because the sun is about four hundred times larger than the moon and four hundred times as far away—a most curious coincidence. But this kind of eclipse is visible only within that narrow part of the Earth's surface located in the moon's shadow. In contrast, a lunar eclipse, which occurs when the Earth passes between the sun and the moon, lasts for several hours and can be seen from any point on Earth where the moon is above the horizon at the time.

The abrupt turn to blood-red, or "dying," of a full moon must have alarmed early humans, often provoking fear even when it could be predicted. An eclipse has often been interpreted as a suspension of the natural order. The Maasai people of east Africa are reported to have thrown sand into the air during an eclipse. Some North American Indians are said to have banged and rattled pots and pans (and probably drums before pans became available) or shot flaming arrows in the direction of the moon to kill the predator consuming its light. People from the Orinoco region of Venezuela buried their fire under the ground in case the moon's fire went out.

During the minutes before a total solar eclipse, with sunlight arriving only from the edge of the sun, colors turn more intense and shadows more distinct. As the temperature drops, an eclipse wind is produced, and shortly before the total eclipse so-called shadow bands appear: shimmering dark lines produced by atmospheric temperature cells produced by the remaining rays of the sun. During a solar eclipse all objects on Earth assume an unforgettable pallor, often described as mainly olive green with tinges of copper.

The French Catalan astronomer François Arago (1786–1853) described the emotions of the witnesses to a total eclipse of the sun that he observed in the eastern Pyrenees on July 8, 1842. Nearly twenty thousand people had assembled with smoked glasses, only the sick staying inside, and

> when the Sun, reduced to a narrow thread, commenced to throw on our horizon a much-enfeebled light, a sort of uneasiness

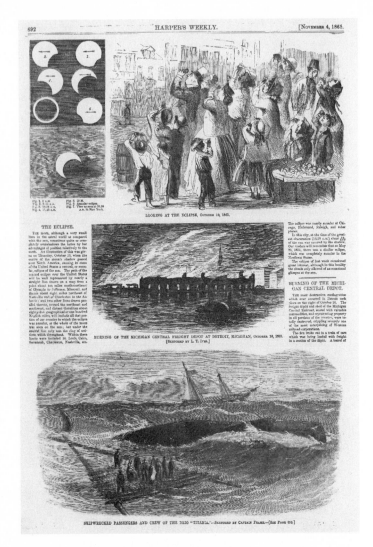

A solar eclipse that passed over the North American continent on October 19, 1865, shared newspaper space with a fire and a shipwreck.

took possession of everyone. Each felt the need of communicating his impressions to those who surrounded him: hence a murmuring sound like that of a distant sea after a storm. The noise became louder as the solar crescent was reduced. The crescent at last disappeared, darkness suddenly succeeded the light, and an absolute silence marked this phase of the eclipse, so that we clearly heard the pendulum of our astronomical clock. . . . A profound calm reigned in the air; the birds sang no more. After a solemn waiting of about two minutes, transports of joy, frantic applause, salute, with the same accord, the same spontaneity, the reappearance of the first solar rays.

According to the French science writer Camille Flammarion (1842–1925), in Africa on July 18, 1860, men and women were seen either praying or fleeing to their dwellings. "We also saw animals proceeding towards the villages as at the approach of night, ducks collected into crowded groups, swallows hurling themselves against the houses, butterflies hiding, flowers—and notably those of the Hibiscus Africanus—closing their corollas," Flammarion writes. On July 28, 1851, the sky was too cloudy to yield an astronomical observation of any kind in the town of Brest in Belarus, so the astronomer Johann Heinrich von Mädler focused his attention on the fauna. Except for the horses, all animals showed some kind of restlessness, beginning fifteen minutes before the actual eclipse. Geese and ducks fell asleep, and chickens scurried to their roosts. The aurochs, one of the last of its kind, became uneasy, soon hiding in the thicket and uttering a call that had rarely been heard. Before the total solar eclipse passed over north India on October 24, 1995, where it happened to coincide with the annual Festival of Light or Divali, astrologers recommended avoiding the shadow of the moon while it crossed the sun to circumvent misfortunes. The eclipse lasted less than a minute and was visible on a path less than twenty-seven miles wide, but many observers fasted or jumped into rivers to be cleansed. A Western traveler observed the eclipse from Khanua, Rajasthan: "I noticed it had begun to darken in the West. The clear sky turned a deeper blue than normal. As more and more of the Sun disappeared, the colors changed and gave the landscape an evening glow. Somebody described it as 'spooky.' Even the villagers quietened down as a chill filled the air. The light was very eerie and it continued to darken, the light changing as I watched. Two dogs sped across the fields heading for home, their tails between their legs. Parakeets squawked and circled uncertainly. The monkeys had disappeared. I could see the black lunar shadow approaching rapidly. I looked up to see the thin crescent of the Sun condense into a point, flicker, and go out."

In some instances anticipation of an eclipse has been used to increase one's credibility and ad-

vance an agenda. When Christopher Columbus was on his fourth trip to the New World, worms in the timbers of his ship caused it to leak and become unstable. He had to land in St. Anne's Bay, Jamaica, to repair it. Ultimately, he and his crew had to spend more than a year on the island, whose indigenous people refused to defer to the Europeans. Finally, Columbus, having calculated that a total eclipse of the moon would occur on February 29, 1504, gained an advantage. The night before this event, he called a meeting of the tribal leaders, invoked the Almighty, and warned that if the natives did not cooperate, the moon would disappear from the sky. When the warning came true, the terrified natives begged Columbus

Christopher Columbus and the Jamaicans at the moment of the lunar eclipse to bring the moon back, providing his crew with food and assistance.

Thanks to precise knowledge of the motion of the sun, moon, and Earth, it is possible to calculate eclipses for thousands of years, in the past as in the future. The Austrian astronomer and mathematician Theodor von Oppolzer (1841–1886), in his *Canon of eclipses* (1887) accomplished the feat of compiling eight thousand solar and fifty-two hundred lunar eclipses covering the time span from 1200 B.C. until A.D. 2161.

Eclipses are rare, but the very notion of rareness evokes another lunar phenomenon. To say that something happens "once in a blue moon" is to say, literally, that it happens as often as a second full moon sneaks in during a calendar month. In an unrelated phenomenon, the moon may in fact appear bluish when the Earth's atmosphere contains a high concentration of very small smoke or dust particles. Particles measuring about one micron in diameter—thus wider than the wavelength of red light—strongly scatter red light, while allowing other colors to pass, so that white moonbeams shining through the clouds emerge as blue and sometimes as green. This phenomenon has also been described as resembling electric glimmer. When the volcano Krakatoa erupted in 1883, plumes of ash rose to the top of Earth's atmosphere; as a result, the moon appeared to be blue for about two years. Forest fires can have the same effect.

Among the rarer optical effects associated with the moon are also moonbows or lunar rainbows.

Although the cause is the same as for a sun's rainbow—light refracted through water droplets—the moon's bow is much weaker, usually appearing as less brightly colored and less discernible in the sky. After crossing from Nassau to Miami on June 16, 1938, the geographer Armin Kohl Lobeck of Columbia University delivered a dramatic account in *Time* magazine. "Tumultuous trade wind clouds towered to gigantic heights and there were occasional squalls of rain. About 11 o'clock, when the moon was well up in the southeast sky, the rainbow appeared in the northwest, where a thunderstorm was in progress. The prismatic colors were fairly distinguishable. The arc was complete, the two ends dipping into the sea." There are also so-called moon coronas produced by high, thin clouds, forming a close fringe around the moon. A lunar halo, a large colored ring surrounding the moon, is sometimes visible when there are ice crystals in the upper atmosphere. Moon pillars, pale shafts of light that extend out either above or below the moon, can be seen when the moon is rising or setting near the horizon. They appear when ice crystals reflect light forward from a strong light source such as the moon.

A number of people claim to have seen red glows, flashes, glows, mists, obscurations, temporary colorations of the lunar soil, and shadow effects when looking at the moon. Claims of such sightings, typically discernible only during short intervals, go far back in time, with some having been observed independently by several

A moonbow   witnesses or reputable scientists. In fact, lunar
transient phenomena or LTPs, as they are now
called, continue to be seen. Some topographical
formations, such as the surface of the lunar impact crater Aristarchus and its satellite craters,
seem to be particularly prone to the phenomenon, accounting for a third of all such observations. Lunar missions have established a higher
occurrence of alpha particles caused by the emission of radon-222 from Aristarchus. This may be
the source of the transient light phenomena observed.

LTPs are hard to analyze, though, in part because they are irreproducible. Reports of LTPs
rarely make it into scientific publications, and
many events probably arise from causes in the

terrestrial atmosphere. To be considered as a genuine lunar phenomenon, an event would have to be observed from two places on Earth at the same time—a hard task for something no one can predict.

A lunar halo

If LTPs are somewhat dubious and irregular, impacts are undeniable, and they occur continually on the lunar surface. The most common impacts are those associated with micrometeorites. Impact flashes from such events have been detected simultaneously at multiple Earth locations. Most lunar scientists acknowledge that such transient events such as the emission of gases from the surface and impact cratering do occur. The controversy lies in the frequency of such events.

A category of another order are "observations" of life forms on the moon. A digression is in order. The idea of lunar life has been part of the human imagination for millennia, long before the invention of the telescope, but the English clergyman John Wilkins (1614–1672) was among the first modern scientists to purport such a view. "'Tis probable there may be inhabitants in this other World," Wilkins wrote in *The Discovery of a World in the Moone* (1638), "but of what kinde they are is uncertaine." Johann Hieronymus Schröter (1745–1818), a German astronomer known for his elaborate drawings of Mars was, "fully convinced that every celestial body may be so arranged physically by the Almighty as to be filled with living creatures." He attributed color changes he observed to cultivation of those areas, and he speculated that signs might indicate "some industrial origin—furnaces or factories of the inhabitants of the moon." The "pro-selenites" also included the eccentric German refugee Sir William (Wilhelm) Herschel (1738–1822), who, in his paper *Observations on the Mountains of the Moon,* published in 1780 in the journal of the Royal Society in London, claimed to have seen "forests" on the surface through his giant telescope and insisted that the habitability of the moon was almost certain. Furthermore, Herschel expressed what could count as one of the most daring statements about the relationship between moon and Earth: "Perhaps—and not unlikely—the Moon is the planet and the Earth the satellite! Are we not a larger moon to the Moon, than she is to us?" He also left no doubt that he would prefer the moon as his preferred habitation.

The absence of a lunar atmosphere has long cast doubt about the possibility of life on the moon. Even early observers noted that stars eclipsed by the moon don't suffer any visible change in their shape or color as they pass into or out of the line of sight to the moon. Still, the notion of life on the moon remained seductive even to scientific minds. A notable figure among the ranks of the moon gazers with particularly vivid imagination is Franz von Paula Gruithuisen (1774–1852). A Bavarian physician who first became known as a pioneer of a less invasive surgical method to remove bladder stones, Gruithuisen later became a professor of astronomy.

Although he was aware of the different temperatures and gravitational conditions on the moon, Gruithuisen claimed in a lengthy article, *On Lunar Inhabitants and Their Colossal Artistic Artifacts* (1824), to have observed manifold changes in color in the thin "lunar atmosphere" — "clouds and mists" not only diffusing on the lunar surface, but also keeping it warm and enabling floral growth. "Some plants such as cress would bear fruit under these circumstances even on the Moon," he wrote, believing that fruits ripened much faster there. He sternly postulated advanced life on the moon, meticulously hunting down traces of "understanding inhabitants." Since heat develops poorly on the lunar surface, he thought it would be impossible for the inhabitants to heat freestanding houses and imagined that they lived underground, where they could go without heating, "even though one cannot deny [their] ability to make fire or to heat in the case of neediness." Gruithuisen even claimed that he had seen a lunar city and a star-shaped "temple." The ashen light he discerned, so he assumed, might be caused by fire festivals celebrating either changes in government or religious periods. He speculated on the character of the lunar surface and said that scientists would discern many more details once they had the technical means to do so. He concluded that "dwarf-like instruments will only produce dwarf-like steps; only with the help of giant telescopes will it be possible to make giant steps, sufficient diligence assumed." Gruithuisen's "discoveries" were clearly over the top,

Franz von Paula Gruithuisen

even for those contemporaries who had not entirely given up on the idea of life on the moon, but they helped him gain an appointment as a professor of astronomy at the University of Munich. They also caused him to develop a strange ambition to make contact with the selenites. Assuming that moon dwellers share our understanding of mathematics, he advocated building a vast geometrical structure in Siberia to attract their attention. Sadly, this project never became a reality.

Even though more and more respectable astronomers clearly declared the impossibility of life on the moon, speculation that there could be *some* sort of life didn't simply vanish overnight. The popular imagination continued to follow its own laws. Even as the twentieth century approached, Camille Flammarion clung to his belief in life on the moon. Well aware of the limitations of telescopes, he wrote: "Now, I ask, what can be distinguished and recognized at such a distance? The appearance or disappearance of the pyramids of Egypt would probably pass unnoticed." From his travels in balloons, he was aware that from a distance, a visitor might not believe in life on Earth. "If, then, the Earth seems like a dead world when seen from only a few miles' distance, what is it but illusion to assert that the Moon is truly a dead world, because we view it at 120 miles or more?" he wrote in *Popular Astronomy* (1894). Even using the highest magnification available at the time, he assumed, signs of life would not be visible. Although Flammarion knew that the

Earth and moon have very different climates, he attributed the occasional "fogs, mists, vapours, or smoke" he observed to human origin. Eventually, he began to content himself with the idea that simpler forms of life existed on the moon. "Why should we suppose that there is not, on this little globe, a vegetation more or less comparable with that which decorates ours?" he asked. "Thick forests, like those of Africa and South America, may cover vast extents of land without our being able to recognize them. On the Moon they have neither spring nor autumn, and we cannot trust to the variations of tint of our northern plants, to the verdure of May and the fall of the yellow leaves in October, to strictly typify that the lunar vegetation should show the same aspects, or should not exist. . . . Do there exist on the Moon passive beings analogous to our vegetation?"

Later on, speculations about life on the moon remained a specialty for a small caste of stubborn moon gazers. The American astronomer William Henry Pickering (1858–1938) believed the more or less sharply defined white spots scattered over the lunar surface to be ice fields, and he claimed to have seen snowstorms on a peak of the well-known pinnacle Pico and even blizzards north of the crater Conon, near the highest part of the moon's Apennines. During some phases of the lunar day he observed a greenish coloring in the center of the crater Grimaldi that he mistook for vegetation. He hypothesized that changes in the appearance of the lunar surface—movement of small dark areas—were due to "lunar insects." He

Camille Flammarion

even went so far as to define their size: similar to tropical red ants, but not exceeding the size of the locusts devastating crops in Africa. Given the difference of the environment, Pickering didn't expect them to resemble any animals found on Earth.

As late as 1960 the astronomy writer Valdemar Firsoff asked, "Is the Moon a museum piece from a geologically remote past, preserved in the vacuum of space as though in a labeled glass case? . . . Has perhaps life secured a foothold on our companion world?" With an enthusiasm bordering on obsession, he tried to revive the idea that there might be life on the moon. He regarded as likely that the moon has an atmosphere made up of water vapor, carbon dioxide, and "heavy vapours" of "volatile substances" that develop "like some sealed aquaria, within walled enclosures, clefts, and hollows, in the depressed portions of the maria, and other, less-well-defined, volcanic regions."

Firsoff reminds us that, in earlier times, Earth's atmosphere had very different constituents. Its animals would have had respiratory organs adapted to these conditions, as still do, for example, deep-sea fish that can survive in complete darkness and under high pressure. Firsoff recognized the extent of temperature fluctuations on the lunar surface, but he didn't see these factors as standing in the way of life. He pointed to the fact that on Earth, lichens and rotifers, distant relatives of the spiders, survive well below the freezing point, and that, at the other extreme, the dry sand of the Sahara Desert is full of microbes, while some protozoa survive even in boiling hot springs. Firsoff theorized that "vital parts of perennial plants may be hidden underground" and assumed that "lunar plants may not depend on the supply of gas from the atmosphere at all, or only to a small extent, obtaining all or most of their needs from the gas marsh below and reaching out beyond it only for the energy of the sunrays required for photosynthesis or similar processes."

As late as 1968 Arthur C. Clarke (1917–2008), one of the great visionaries of science fiction, speculated, "If life ever got started on the Moon — perhaps in some long-vanished lunar sea — it may still be there. Any biologist worth his salt could design a whole menagerie of plausible Selenites, granted the existence of a few common chemicals on or below the lunar surface." For skeptics Clarke invoked the example of the deserts of the American Southwest, which appear to be barren from the air but are really seething with life, as Walt Disney's documentary *The Living Desert* (1953) had impressively shown.

While some eager believers may have seized on these remarks when they were written, such fantasies finally came to a sudden end in 1969 . . . when humans set foot on the moon.

CHAPTER TWO

Moon of the Mind

Like an ongoing discussion, our understanding of the moon is constantly overlaid by new discoveries and events. Each generation has a collective perception of what the moon is, means, and symbolizes. Any effort to reconstruct this shared field of meaning from a past time reminds us that, as the historians of science Stephen Toulmin and June Goodfield put it in *The Fabric of the Heavens* (1961), we are "confronted not by unanswered questions, but by problems as yet unformulated, by objects and happenings which had not yet been set in order, far less understood." In other words, there is not one moon but many, each particular to a different culture. So to see the moon from the perspective of earlier peoples, we must try to shed all we have learned about it — at least for now. In so doing we may better discover the place of the moon in our forebears' myths, in their stories about themselves, and in their attempts to order the flow of time.

Where to begin? We may want to start with the moon and the sun, the two brightest objects in the sky. If we follow their peregrinations over the course of the day and night, we can see one as the death of the other.

It hardly comes as a surprise that both the

In the Islamic world, the appearance of the lunar crescent was trumpeted to the people.

Caricatures of
sun and moon

moon and the sun are core elements of the world's ancient religions—in fact, in many instances they were held to be the two principal sky gods. How has the relationship between these two most conspicuous celestial actors been understood? In old stories across cultures they were often characterized in human terms. They were brother and sister; a "mismatched" couple exhibiting very different personalities; or married souls forever fighting and unable to reconcile their differences. In the Rig Veda, a sacred Hindu text that dates back some four thousand years, a hymn celebrates the marriage of the Moon God with the Sun Goddess. In medieval Provence, among the Jews of Avignon, the moon was considered to be a bedraggled or evil sun—the sun of the wolves or of the hares.

In some cultures—in warmer climates, for example, like India, Mesopotamia, and Egypt, where the sun was perceived as an enemy rather than a life-giving force—the moon took precedence over the sun as an object of veneration. In temperate zones, humans seem to have realized early on that the sun's warmth was the major factor both in plant growth and in determining the seasons. Here the moon began to be associated more often with cold and darkness. Both science and religion contributed to the increasing status of the sun: astronomers revealed that moonlight is merely reflected sunlight, and the emergence of Christianity brought the belief in a transcendent God who could be symbolized by the sun.

Despite this shift, the moon remained more accessible, better suited to human measure than the sun, with its full and radiant energy. If the sun is a metaphor for unfathomable God, the moon can be interpreted as a god in a form intelligible to humans.

We have seen how the "face" of the moon was interpreted. Only a small step was required for stories to evolve from these images. A German tale tells of a man who went to collect wood in the forest, even though it was a Sunday, the day of rest. Immediately sent to the moon for his sin, he has been visible there ever since, a warning to other humans that they too could end up imprinted on the moon's face forever should they dare commit some unlawful act here on Earth.

The native people of New Zealand, the Maori, have a story that symbolizes the influence of the moon on the rain and on the waters of the Earth. In the patterns on the moon they see a woman with a bucket. This woman, Rona, was the daughter of the sea god Tangaroa, whose task it was to control the tides. One night she was carrying a bucket with stream water home to her children when the path darkened—the moon had disappeared behind the clouds. Rona, continuing her walk in the dark, tripped over a root and made some unkind remarks about the moon. From this point on Rona's people were cursed. The moon also retaliated by snatching away Rona, who continues to live in the sky. When Rona upsets her bucket, they say, there will be rain.

There is some evidence that the moon not only

An engraving in the cave of La Mouthe

had a firm place in the life and mind of certain peoples but may also have been understood as a living being itself, or regarded as intimately connected with certain animals. Given the absence of written documents, we must depend on the association of symbols and imagery found in certain artifacts that only suggest what significance the moon had. For example, the Venus of Laussel, which dates back to 25,000 B.C., is an engraving on limestone found in the entrance of a cave in the Dordogne area of France. In its right hand the nude female figure holds something which could be the horn of a bull but could also be a half-moon. It is carved with thirteen notches, which may refer to the number of lunar months (and menstrual cycles) in a year, especially since the figure's left hand points toward her womb. Farther east an association of the bull, pregnancy, and the moon is evident in religious shrines at Çatal Höyük, a central Anatolian settlement from the Neolithic Age, approximately 8,000 years ago. Several bulls' heads are depicted in plaster relief on the walls, along with a mother goddess, whose horns resemble the lunar crescent. In an engraving of a running bull found in the cave of La Mouthe in southwestern France, the two horns not only are larger than life but are contorted, as if to invoke more realistically the waning and waxing moon.

Sculptures of clay and bronze symbolizing bulls and the moon have been found dating back to the Bronze Age, which indicates that even then humans concerned themselves with the sky. These include the beautiful Nebra sky disk, thirty-two centimeters in diameter, which was found in Germany and dates back to ca. 1600 B.C. Among the gold ornaments on it is one that can be interpreted as a lunar crescent or an eclipsed sun. Some researchers go so far as to see the disk as evidence that the Bronze Age people already used a combination of solar and lunar calendars.

In Mesopotamian and Assyrian mythology, the god of the moon was called Sin, symbolized both by the crescent and by a strong bull with perfectly shaped limbs and distinct horns. The double axe with its concentric rings, found not only in Greece but also in the Middle East and at pre-Columbian sites, also may symbolize half-moons.

According to the poet William Butler Yeats (1865–1939), the moon "is the most changeable of symbols, and not just because it is the symbol of change." It is not difficult to see how the moon could have become a symbol for both transience and rebirth in many early cultures. References to the moon as "new," "young," and "old" reflect this pattern. Myths related to the moon are often characterized by paradoxes. As often as the moon is seen as the source of renewal, some cultures saw it as the possible source of death. For the Maori, the moon was the "man eater"; the Tatars of central Asia thought that a man-devouring giant lived in the moon; and the Tupí, at the time of colonization by the Portuguese one of the largest ethnic groups in what is now Brazil, are said to have believed that "all baleful influences, thunder and floods proceed from the Moon." Many cultures share the idea that we fly to the heavens at the moment of death. In this context, the moon often becomes the first stage of such a journey. According to the Upanishads, the ancient Hindu scriptures, for example, the deceased could take two different paths. Both passed by the moon, but one returned to Earth, while the other continued to the sun, where the union with Brahman was achieved. Arriving at the sun thus signified the end of the cycle of reincarnation. Often, the moon was imagined as a gate to another world, a mediator between the Earth and the sun, or a transitional place to an eternal world. Some Buddhist monasteries have "moon gates," thresholds of passage to an altogether different reality.

And what of the moon's gender? The list of peoples for whom the moon is considered male is long indeed; it includes, among others, the Ainu, Anatolians, Armenians, Australian aborigines, Balts, Basques, Finns, Germans, Hindus, Japanese, Melanesians, Mongolians, Native Americans of the Pacific Northwest, Persians, Poles, and Scandinavians. In some Slavic folklore traditions, such as those of Russia and Bulgaria, the moon was called "father" or "grandfather." In most cases in which the moon was male, the sun was female.

The gender of the moon has not been a constant in all cultures. In the English language the moon was masculine up until the sixteenth

century, an idea retained with "the man in the moon." In other cultures, though, the moon has been assigned feminine attributes—though it is useful to remember that differences between "feminine" and "masculine" have not always been clear-cut and immutable. In India, for example, the word for the moon was feminine, coming from the same root as the words for mother and spirit, but the moon was brought to life as the god Chandra (both a boy's and a girl's name today, the difference only in the way it is pronounced), and sometimes Chandrasasin— the god of the moon with the hare. Today on the Indian subcontinent the moon is associated with calmness and softness, and it is common for adolescent girls to display their love for the moon. In the myths of some cultures the moon can even be male in the waxing and female in the waning.

The Esala Perahera, a procession in the honor of Buddha at Kandy in ancient Ceylon, in tune with the moon

It is male for some North American Indian tribes and female for neighboring tribes. In China, on the other hand, it is difficult if not impossible to operate with these adjectives, because sun and moon are perceived as sexless. However, they can be associated with the opposing principles yang (luminous and warm — Sun) and yin (shadowy and cold — moon).

As far as we can reconstruct beliefs today, in the Paleolithic and Neolithic periods mother goddesses were related to both the moon and the Earth. The Egyptians had various lunar myths, often connected with fertility. They venerated Isis, who gave birth to Horus, the god of the sun. Their god Thot was also connected with the moon. Sin, the Sumerian moon god and patron of the sacred city of Ur, combined both womb and bull. In ancient Greek mythology, the moon is associated with the goddess Selene, but also with Artemis and Hecate. Selene is Greek for "moon" and also the source for the word *selenology,* the encompassing term for lunar study, and for the name of the element selenium. Later, Luna and Diana were the goddesses in Roman mythology representing the moon, which explains why in Romance languages the article for the moon has traditionally been feminine (*la lune* in French, *la luna* in Spanish and Italian, *a lua* in Portuguese, and so on). In contrast, the Germanic peoples characterized the moon — *der Mond* in modern German — as a god, Mani.

A few centuries before the Christian era, Greek philosophers took a greater interest in the moon. In fact, as Dana Mackenzie has remarked, it became "a very important test case for philosophers and their competing cosmologies." The astronomer Anaxagoras (500–428 B.C.) was the first Greek philosopher to explain the origin of solar eclipses. He thought that the sun was a red-hot stone and the moon "a stone star" made of earth, and was banished from Athens for this impious belief. The philosopher Democritus (ca. 460–370 B.C.) not only believed that there are many worlds and that all matter is composed of atoms but also proposed the idea that the lunar markings are caused by valleys and mountains. For Aristotle (384–322 B.C.) the moon marked an important cosmic boundary; he envisioned a distinction between the sublunar and supralunar worlds divided by a spherical shell in which the moon is located. According to this concept, the sphere extending from the Earth to the moon was made up of the four elements and subject to generation and corruption—in other words, it was characterized by birth, death, and change of various kinds. In contrast, the spheres above the moon are marked by the regular movements of stars, sun, and planets moving around the Earth, following the eternal laws of the divine order. They are fixed, and their motion is circular and perfect. Aristotle also deduced that the moon must be a sphere which always shows the same face to the Earth and that the moon must be nearer to Earth than Mars, because this planet is sometimes hidden by the moon.

Aristarchus of Samos (320–230 B.C.) pro-

circulum 28. gratis Lunaris: habebis 101 | numero ætatis Lunaris designata: habebis horologia Lunaria separata, quot circuli in | systema horarum Lunarum, ex quo, quo- præcedentibus duobus continentur. His fic | modo educit paulo ante dix imos, facillimè delineati,horum singula suo stylo monies, | noctiurnam horam dicas; illud enim horo- phasique Lunati circa centrum, una cum | logium illa nocte horam demonstrabit. cui artes

"The Selenic Shadowdial or the Process of the Lunation," from Athanasius Kircher's *Ars Magna Lucis et Umbrae* (1646), which gives the moon twenty-eight phases

posed—eighteen centuries before Copernicus—a heliocentric theory of the solar system and undertook serious efforts to measure the distances between Earth, moon, and sun with the help of geometry. Given the imperfect nature of his operations, most of his results deviate considerably from what is known today, but he established the relative distance to the moon with only a minor error. His estimate that about sixty Earth radii separate the moon and the Earth is within the range of the moon's elliptical orbit: the distance varies from fifty-five to sixty-three radii. (It was Hipparchus [190–120 B.C.] whose additional observations determined that the moon's motion could best be described as an oval rather than a circle.) But Aristarchus's model was far superseded as the dominant one during the centuries to come: the geocentric cosmology of the Greek Ptolemy (A.D. 90–168)—in which the moon was Earth's closest neighbor and, like the sun, orbited Earth—remained the most widely accepted view. Greek observers were probably the first to come up with the idea that the dark patches on the lunar surface were seas and the bright regions land. Somewhat later, the Latin term *mare* (plural *maria*) came to designate the dark areas, and *terrae* was used for the highlands.

In *De Facie in Orbe Lunae* (On the face in the orb of the moon), a rich dialogue and catalogue of competing ideas about the moon, the Greek biographer Plutarch (ca. A.D. 46–120) wrote that the Arcadians of pre-Hellenic Greece imagined themselves as *Proselenes,* or people whose origin reaches back to the time "before the Moon," when the moon was not yet in the Earth's sky. In claiming this history, the Arcadians attributed a special sense of nobility to themselves. The Mozca (or Muysca) Indians of the Bogotá Highlands in the eastern Cordilleras of Colombia also relate some of their tribal reminiscences to the time before there was a moon. But these myths linking human history to the moon are exceptions. For most the moon was always there, and always would be. But sometimes, the moon's position was believed to be subject to change. With the help of magic one could make the moon descend from the sky. Aristophanes (ca. 446–ca. 386 B.C.) speculated in his play *The Clouds:* "What if I were to hire a sorceress from Thessaly, and if I made the Moon sink out of the night; if I then enclosed it in a round frame, like a mirror; and then if I kept it closely guarded?"

Working during what the Western world calls the Middle Ages, Arab astronomers such as Abu Abdullah Al-Battani (858–929) should be recognized for improving many of the calculations of the classic astronomers, including that of the moon's orbit. Arabian astronomers also invented the astrolabe, which helped to perform measurements in the sky before the telescope became available. Al-Hasan Ibn al-Haytham (965–1039), who is best known for discovering the mechanism of how light enters our eyes and enables us to see, also commented on the constancy of the lunar markings and the character of the moon reflecting sunlight.

How was the distance between moon and Earth imagined in this era? Three-quarters of a millennium ago, Roger Bacon (ca. 1214–1294), the legendary English Franciscan monk and one of the first modern scientists, calculated that if a person walks twenty miles a day, he would reach the moon in fourteen years, seven months, twenty-nine days and some hours. As we know now, this space hiker would have needed more than twice as long, but who is to blame Bacon? He used the best data at his disposal.

According to the teachings of the Roman Catholic Church, based on both the book of Genesis and the works of Aristotle, the cosmos was divided into heavenly and earthly spheres. The moon—"the lesser light to rule the night" (Genesis)—which is chronicled to have come into existence on the fourth day of creation, belonged to the heavenly realm and was thus divine and composed of ether, the fifth element, the quintessence of everything. Within Christian belief systems, ancient myths connected with natural appearances often survived under different forms. For the Lusitanians, for example, an ancient people living in the western Iberian Penin-

sula before and during the time of the Roman
Empire, the moon was a goddess responsible for
fecundity in the human, vegetable, and animal
worlds. With the advent of Christianity an important part of the lunar symbolism of this pre-
Christian goddess passed to Mary. Although the
ancient faith was maintained, the external appearance followed the requirements of the new
times, and in the wake of this transformation the
influence of the moon goddess was kept alive.
The association of Mary with the moon had far-
reaching consequences—like the Mother of God,
the moon could not be the source of light herself,
and she had to be immaculate. Although the first
requirement was clearly confirmed as astronomical science progressed, the second became a challenge to the Christian worldview when Galileo

Madonna
and child by
Albrecht Dürer

demonstrated that the lunar surface was not in fact smooth, but rough and uneven. But this discovery was not the death knell for the connection between Mary and the moon, as the rich legacy of such images in modern times testifies.

The Catholic Church also undercut the moon's long-standing influence on reckoning of time in months and years. When Pope Gregory XIII in 1582 established the calendar that was to take his name and eventually gain international acceptance, the lunar calendar, seen by the church as profane, bowed to a system based on the sun, which symbolized the risen Christ. The Gregorian calendar replaced the Julian, credited to Julius Caesar. Before Caesar, the start and finish of the month had been determined by the phase of the moon. The Roman dictator's hybrid "luni-solar" system counted the days of the year on the basis of both moon and sun and regularized the leap year to keep the nominal and seasonal years in sync.

But how did the moon's role in early time-keeping systems develop in the first place? Even if it is hard for us to imagine, early humans many not have assumed a continuity between the crescent moon disappearing in the morning sky and the body appearing in the evening sky some days later. Over time they must have begun to understand the moon as an entity of light that always changed shape in the same way—a continuity that resulted in a recognizable pattern and, perhaps, a story.

Stone structures and megaliths grouped in

circles seem to have been connected with the sky, because their alignments can be associated with the direction in which particularly visible celestial bodies rose and set. Stonehenge, a megalithic structure in the south of England, built over a time span of a thousand years, is the best known representative of such architecture. Some astronomers have speculated that Stonehenge served as an observatory to the people who erected it and was used to predict lunar eclipses. They may have understood these occurrences as affirmations of a cosmic order, not necessarily as omens of impending disaster. There are many other such monuments. In a quest to explain the multiple rows of posts at Carnac in Brittany, the archaeoastronomer Alexander Thom showed in the early 1970s how the sight lines of various tall upright stones (menhirs) and burial mounds correspond with extreme positions of the rising and setting moon.

The erection of such structures must have been very costly, suggesting that they may have had social and perhaps economic significance beyond the religious one. Humans have inevitably thought about the cyclical nature of life on Earth: recurring periods of heat and cold, drought and rain, the life stages of the animals on which they relied, the seeds and grain they gathered, and, above all, their own living and dying. At some point it became clear that the moon's peregrinations provided a means of structuring time. Once the regular disappearance and reappearance of the moon was understood, people could count the days and group them into months and smaller periods related to the phases. Martin F. Nilsson (1874–1967), in his seminal study *Primitive Time-Reckoning* (1918), reminds us that time was first counted only when the moon was visible, omitting days when the moon was dark. In ancient Greece, for example, the day of the new moon marked the start of the month and, just like the day of the full moon, it was considered a holiday.

In some instances the moon's position in the sky was taken into account in telling time, a fact reflected in sayings among central African people such as "the moon falls upon the forest," which means that it sits low on the horizon, or "it sleeps in the open air"—is present in the sky at daybreak. We also have to remember that the time between the quarter phases is a bit more than seven days—a week in the modern sense. Nilsson proposed that the moon easily fulfilled the requirements of a timekeeping device because its monthly cycle is short enough that each night can easily be distinguished by the shape of the moon and its position in the sky at sunrise and sunset. Alas, with the exception of pregnancy, the principle of counting time in moon months has not prevailed, since there are, as Nilsson has noted, "too many months in the course of one human life."

In ancient Rome, the month was likewise divided according to the lunar phases, with the first day of the month called *kalendae* (from *calere,* to be warm or glow). Farther back, the ancient Greeks developed the Metonic cycle in an

attempt to determine the correspondence between the lunar phases and particular days of
the year. Meton, a mathematician and astronomer who lived in the fifth century B.C., calculated
that nineteen solar years correlate to 235 lunar
months or 6,940 days. After nineteen years, the
same lunar phase returns on the same date of the
solar year. Just once in 312 years is this cycle off
for one day.

Astronomers of ancient Egypt tried to reconcile the timekeeping systems based on the stars,
moon, and sun. At some point, around 3000 B.C.,
they defined the year as having 365 days and the
months 30 days, which were further divided into
periods or long "weeks" of ten days. Such a system is practical, especially in the context of book-
keeping, and Copernicus used the Egyptian year
for his calculations. But because it doesn't reflect
the development of the visible moon, conflicts
with religious-minded people arose.

Centuries later, the Babylonians used the term
shabbatum for the day of the full moon, and the
exiled Jews in Babylon probably adopted the
word for their Shabbat or Sabbath. As Peter Watson has noted, "the Sabbath was originally the
unlucky day dedicated to the malign god Kewan
or Saturn, when it was undesirable to do any
kind of work." The Jewish Passover, in turn, falls
on the day of the full moon following the vernal
equinox (March 21).

In the Christian tradition the moon is used
to set the date of Easter, the commemoration of
Christ's resurrection. Western Christians cele-

brate Easter on the first Sunday after the first full moon following the vernal equinox, which can fall from March 22 to April 25. The death of Jesus, his descent to hell, and his resurrection on the third day—a Sunday—is traditionally associated with the setting and rising of the sun, but some see the older imagery of the rising moon lurking behind it. As the theologian Saint Augustine (354–430) has written: "The Moon is born every month, increases, is perfected, diminishes, is consumed, is renewed. As in the Moon every month, so in resurrection once for all time."

In parts of Asia the moon is deeply connected with regularly occurring festivals. The Kumbh Mela, which involves tens of millions of pilgrims, takes place every twelve years at four different locations in India. The climax occurs on the night of the full moon, when the assembled crowd bathes in holy rivers, such as the Ganges and Yamuna. In China and among scattered Chinese communities around the world, the fifteenth moon day of the eighth Chinese lunar month brings the moon festival or midautumn day, a traditional holiday when families gather to eat moon cakes—pastries with a filling of lotus seeds—and watch the moon. If families are separated, they at least know that their kinspeople are gazing at the same moon. Legend has it that Chang'e, the goddess of the moon and wife of the hero who shot the sun, flew to the moon and has lived there ever since. During the festival her devotees hope to catch a glimpse of her dancing on the moon.

Dividing the seasons with the help of the moon is a cross-cultural phenomenon. The Hopi Indians, who, unlike many Native American groups, resisted influence by missionaries and settlers until the 1870s, clinging to their traditional ways, provide a useful non-Western example. According to a careful reconstruction by Stephen C. McCluskey, the Hopi based their calendar on observations of the respective positions of the sun and the moon, and their ceremonies were closely related to the agricultural cycles. In contrast to the science-based astronomy of the Greeks and Babylonians, the Hopi system was

Chang'e, the goddess of the moon, from a box for Chinese moon cakes directed toward practical ends, serving "to mark time for planting and harvesting crops or indirectly as it regulates religious festivals explicitly concerned with the success of crops." In fact, the Hopi used various names for the new moon at different times of the year: moisture moon (*Pa-muya*), big feast moon (*Nashan-muya*), basket moon (*Tühóosh-muya*), harvest moon (*Angok-muya*), and cactus blossom moon (*Isu-muya*).

The prophet Mohammed is said to have declared the moon to be the true timekeeper and its calendar the only valid one. The Islamic calendar therefore follows the lunar year (six months with 30 days and six months with 29 days, for a total of 354 days), which explains why Ramadan, the traditional month of fasting and spiritual cleansing during the ninth month of the Islamic calendar, can occur at different seasons. According to a centuries-old rule, Ramadan begins as soon as the sliver of the moon is spotted, a signal that can be difficult to spot when clouds obstruct the view. Because not all Muslims consider it lawful to collaborate with astronomers in order to predict the beginning of Ramadan, the most important Islamic holiday doesn't start at exactly the same time in all Islamic communities.

While in the West, imprecise lunar calendars have come to be associated with superstition and backwardness and are thus out of fashion today, the moon is still honored here in the calendars, if only in the word *Monday* and its equivalent in other languages. The name goes back to the Latin *dies lunae,* which conquered German tribes translated directly into their languages. In Japan and Korea the first day of the week is also associated with the moon. According to the ancient geocentric worldview, all objects moving in the sky, as opposed to the fixed stars, were called "planets." They included the sun, the moon, Mars, Mercury, Jupiter, Venus, and Saturn, and each came to stand for one particular day of the week. The moon may have declined in timekeeping power, but Monday reminds us of this older order.

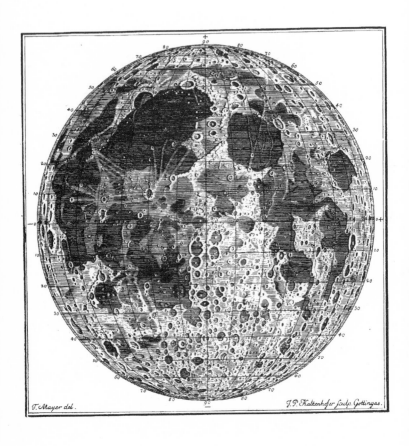

T. Mayer del.

J. P. Kaltenhofer sculp. Gottingae.

CHAPTER THREE

Charting the Moonscape

We have the ability to imagine places and spaces that don't even exist, but sometimes reality surpasses the imagination. The discovery of America not only was a complete surprise, it also opened up the world in geographical terms and brought about a complete change in perspective. In the case of the moon the starting point was different. Its existence had long been evident, but the first glimpse of the moon through a telescope—just over one hundred years after Columbus set foot on soil in the New World—triggered a dramatic conceptual shift. No longer a mythic figure, the moon became an object whose surface could be studied in detail. Although the physical distance between Earth and moon remained the same, the moon did not seem as out of reach anymore. Suddenly an extension of one of the human senses created an illusion of much greater proximity. This was a revolution. Humankind's progressive ability to explore the moon visually surely paved the way for exploration of the moon up close with the other senses.

Just a couple of years before the telescope came into use, around the year 1600, William Gilbert (1540–1603), physician to Queen Elizabeth I, produced a rather detailed pen-and-ink sketch

Tobias Mayer's lunar map

47

Moon map by
William Gilbert

of the moon. This would not be noteworthy in itself, but Gilbert also introduced the first brief nomenclature for features of the moon. Gilbert, who assumed the dark spots to be land rather than seas, coined thirteen terms. What is called Mare Crisium today was "Brittannia" (Britain) for him, and his "Regio Magna Orientalis" (large eastern region) corresponds relatively well to the current Mare Imbrium. "Regio Magna Occidentalis" is a conglomerate of what are now Mares Serenitatis, Tranquillitatis, and Foecunditatis. "Continens Meridionalis" and "Insula Longa" are segments of the current Oceanus Procellarum. Gilbert's names never came to be widely used, certainly in part because they weren't published until 1651, by which time two other major schemes had been made public.

A Dutch spectacle maker, Hans Lippershey, introduced the earliest type of telescope, using two lenses, in 1608. Soon, unseen worlds appeared, but the new device first had to overcome considerable prejudice. As the art critic Martin Kemp reminds us, "strange things were becoming visible for which no ready frame of interpretative seeing existed," and the telescope "was vulnerable to the charge that what was seen through it was in whole or in part produced by the instrument itself."

Although Galileo is commonly believed to

have been the first to examine the moon with the help of a telescope, the English scientist and mathematician Thomas Harriot (ca. 1560–ca. 1621) drew the first known sketch of the moon based on a telescope viewing. Harriot made his sketch on July 26, 1609, four months before

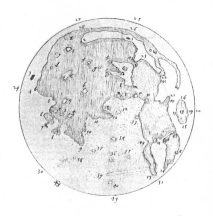

Thomas Harriot's map of the moon

Galileo's observations, and created a map about two years later. He used letters and numbers to help identify certain spots. His endeavors seem to have been known to some scholars in his field, but his renown was limited because he did not publish his lunar drawings.

Galileo Galilei (1564–1642) soon heard about the groundbreaking invention and set about improving upon it. Directing his telescope toward the Adriatic Sea, he was able to discern—as somewhat blurry images fringed with color—ships some hours earlier than with the naked eye. When he steered his device toward the moon on the clear night of November 30, 1609, he saw that its surface was not "uniformly smooth and perfectly spherical, as countless philosophers have claimed about it and other celestial bodies, but rather, uneven, rough, and full of sunken and raised areas like the valleys and mountains that cover the Earth." He wrote that the large dark spots "are not seen to be at all similarly broken, or full of depressions and prominences, but rather to be even

Galileo Galilei presents his telescope to the Doge Leonardo Donato and the Senate of Venice.

and uniform; for only here and there some spaces, rather brighter than the rest, crop up."

In his book *Sidereus Nuncius* (Starry messenger), published in March 1610, Galileo showed that characteristics of the Earth were not unique in the universe and that the celestial bodies didn't display the absolute perfection older traditions of thought ascribed to them. But despite the noted similarities between the two orbs, Galileo didn't share the view of his contemporaries that the moon was just another Earth composed of soil

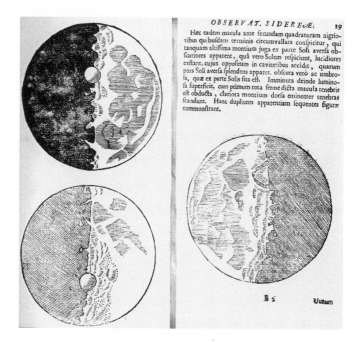

Hæc eadem macula ante secundam quadraturam nigrio-
ribus quibusdam terminis circumvallata conspicitur, qui
tanquam altissima montium juga ex parte Soli aversa ob-
scuriores apparent, quà verò Solem respiciunt, lucidiores
exstant, cujus oppositum in cavitatibus accidit, quarum
pars Soli aversa splendens apparet, obscura verò ac umbro-
sa, quæ ex parte Solis sita est. Imminuta deinde lumino-
sa superficie, cum primum tota ferme dicta macula tenebris
est obducta, clariora montium dorsa eminenter tenebras
scandunt. Hanc duplicem apparentiam sequentes figuræ
commonstrant.

and water. He conceded that other factors than land and water might account for the differences of brightness. Galileo's great accomplishment was to significantly narrow the gap between speculation and knowledge, to bring the moon almost within grasp. He was able to see what others had not seen, but the capacity of his telescope was limited. He saw neither mountains nor valleys but, as Scott L. Montgomery has remarked, "a jagged outer edge, evolving and irregular patterns of light and shadow." In his paintings, then, he has altered various aspects for effect: "The terminator is far more irregular than in reality, and the craters are enlarged to almost double their size."

Fantasies are often more attractive than reality,

Galileo's images from *Sidereus Nuncius*

and close examination of a beloved object can reveal things we'd rather had remained hidden. The power of the telescope soon made it apparent that other satellites existed; the uniqueness of Earth was put into question. In fact, the Chinese astronomer Gan De, who lived in the fourth century B.C., had discerned Jupiter's satellites with the naked eye, but in the West Galileo is credited with this discovery in 1609. He called them *planetae*. In the following year he discovered the largest four satellites of Jupiter, namely Callisto, Io, Europa, and Ganymede, later called the Galilean moons.

Soon the development of increasingly powerful telescopes led to fierce competition to create the most accurate map of the moon—which led, ironically, to further inaccuracies. To begin with, drawing an image seen in the telescope was a challenge, requiring frequent repositioning to compensate for the movement of the moon. Competing mapmakers in different countries published their works at different stages of the production process, and on some copied maps the same feature on the surface might be perceived differently and given different names.

After Galileo, a line of scientists produced or commissioned engravings on the lunar theme. The Parisian mathematics professor Pierre Gassendi (1591–1655) provided names for some features revealed with the help of a telescope, such as *umbilicus lunaris* (the navel of the moon, now called the Tycho and ray system). Gassendi's beautiful map was executed by Claud Mellan and published in 1637.

Michael van Langren (1600–1675), also called Langrenus, came from a well-known Flemish globe and mapmaking family. His moon project was driven by the problem of determining longitude to orient ships on the oceans. He came up with the idea of timing sunrise and sunsets on "islands and the peaks of mountains most often isolated from the main continuum which in an instant appear on the face of the waxing Moon, and also those which suddenly vanish on the waning Moon, which first instant and duration is a help in finding longitude."

Mapping, of course, involves more than just creating an accurate visual representation. The identified bodies on a map need names that will resonate with those who use the map. Some artists and users were more able than others to enforce nomenclature. King Philip IV of Spain, for example, demanded that van Langren's map, which was published in 1645 with a moon measuring more than thirteen inches across, feature large formations named after the Catholic king and both living and deceased members of the royal family, such as the Oceanus Philippicus, now the Oceanus Procellarum. The royal hoped this act of naming would enhance his importance and even endow him with immortality. Scientific nomenclature played only a limited role in van Langren's map, and use of contemporaries' names for so many features partly explains why the map could not stand the test of time. Van Langren also used the names of thirteen saints, gave "water" features such names as Mare Venetum (Venetian Sea) or Portus Gallicus (French harbor), and dubbed the main highlands in honor of human virtues, resulting in features named Terra Pacis and Terra Dignitatis. Scott Montgomery recognizes the effort "to claim the Moon as Catholic territory." "The lunar surface, according to this scheme, did not belong to astronomy; astronomy, however, belonged entirely under royal power." None of the place names mentioned above survives today. On the other hand, vanity inspired van Langren to name a prominent crater Langrenus, and this name has stuck—a charming reminder of these early mapping efforts. Both Gassendi and van Langren used Caspian Sea— the name of a sea here on Earth—to designate the large dark oval spot now known as Mare Crisium. Ewen A. Whitaker assumes that this feature got its name "because it occupies roughly the same position with respect to the Moon's face that the Caspian Sea does with respect to a map of Europe, N. Africa, and the Middle East."

It is not clear whether the German-Polish politician and astronomer Johannes Hevelius (1611–1687) was familiar with van Langren's map. The hills, mountains, and crater rims in Hevelius's *Selenographia* (1647) resemble rows of termite hills, which was the usual geographic practice at this time for terrestrial maps. After considering the option of naming the lunar features for mathematicians credited with advances in astronomical science, Hevelius chose instead a profusion of geographical terms for newly discovered lunar features. As he put it: "I found to my perfect delight that a certain part of the terrestrial globe and the places indicated therein are very comparable with the visible face of the Moon and its regions, and therefore names could be transferred from here to there with no trouble and most conveniently; namely, think of the part of Europe,

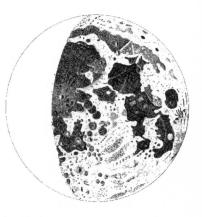

Drawing of the moon by Johannes Hevelius

Asia, and Africa that surround the Mediterranean Sea, Black Sea, and Caspian Sea." His concept did not stand up. The names he chose were too awkward and sounded too archaic, and only ten have survived to modern times. Furthermore, his map created confusion because he accorded several groups of craters a single name and saw mountain features where none existed. Elsewhere, he mistook peaks for craters. Still, despite its shortcomings, Hevelius's map retained its primacy for almost one and a half centuries.

The Italian Giovanni Battista Riccioli (1598–1671), who competed with Hevelius, used 63 of van Langren's names in his 1651 *Almagestum novum* (New almagest), but applied three of them to different lunar features and added 147 new names taken from persons connected to astronomy. Doing away with the catalogue of moral qualities, he chose names related to the weather on Earth, such as Terra Caloris (land of heat) and Terra Nivium (land of snows), none of which are used on maps of today. Riccioli was convinced that there were no humans and no water on the moon, and, as a Jesuit, he was obliged to reject the Copernican system—though it is revealing that he named three prominent craters after Copernicus, Kepler, and Aristarchus. The basic nomenclature devised by Riccioli—assigning the names of famous scientists to craters and classical Latin names characterizing weather or states of mind to maria (such as Mare Crisium and Mare Serenitatis)—has survived to the present day, though the names themselves have changed.

Beginning in 1748, the German astronomer Tobias Mayer (1723–1762) made more than forty detailed drawings of various lunar regions in an attempt to create a more accurate map. Although the moon globe he envisioned was never constructed, two lunar maps and some drawings eventually emerged. The engraver reversed one drawing by mistake, so the plate had to be accompanied by the following caption: "If you want to view the print correctly, you must hold it up to a mirror."

Johann Hieronymus Schröter (1745–1816) based his map on the measurements Mayer had performed. He surpassed his contemporaries in not only creating a vast number of drawings of many smaller areas of the moon at various states of illumination but also approximating both mountain heights and crater depths by measuring shadow lengths. Since there was now a consensus that water could not exist on the moon's surface, Schröter had to face the problem of outdated terms like *peninsula, river,* and *swamp* in

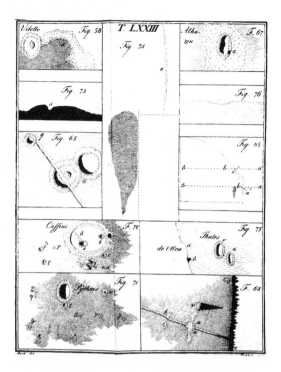

older maps. As Latin no longer was the sole language of scholarship, he also introduced the German word *Rille* for "groove."

Some details from Schröter's map

The most influential contribution to lunar topography in the nineteenth century—a highly detailed large-scale lithographic map that surpassed all predecessors in refining the system of feature names and positions based on micrometric measurements—was created in Berlin by Johann Heinrich von Mädler (1794–1874) and his friend Wilhelm Beer (1797–1850). Beer was a banker who provided the observatory and a four-inch refracting telescope and Mädler was the observer, artist, and scientist who drew the map.

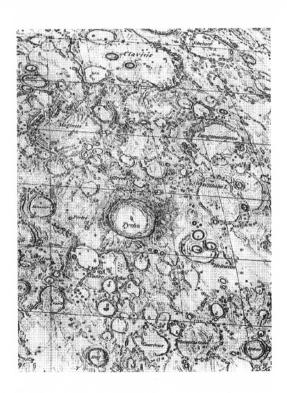

A detail from the map by Mädler and Beer

Studying the moon for about six hundred nights, Mädler created a *Mappa Selenographica* more than three feet square that depicted the moon not in spatial lunar coordinates but as a disk with contortions on its fringes. He designated secondary craters by assigning each a letter associated with a nearby "parent" crater, and he assigned Greek letters to elevations, ridges, and rilles. Mädler concluded that the moon has no atmosphere or water and does not change: "The Moon is no copy of the Earth." Part of the publication was a bulky volume titled *Der Mond,* which was a resource of knowledge about the moon, "our satel-

lite which in front of the eyes of all inhabitants of the Earth repeats its eternal cycle."

In 1840 Mädler became the director of the Dorpat Observatory in Estonia. Based on the work with a larger refracting telescope available there, he published a revised version of his map in 1869. By this time, photography was starting to decrease the interest in lunar cartography, but illustration had the advantage of allowing emphasis of detail. The revised edition of Mädler's book seemed to answer conclusively the most important questions about the physical condition of the moon; the book had the effect of effectively paralyzing the study of the moon for decades.

As lunar maps evolved, their functions became

Johann Heinrich von Mädler and a moon globe

more firmly fixed. Certainly they helped to serve the vanity of rulers who commissioned them, but otherwise they served no obvious political function, such as the territorial claims that terrestrial maps articulate. And unlike maps of Earth, visual representations of the moon didn't help travelers orient themselves in little-known terrain. Their function was largely symbolic—even though the moon was out of physical reach, mapping it had an imaginative value. More than at any previous era, the nineteenth century was obsessed with counting and measuring—an activity that engaged the many blank spaces and uncertainties related to the cosmos without quite solving them. Precise maps also had the effect of elevating science at the expense of ancient myth.

The degree of concentration required to chart a world so utterly out of reach was immense, and before the advent of photography it was a tedious task to observe the details and then transfer these impressions with a pencil. Mädler's obsession with his project was such that he is said to have met his wife in one of the rare moments when he was not looking through the telescope. The German astronomer Johann Friedrich Julius Schmidt (1824–1884) is credited with the most detailed lunar map of the nineteenth century, featuring no fewer than 32,856 craters. Such an obsession with ever-smaller features, as Paul D. Spudis has stressed, was inextricably linked with the idea common during this time "that the secrets of lunar history were in the details" rather than in the broader features.

The invention of photography had an impact on the depiction of the ·moon, just as it did on all forms of visual representation. In 1840 John William Draper (1811–1882), then a professor of medicine at New York University, created a few crude daguerreotypes of the moon. They were soon superseded in quality by the photographs by his son Henry (1837–1882), and those of William Cranch Bond (1789–1859), the first director of the Harvard College Observatory, who used a refracting telescope of fifteen inches, then the largest in the world. A particular challenge facing photographers was the need for long exposure times combined with the movement of the moon in the sky, which led to blurred images. But the complication could be resolved. A clockwork mechanism was attached to the camera to compensate for the rotation of the Earth and moon. Another pioneer of lunar photography was Lewis Morris Rutherfurd (1816–1892). A lawyer by education, he started to devote himself to astronomical photography in the 1850s and soon maintained a little observatory in Manhattan. Rutherfurd also designed a special telescope for this kind of photography. The quality of photographic emulsions steadily improved, but well into the twentieth century scientists continued to rely more on illustrations of the moon, which were richer in detail, than on photographs.

How did the progressive discovery of the moon fit into the larger development of astronomy? The observation and mapping of the moon in the centuries after Galileo continued to be-

PHOTOGRAPH OF THE MOON.

Photograph taken by Henry Draper in 1863

come more precise, and astronomers grew ever more interested in the planets. Some of the scientists we have met in this chapter also began to study Mars and other planets. In 1830, for example, Wilhelm Beer published the first map of Mars. In 1869 Richard Proctor produced an even more detailed map. Cassini also devoted much of his attention to the Red Planet.

By the end of the nineteenth century, the telescope, now much improved and widely used, had rendered the moon rather mundane. Many researchers now devoted their energies to more promising fields of stellar enquiry. Some lamented that the profession had abandoned the moon to amateurs. What attention it did receive was mostly in the context of its surface and of larger issues of space exploration. But in the 1960s, some geologists looked up from Earth and began to take an interest in the surface and makeup of the moon. They began also to transform their methods to accommodate lunar exploration.

The scientist connected most with this major transition is the American geologist Eugene Shoemaker (1928–1997), who had devoted himself to the study of cosmic impacts, examining Barringer Meteor crater in Arizona and craters blasted by atomic bombs tested in Nevada. Realizing that geologists would eventually be needed to investigate the composition of the moon, Shoemaker founded the astrogeological branch of the U.S. Geological Survey in 1961, and the group began using telescopes and photographs to map lunar geological features. One result of this work, a few years before the moon could actually be scrutinized and mapped up close, was the *Photographic Lunar Atlas* (1960, 1967), published by the U.S. Air Force and the University of Arizona, which depicts the moon's near side under various light conditions. Originally a potential Apollo astronaut, Shoemaker had to abstain from a lunar trip for health reasons. But in another sense he still made it to the moon—his cremated remains were deposited there by the mission Lunar Prospector in the year after his death. So far, he is the first and only human interred on the moon.

Lunar features continued to be named and re-named throughout the twentieth century. Lunar nomenclature is a discipline both esoteric and complex, often resulting in confusion, if not chaos. Ewen A. Whitaker's *Mapping and Naming the Moon* (1999) deals with some of these "nomenclatural nightmares." When new features were discovered—for instance, after the photographs were made of the moon's far side—the lack of specific rules became apparent. Special committees were formed and international resolutions adopted. Thus it was determined that craters can be named only after deceased people and that specific types of lunar features, such as mountains, rilles, and valleys, must named in Latin: mons, rima, vallis.

Over time, improved resolution of images—as with the metric camera photography from Apollo missions 15, 16, and 17—brought new challenges in nomenclature: In how much detail should maps represent the surface now that even much smaller formations became discernible? And while few of the fanciful names given centuries ago to lunar features have survived, the dark patches continue to be called maria. Lunar nomenclature continues to reflect the advancement of scientific knowledge—with some erratic leftovers from older times.

Visual exploration of the moon only fore-shadowed a much more complex scientific enterprise. Twenty-first-century study of the moon extends well beyond sight. Spectrometers allow for ever more precise analyses of the moon's mineralogy and composition. Measurement of gravitational fields, temperature, and radiation are becoming increasingly accurate. Navigation data of lunar spacecrafts revealed the existence of mascons: patches on the surface, mainly on the near side, with particularly high gravity. Efficient camera systems are able to detect the details of rock formations hardly three feet high. The Japanese lunar explorer Kaguya (Selene) has, since 2007, circled the surface at a distance of roughly sixty miles, using so-called laser altimeters that constantly send impulses to provide highly detailed three-dimensional information about the moon's topography.

These efforts to map the moon and name its features reveal that conquest no longer is the exclusive province of kings and gunships, but includes detailed scientific investigation. The moon is more than an object of study and contemplation; it is also a screen upon which our desires and aspirations can be projected. It is fascinating to note that the Italian philosopher and optician Giambattista della Porta (ca. 1535–1615) took this idea literally and proposed precisely such a project. In his book *Magia naturalis* (1589), he suggested that the moon could become a news medium. According to his proposal, a parabolic mirror with a wide focus would project letters onto the lunar surface that could then be read by people on Earth. Della Porta's idea was never put into practice, and the stories that shake the world today are to be found on computer screens, not on a lunar screen.

CHAPTER FOUR

Pale Sun of the Night

That the moon shines is irrefutable, but just how bright is it? There is no easy answer, because the intensity of moonlight—measured in lux—varies considerably and is influenced by several factors. During a full moon the luminance is about 25 times as great as at the time of the quarter moon and 250 times as great as on a clear, moonless night. The distance between Earth, moon, and sun, which changes because the terrestrial and lunar orbits are not perfect circles, also has a bearing on the intensity of light. The moon's lighter lunar highlands reflect much more sunlight than do the darker areas—12 to 18 percent compared with 5 to 10 percent. Also affecting the moon's brightness is the phenomenon of atmospheric extinction—that is, the loss of light as it passes through the Earth's atmosphere. In dry, clear air, the atmosphere has relatively little impact, but moist or dusty air mutes the intensity of moonlight; thus moonshine can appear particularly intense in dryer climates. Apart from the air quality, the very amount of air through which the light beams pass—measured in units of "air mass"—can vary. Moonlight passes through one air mass when the moon is directly above us, forty when it is on the horizon. These two fac-

Moonlight as atmosphere

tors can combine to produce enormous variations in how bright the moon appears. But even at its most intense, the moon's light is of dramatically low intensity compared with the sun.

But a question remains. A piece of jet-black coal exposed to sunlight here on Earth is black no matter what, so why do we perceive the full moon in the sky as white even though the material that makes up its surface is gray or black? The standard explanation used to be that the diffuse reflection of all the wavelengths of sunlight causes the perception, but recently scientists from different disciplines have suggested a different hypothetical source for the illusion. According to Sobha Sivaprasad and George M. Saleh of the Princess Royal University Hospital in Orpington, Kent, "the eye computes the colour based on relative inputs." Compared with its surroundings, the black of space, the moon undergoes "an artificial brightening," and as a result, we visualize the gray moon as white.

Robert W. Kentridge of the University of Durham offers an alternative explanation. He believes that the moon "is most likely seen as white because it is mistakenly perceived as being lit by the same illuminant as its surroundings when, in fact, its direct scattered illumination is far more intense than the scattered illumination of its apparent surroundings." Moon rock viewed up close on Earth appears dark "because there is no confusion about its illumination." As Paola Bressan of the University of Padua reminds us, the understanding of this mechanism can be credited to Adhémar Gelb. Gelb, in 1929, suspended a disc made of black paper in a dark room, then illuminated it with white paper. "The fact that it was actually black became evident only when a larger surface of higher luminance, such as a sheet of white paper, was brought into the beam of light and placed behind the black disc," Bressan explains. "But as soon as the white paper was taken away, the black surface went precipitously back to white, demonstrating that perception was impermeable to the knowledge of the 'real' colour of the disc." The so-called Gelb effect shows "that the highest luminance in a scene appears white" and confirms that "the moon looks white simply because it is the brightest region in the nocturnal sky."

Our perception of how sunlight and moonlight affect the world around us is a study in opposites. The intensity of sunlight makes everything appear etched in sharp contours and gleaming in its brightest colors. Sunlight is a metaphor for clarity and truth, making misperception all but impossible. Against the dark sky, the moon appears even whiter than the sun, but its light seems to drain the world around us of its color and veil everything in shades of gray. Nearby objects appear far away and vice versa, tricking us into believing what is untrue. By the light of the full moon, objects motionless during the day become eerily animated and changeable—the dim light leaves us room to interpret the evidence of our senses.

If we consider physical properties alone,

moonlight is practically the same as sunlight, despite how different they appear to the human eye. Objects look gray by moonlight only because moonlight falls just under the threshold at which color-sensitive receptors in the retina function effectively, but well above the much higher threshold for simple light receptors. Long photographic exposures under moonlight with color film have demonstrated that the colors are in fact there, even if we are not able to discern them. If we look at a landscape illuminated by moonshine long enough for our eyes to adapt to the dark, the gray appears blue. This phenomenon of blueshift is the result of a particular response of the rods, the photoreceptor cells in the retina.

Some philosophers of ancient Greece, such as Thales of Miletus, Anaximander, Pythagoras, Parmenidis, and Empedocles, theorized that the moon shines because it reflects the light of the sun. The Egyptians had also made this discovery, but this news reached western Europe only much later, in the twelfth century. Even then, uncertainty long persisted about whether the moon generates its own light. Among his many more celebrated pursuits, Leonardo da Vinci (1452–1519) devoted some attention to this question:

> Either the Moon has or has not its own light. If it has its own light, why does it not shine without the help of the Sun? And if it has not its own light, it must be a spherical mirror. The Moon does not have its own light but shines only as long as the

Sun makes it shine; and we see exactly as much of its light as it sees of ours. Its night receives as much brightness as our waters lend to it as they reflect the image of the Sun on it, for in all the waters which they look down upon, the Sun and the Moon are reflected. The Earth and the Moon lend each other light.

Leonardo didn't have it quite right: Earth shines on the moon with much greater intensity than the other way around. In fact, this light is so bright that it is perceptible to us by means of a secondary reflection. Some days before and after a new moon, though only a crescent is illuminated by the sun, the rest of the disk is clearly visible as well—"the young moon with the old moon in her arms." It faintly marks out the night hemisphere and grows in intensity with the narrowing of the crescent or vice versa. This is the earthlight or earthshine of the moon: a reflection of a reflection or secondary moonlight, also called *clair de terre.* The light has traveled three times— from the sun to Earth, from Earth to moon, and from moon back to the eyes of the observer on Earth. When the moon is full, this shine is no longer noticeable from Earth.

The explanation of earthshine historically presented some problems. Sometimes it was attributed to the moon itself, thought to be transparent, or to the reflection of Venus. The Aristotelian notion that the moon mirrors the earth was developed further by some eighteenth- and nineteenth-century theorists. The German physicist and astronomer Johann Heinrich Lambert (1728–1777) may have taken the theory a step too far. He claimed to have observed, on February 14, 1774, that the ash color of the moonlight changed into an olive green bordering on yellow. He explained the phenomenon as follows: "The moon, which then stood vertically over the Atlantic Ocean, received upon its night side the green terrestrial light, which is reflected toward her when the sky is clear by the forest districts of South America." Alexander von Humboldt (1769–1859) was probably on solider ground when he observed that "the extremely variable intensity of the ash-gray light of the moon depends upon the greater or less degree of reflection of the sunlight which falls upon the earth, according as it is reflected from continuous continental masses, full of sandy deserts, sandy steppes, tropical forests, and barren rocky ground, or from large ocean surfaces." And Camille Flammarion (1842–1925), the great French popularizer of astronomy, wrote, "This ashy light, the reflection of a reflection, resembles a mirror in which we may see the luminous state of the Earth. In winter, when a great part of the terrestrial hemisphere is covered with snow, it is perceptibly lighter." He also cites a very curious case: "Before the geographical discovery of Australia, astronomers suspected the existence of that continent from the ashy light, which was very much brighter than could be produced by the dark reflection from the ocean."

In ancient India, the moon was called "the king of the stars of cold" and was thought to radiate coolness. Plutarch came closer to the truth, complaining "that the sunlight reflected from the moon should lose all heat, so that only feeble remains of it were transmitted by her." The moon does emit some heat, but it's hardly measurable. In 1725 the French geophysicist and astronomer Pierre Bouguer (1698–1758), the founder of the science of photometry, studied the illumination cast on a screen by the image of a full moon in a concave mirror. After repeating the experiment during four nights, he wrote, "The sun illuminates us about 300,000 times more than the moon." In 1860 the director of the Harvard University Observatory, George Bond (1825–1865), used a large glass sphere with a silvered reflecting surface to determine that the sun is 470,980 times brighter than the full moonlight. The ratio of brightness as determined by modern analytical methods is about 400,000 to 1.

In 1846 the Italian physicist Macedonio Melloni (1798–1854) tried to determine the temperature of lunar rays reaching the atmosphere during the different phases of the moon from his meteorological station high on Mount Vesuvius. The experiment was repeated ten years later by Charles Piazzi Smyth (1819–1900), the Royal Astronomer for Scotland, who ascended Mount Guajara in Tenerife armed with a state-of-the-art temperature gauge. Although Melloni could find only the smallest elevations of temperature,

Smyth compared the heat to one-third that of a wax candle placed at a distance of about sixteen feet.

In the industrialized parts of the world, the moon's luminescence is no longer what it used to be. Thousands of generations before us lived in a world with no artificial light and had a much different experience of the night sky. Back then day and night were regarded as more different than they are today—*were,* in fact, more different. If you happen to live in a large city or metropolitan area, you never experience pitch dark. The moon and some stars and planets may be visible, but they compete with a vast number of street lamps, lighted billboards, and other sources of illumination. The English essayist Alfred Alvarez goes so far as to say that if you live in a large city, you forget the night—and the moon, we might add. Only during a power blackout or when we walk in a forest at night do we have a chance to taste what it must have been like in earlier times.

How was the experience of moonlight different? In earlier times people were used to spending a great deal of time outside, where their bodies and senses experienced bright days and dark nights, and the rhythms of their bodies, including their sleep patterns, adapted accordingly. Nights without the moon in the sky had a dramatic quality. As Jérôme Carcopino wrote about ancient Rome:

> When there was no moon, its streets were plunged into impenetrable darkness. No oil

lamps lightened them, no candles were af-
fixed to the walls; no lanterns were hung
over the lintels of the doors, save on fes-
tive occasions when Rome was resplendent
with exceptional illuminations to demonstrate her collective joy. . . . In normal times
night fell over the city like the shadow of
a great danger, diffused, sinister, and men-
acing. Everyone fled to his home, shut himself in, and barricaded the entrance. The
shops fell silent, safety chains were drawn
across behind the leaves of the doors; the
shutters of the flats were closed and the
pots of flowers withdrawn from the windows they had adorned.

Of course, there were fires and eventually fire-
places and—only much later—candles and oil
lamps. But candlelight was introduced around
the early sixteenth century and was at first popular only among privileged people. For the vast majority of people nighttime darkness was the rule,
interrupted only by nights with intense moonlight. And even the early sources of artificial light
produced only a fraction of the illumination of
the ubiquitous electric lights of today, and their
disruption of the natural rhythm of darkness and
light was therefore minimal.

In a world dominated by natural illumination,
the moonlight visible on a night with a clear or
only partly cloudy sky took on a special significance. People stayed up longer and slept less during such a night; they lived by the moon—their

pale sun of the night. Farmers often could perform chores by moonlight that they were not able to finish during the day, especially when all the fruit was suddenly ripe. In the fall, under the light of the harvest moon—the full moon rising at around sunset for several evenings in a row—they could continue to gather their crops long after the sun had set. Centuries earlier, hunters and gatherers had to find ways to make the best use of limited light; during moonlit nights they were able to extend their pursuits into the night as well. When we say today that somebody is "moonlighting," undertaking some unusual or extracurricular activity for profit, we invoke an earlier time when people fulfilled their duties after dark.

Contemporary nomadic tribes in the Sahara Desert seek protected spaces during the day when it's too hot to travel and take advantage of the moon as a gentle guardian to guide them as they journey through the cool dark. I have seen fishermen searching the shallow waters of the Red Sea by moonlight, only occasionally using an additional lamp to attract their large and colorful prey. And in our high-tech culture, some hunters take a special thrill from catching prey under adverse circumstances; special night-vision lenses aid such hunters by gathering light from the moon and stars and directing it into a photocathode tube. By converting photons to electrons, these devices increase the hunter's visual acuity.

But even if the only activity was a stroll, before the competition of artificial light, the moon

showed the wanderer the way to the well and back to his house. Old clocks mark not only the hours of the day but also the phases of the moon, so that travelers could plan their journeys for full moons, to take advantage of the extra hours of visibility. A moonlit night was never as bright as day, but the moon's bright glow could at least allow nocturnal wanderers to see the contours of the landscape in some detail. Henry David Thoreau, an astute observer of many natural phenomena, often deliberately sought refuge from civilization by rambling at night in unpopulated areas. He wrote in his journal, "I saw by the shadows cast by the inequalities of the clayey sandbank in the Deep Cut, that it was necessary to see objects by moonlight as well as sunlight, to get a complete notion of them. This bank had looked

The moon gives the icy landscape along the northern Siberian coast a particular intensity.

much more flat by day, when the light was stronger, but now the heavy shadows revealed its prominences." He was also led to think of the moon being in a "continual war with the clouds," a war that had implications for the orientation of anyone crossing the landscape at night. When the moon was absent from the sky, travelers risked losing their way. Perhaps this disorienting effect explains why Julius Caesar supposedly preferred to start his battles when the moon was dark.

Moonlight is often most spectacular in winter, when it can produce dramatic reflections in the snow. A similar effect can be observed on long stretches of beach or dunes. In the mountains, when the moon is close to the horizon, its surface resembles a faraway rocky peak, which led some early observers to conclude that the moon is made of the same matter as the Earth. Along these lines, Sir William Herschel's son Sir John Herschel (1792–1871), in his *Outlines of Astronomy* (1849), compared the illumination of the moon's surface he had observed during his trip to South Africa with that of weathered sandstone rock in full sunshine: "I have frequently compared the moon setting behind the gray perpendicular façade of the Table Mountain illuminated by the sun just risen in the opposite quarter of the horizon, when it has scarcely been distinguishable in brightness from the rock in contact with it. The sun and moon being nearly at equal altitudes and the atmosphere perfectly free from cloud or vapor, its effect is alike on both luminaries." On the other hand, moonlight can convey a remarkable effect when reflected by the surfaces of mountaintops. When the German writer Johann Wolfgang von Goethe (1749–1832) watched Mont Blanc, the highest mountain of the Alps at the border of France and Switzerland, at night, he saw "a broad radiant body belonging to a higher sphere; it was difficult to believe that it had its roots in earth." Even buildings can develop a particular magic under the moon's light. Jacqueline Kennedy Onassis is known to have returned to the Taj Mahal, the famous Indian mausoleum made of white marble, to see it again when veiled in moonlight.

When the moon is visible during the day, the quality of its light is much different. No longer even slightly yellow, it appears white under the influence of the blue light of the atmosphere, its complementary color. In *Cosmos: A Sketch of a Physical Description of the Universe* (1868), Alexander von Humboldt compared the sunlight reflected from the moon with that reflected by a white cloud during the day and found it "not infrequently difficult to distinguish the moon between the more intensely luminous masses of clouds." High in the mountains, Humboldt found that detecting the lunar disk was much easier because "in the clear mountain air, only feathery cirri are to be seen in the sky." Due to their loose texture, these clouds reflect sunlight, "and the moonlight is less weakened by its passage through the rarer strata of air."

Moonlight once facilitated social gatherings and helped slaves to find the path away from their oppressors, but on the other hand a night with the full moon in the sky encouraged thieves and criminals in their pursuits, since they didn't have to risk discovery by using artificial light. Despite a certain danger involved in crossing the landscape at night, the moon's presence makes solitude more bearable to the lonely wanderer. And if he needs to rest or sleep in the open, the moon serves as a watchlight, providing a sense of protection. Today's Zande people, living along the Nile-Congo watershed in central Africa, use moonlight ritually, sharing trickster tales during moonlit nights, woven together a little bit differently at each telling.

The moon's significance must have been immeasurable for travelers onboard ships crossing the oceans. During a tempest, when everything seemed on the verge of vanishing into nothingness, a full moon, often just lurking from between the clouds under these conditions, had the dual role of illuminating the ghastly scene and being the only beacon offering a fixed position to the people in peril. More recently, it has been speculated that had there been a full moon, the lookouts on the *Titanic* might have been able to discern the iceberg in time to change the course of the ship. But the moon's value for travelers is not only practical. Henry Matthews, who in 1817 traveled from England to Lisbon for his health, began his reflections on deck one night with an interesting comparison: "If the sunrise be best seen on shore, the moonlight has the advantage at sea. At this season of repose, the absence of *living* beings is not felt. A lovely night.—The moon, in this latitude, has a silvery brightness we never see in England.—It was a night for romance." Even today, when vast distances are usually covered by airplane, memorable moments sometimes surprise us, as anyone knows who, during a long intercontinental night flight, has been lucky enough to have seen the moon lighting the ocean surface below.

Of course, moonlight has not only served to extend typical daytime activities into the night. The array of activities for which moonlight is preferable can never be complete. The poet Rupert Brooke (1887–1915) frequently bathed in the river by moonlight with his childhood friend Virginia Woolf. Common lore across cultural barriers often associates the moon with love and other strong human emotions. "You are as beautiful as the full moon" is a common compliment in Arabic, whose alphabet has a range of lunar letters. The moon, often personified with a gentle and comforting smile, is deemed able to observe humankind from afar, possessing secret wisdom or wielding supernatural control over events. It's always up there, and it sees everything. When a loving couple secretly meets under its glow, the moon is the only other witness—perhaps the first to become aware of this bond. The lovers' awareness of the moon intensifies their emotions, and the moon becomes a recurring symbol of, even a witness to, their union. The moon assumes a similarly watchful role in many children's stories, where it sends its light through the night as it protects sleeping children. This resonates with ancient myths in which the moon acts as a judge.

In fact, as Doreen Valente has remarked in her classic account *Where Witchcraft Lives* (1962), it is generally acknowledged "that moonlight has a sexually stimulating and exciting effect upon human beings." But it is difficult—sometimes impossible—to distinguish the physical impact the moon has on us from the impact of our cultural associations. If the circumstances are right, moonlight may induce a romantic or sexual mood. And because this mood immediately calls forth certain contexts, memories may be evoked that give rise to such associations more or less automatically. Making love under a full moon in an open sky has an honored place right next to the candlelight rendezvous. Yet for all its romantic power, such a scene could have prosaically unpleasant consequences. Intense moonlight, by overwhelming the senses via sheer beauty, exalted feelings, or the stimulating company of an adored person, can make those experiencing it forget the risks involved in exposing the nude or partially clad body to the elements. Marcel Proust must have had something of this sort in mind when he wrote about the "man who has forgotten the glorious nights spent by moonlight in the woods" but still suffers "from the rheumatism which he then contracted." Obviously aware of the moon's noxious potential, the French novelist also invoked its diurnal innocence, when the moon is "no more than a tiny white cloud of more definite and fixed shape than other clouds."

Finally irresistible for Proust was "the ancient unalterable splendor of a moon cruelly and mysteriously serene."

The moon has aided less amorous pursuits as well. The Lunar Society, a club of prominent businesspeople, naturalists, and intellectuals, met regularly for some fifty years starting in 1765 in Birmingham, England. Its members, including Erasmus Darwin, the grandfather of Charles Darwin, met to perform experiments and to discuss the latest developments in chemistry, electricity, medicine, and economics. The society's name commemorated the timing of its meetings, which occurred on the first Monday night after the full moon, when the additional light made the journey home safer.

The romantic moon

A particular climate of scientific inquiry and artistic exploration, combined with a Mediterranean sensibility, has made Italy a hotbed of lunar musings. The tradition stretches back at least to Dante Alighieri, an intimate scholar of celestial science, who, in his *Divine Comedy,* meets the inconstant souls in the sphere of the moon. Leonardo, Giordano Bruno (1548–1600), and Galileo carried on the tradition of lunar devotion, which culminated in the melancholic moon poems of Giacomo Leopardi (1798–1837). In Italian culture, the moon is not just something to be contemplated; it amounts to a persona with particular capabilities or an understanding being that can be addressed—a conceit carried into a modern Italian-American community in the film *Moonstruck*. As Italo Calvino (1923–1985) once wrote, "As soon as the moon has risen in the poet's verses, it has always had the power to communicate a sense of weightlessness, of suspension, of mute and calm enchantment."

But Italians scarcely have a monopoly on lunar devotion. In the English-speaking world, literature and other forms of artistic expression abound with similar patterns of emotion. William Shakespeare's *A Midsummer Night's Dream* takes place in a land of fairies under the light of the moon. Here the moon is a powerful force with an intoxicating effect on the characters, inciting bizarre and at times illicit behavior. The moon is connected to dreaming and to a world in which fantasy and reality can blur. As Duke Theseus of Athens looks forward to his marriage—and the tensions and pleasures of the wedding night—he invokes the moon as a witness that

> like to a silver bow
> New-bent in heaven, shall behold the
>   night
> Of our solemnities.

At the same time, the very moon that blesses the lovers joining in matrimony also threatens this social bond with lust and chaos. And in another play by the same author, because the inconstant moon is a threat to unending love, Juliet forbids Romeo to swear by the moon.

The moon and its light have been characterized in many ways, but often as soft, pale, sweet, enchanting, serene, lovely, mysterious, gentle, and calm. It has been seen as warm or cold irrespective of the actual temperature, but according to the accompanying psychological evocations. The moon is sometimes connected to the idea of melancholy, perhaps because of the perceived paleness of its light, perhaps because we are particularly susceptible to this state of mind at night. The English poet Percy Bysshe Shelley (1792–1822) compared the moon to a "dying lady lean and pale,/Who totters forth, wrapp'd in a gauzy veil."

In some cultures the moon stands only for innocence and purity. For example, a saying in Sri Lanka holds that "love has to be as pure as the moonlight." In other traditions, though, the moon has a dark side, both literally and metaphorically: it is a part of the night—a time apart in which the order and continuity of the daylight world are suspended. Night can provide the necessary rest from a busy day, but it is also sometimes associated with insomnia, ghosts, and even death—and so is the moon. On moonlit nights figures skulk, glowing eyes peer from bushes, and crazy laughter echoes; often the boundaries of love and death, madness and sex get confounded. The moon can symbolize darker impulses and was at times associated with immorality. Writers of the Gothic literary tradition in various cultures have exploited the fears associated with the night and the moon. In *Lord Jim,* Joseph Conrad provides a vivid example: "There is something haunting in the light of the moon; it has all the dispassionateness of a disembodied soul, and something of its inconceivable mystery." *Murder by Moonlight* is not only the title of a book and a movie but also a common artistic trope. In this genre, the moon is often counterpointed by crepuscular and nocturnal animals like owls and bats—or even vampires and werewolves, in fantasy and horror novels and films too numerous to catalogue.

But viewed from a different perspective, the moon also brings light to the darkness; it softens the difference between day and night, thus soothing fear and superstition. Paradoxically, the moon thus serves as a harbinger of hope against the very fear and superstition that the darkness often inspires. In this role, the moon can evoke Enlightenment rationalism of the kind embodied by the Lunar Society science—a view that not only rejects superstition but also promises to illuminate all subjects shrouded in darkness.

In paintings, the sight of the moon in the background or overhead lends a dramatic quality, a magical aura, to the proceedings depicted. *Flight into Egypt* (1609/1610) by Adam Elsheimer—a painter who seems to have been familiar with Galileo's work—is remarkable not only for its representation of the moon with all major maria but for the rich evocation of a nocturnal landscape with different sources of light, including the reflection of the full moon in a lake. Romantic artists have created an abundance of paintings featuring the moon. It appears close, but is in fact far away. Often, it takes on mythical

or religious symbolism—the awe of nature, for example, or the power of creation. Artists have portrayed its light reflected in lakes and oceans, peeping between the branches of trees, or shining over plains, mountains, cities, and harbors. Such paintings allow us to uncover an entire taxonomy of lunar moods. By presenting the moon's more appealing aspects, these artists helped to shape a new attitude toward the night as a time to feel at ease, a time to be savored.

A Summer Night on the Rhine (1862) by Christian Eduard Böttcher (1818–1889)

Some of the dark pastoral scenes painted by Joseph Wright of Derby (1734–1797) were lit exclusively by moonlight. One example, located in Naples, is *Virgil's Tomb*. In this work the lunar light is represented with such an intense quality that it illuminates even the inner parts of a tomb. Wright also chronicled meetings of the Lunar Society, taking the opportunity to learn Enlightenment ideas firsthand.

A modern example of the metaphorical use of the moon in literature is *Journey by Moonlight* (1937) by the Hungarian novelist Antal Szerb (1901–1945). Here the moon is a backdrop for the protagonist Mihály's feverish odyssey through Italy, where he breaks free from conformism, drifting into his dreams, his past, and his longing for death. Other artists, though, have rejected the moon's power over our imaginations. Filippo Tommaso Marinetti (1876–1944), an Italian futurist who later became an ardent follower of the dictator Benito Mussolini, expressly rejected traditional artistic concepts and ideals by exclaiming in his manifesto, "Let's murder the moonshine." For Marinetti, the moon was connected with the "sentimentality and lechery" of love, love of the "fragile woman, obsessing and fatal, whose voice, heavy with destiny, and whose dreaming tresses reach out and mingle with the foliage of forests drenched in moonshine."

Classical music, too, offers various references to moonlight. The *Suite bergamasque,* among the most famous piano compositions by Claude Debussy (1862–1918), includes a movement known as *clair de lune* (moonlight), which was probably named for the poem by the same title by Paul Verlaine (1844–1894). Gabriel Fauré (1845–1924) also composed a calm mood song carrying this title. It is a bit odd to realize that the composer of the most famous example of moon music—*Moonlight Sonata* by Ludwig van Beetho-

ven—had nothing to do with the title, which was attached to the piece after the composer's death. A music critic, while listening to the piece, was reminded of moonlight shining upon Switzerland's Lake Lucerne.

The moon also features in much modern music. "Blue Moon," composed by Richard Rogers and Lorenz Hart, has been sung by famous singers of which there are almost too many to name, including Frank Sinatra, Billie Holiday and the Supremes. In *Tintarella di luna* (Moon tan, 1959) the Italian pop singer Mina sings about a girl who chooses to "tan" in the moonlight rather than in the sunlight so that her skin will become the color of milk. This complexion makes her "a beauty among beauties" in a land where the Mediterranean sun makes bronzed skin commonplace. She spends "all night up on the roof" like a cat. In a time before political correctness, the rays of a full moon promise to make her completely white, more white and beautiful than any other girl.

Some of the most impressive works of art relating to the moon and moonlight can be found in Japan. The moon was sacred to Tsukioka Yoshitoshi (1839–1892), and during the 1880s, he produced a formidable series of woodcut prints under the title *One Hundred Aspects of the Moon,* with figures from both Japanese and Chinese mythology at its center. In the Japanese aesthetic, the full moon is an object of beauty often linked with autumn. This season evokes melancholy because of the imminent passing of the year, in contrast to the Western mood of harvest celebration on the one hand, and the association with death, spirits, and the uncanny in the context of Halloween, on the other.

To take advantage of the particular qualities of moonlight, some gardeners plant so-called moon gardens, with evening or night-blooming plants such as the intensely fragrant and twining moonflower, the shrublike four-o'clock or "beauty of the night," and the toxic angel's trumpet. In wintertime, "Harry Lauder's walking stick," a bizarrely twisted white-stemmed shrub with a mysterious air, is also suitable for this kind of

garden. Viewed by the light of the moon, such a collection of plants exudes a special dreamlike atmosphere. The moonlight enhances the sight of the blooms, which stay open only until the morning sun touches them. This odd pattern of display protects the flowers from the heat of the day and allows them to attract by their scent nocturnal insects for pollination. Red roses would make little contribution to such a garden: moonlight would simply turn them gray.

Geisha strolling by moonlight (1887), an illustration from Tsukioka Yoshitoshi's (1839–1892) *One Hundred Aspects of the Moon*

CHAPTER FIVE

Encounters of a Lunar Kind

We have seen the speculation by scientists about the possibility of life on the moon, and their later confrontation of the reality. Writers of fiction took many more liberties with the facts, but their imaginative moon cultures were not formulated arbitrarily. These literary fantasies are products of particular times and circumstances, sometimes incorporating newly emerging technical insights and possibilities, sometimes reflecting philosophical thought about what a perfect world should look like or how a less perfect world might be construed: the fictional moon became utopia or dystopia, or something in between. Or, as Scott L. Montgomery has expressed it, the moon "served as a historical sponge for sensibilities then percolating through European society."

Speculation about life on the moon long precedes the telescope, of course, going back at least to the philosopher Philolaus (ca. 470–ca. 385 B.C.), who believed that the moon had mountains, valleys, humans, animals, and plants that were much taller and more beautiful than those on Earth. Because of the longer duration of the lunar day, Philolaus assumed that the humans there would be fifteen times as large as those on Earth. Plutarch, for his part, not only believed

This is how Gustave Doré imagined Lucian of Samosata's tale about a trip to the moon.

the moon to be of an earthen substance but also thought there were people on the moon and that the souls of the deceased go there.

In 1686 the French mathematician Bernard de Fontenelle (1657–1757) published his *Discourses on the Plurality of Worlds.* Apart from popular explanations of the heliocentric model of the universe, this work contained many speculations about life on other planets in the solar system. Fontenelle thought the lunar air too rarefied to support life, but he nonetheless portrayed the moon as a sort of utopian alter ego to the Earth. For Fontenelle the moon was only the first of a series of second worlds the newly discovered universe must contain. Furthermore, the developing new model of the cosmos, which abandoned the medieval concept of crystalline spheres separated from one another, made space travel a theoretical possibility after all.

How did philosophers and writers imagine these first explorations of the moon? How did they envision the Selenites or Lunarians—as the moon's inhabitants were called—and their world? Would moon dwellers resemble humans? In *True History,* the Greek Lucian of Samosata (ca. A.D. 120–180) describes a whirlwind that lifts a sailing vessel from the ocean and carries it to the moon. He calls the lunar inhabitants Hippogypi and portrays them as riding on three-headed vultures with unusually large feathers. In another work with a similar theme, Lucian's hero Icaromenippos reaches the moon merely by flapping his arms after attaching the wings of a vulture and an eagle, "to make a bird of myself, and fly up to heaven." The feat is possible, in Lucian's scheme, because he assumes an unbroken atmosphere of air, based on Aristotle's dictum that nature abhors a vacuum—an error that colored scientific thought for almost two millennia. Icaromenippos takes off from Mount Olympus and, after three days, comes to rest on the moon. From that vantage point he can observe the crimes on Earth, and he meditates on the great universe in front of him.

The next known imaginary trip to the moon is centuries later, in Persian literature. The epic poem *Shāhnāmeh,* written by Firdausi and published in 1010, incorporates orally transmitted legends into an account of a marvelous flight to the moon and through the heavens. Five centuries later, the Italian Ludovico Ariosto (1474–1533), in *Orlando Furioso* (1516), an epic poem in four cantos, told of four steeds "redder than flames" that transport the British duke Astolpho, accompanied by John the Evangelist, to the moon. It is both a place for contemplation and a fantastic setting, a stopover for events past and still to come. "Swell'd like the Earth, and seem'd an Earth in size," it had cities and castles just like our planet.

The eminent German astronomer Johannes Kepler (1571–1630) knew that there was no atmosphere between the Earth and the moon, so the use of animals for the trip was out of the question. He could not provide a plausible means of travel to the moon without bringing supernatural forces into play. In his tale *Somnium* (The dream, 1634) a demon—abhorring sunlight, but able to travel at night—makes possible the four-hour trip from Volva, the Earth, to the island Levania, the moon. The beings envisioned by Kepler lived in caves and crevices, emerging only briefly during the day.

The Man in the Moone; or, A Discourse of a Voyage thither, written by Bishop Francis Godwin of Hereford (1566–1633), is the earliest surviving moon travel story in the English language. Godwin's hero, a certain Domingo Gonsales, is a good Spaniard of a poor family. Marooned on an island, Saint Helena, he tames a kind of wild swan called a gansa, which has one webbed foot and another with eaglelike talons. Gonsales constructs a contraption with a seat to be carried by the tamed birds. Unaware that the gansa is migratory, and accustomed to wintering on the moon, he becomes an unwitting passenger on the birds' annual migration. As they ascend, he feels the Earth's gravity diminishing. He is surprised to find his birds flying "as easily and quietly as a fish in the middle of the water . . . whether it were upward, downward or sidelong, all was one." En route to a safe landing atop a lunar hill, Gonsales observes that the Earth rotates around its own axis—just as he had been taught as a young student in Salamanca, the first university to introduce the heliocentric cosmology of the Polish astronomer Nicolaus Copernicus (1473–1543). But Gonsales doesn't go as far as the famous astronomer and declare that the sun is the center of the universe.

Francis Godwin posits his "second Earth" as a utopia whose natural environment makes "it seemeth to be another Paradise." Godwin portrays a surface covered by a huge sea with trees three times as high as on Earth and more than five times as thick. The moon is inhabited by a race unlike any an earthling has ever encountered. Their skin is "lunar" in color, "having no affinitie with any other that ever I beheld with mine eyes." Happy and satisfied, these people display no feelings of hatred or envy; fighting and murder are unknown to them. They cheerfully take Gonsales by the hand and bring him to their homes. He crosses himself and, amazed by this encounter, invokes the names of Jesus and Mary. The immediate reaction of his hosts fills him with delight: "No sooner was the word Iesus out of my mouth, but young and old, fell all downe upon their knees (at which I not a little rejoyced) holding up their hands on high, and repeating all certain words which I understood not." The Lunars, as he calls them, sleep whenever the sun is shining, and none live on the far side of the moon. Observing their fondness for tobacco, he assumes a connection to the Native Americans; in fact, he believes the Lunars are their descendants.

THE
MAN IN
THE
MOONE.

T is well enough and sufficiently knowne to all the countries of *Andaluzia*, that I *Domingo Gonsales*, was borne of Noble parentage, and that in the renowned City of *Sivill*, to wit in the yeare 1552. my Fathers name being *Therrando Gonsales*, (that was neere kinfman by the mothers fide unto *Don Pedro*

B

The opening sentence of Bishop Francis Godwin's *The Man in the Moone*

Gonsales's
flight to the
moon with the
help of gansas

On Earth, Gonsales had to learn Italian,
French, and German to make himself understood.
The Lunars, though, speak the universal language
from before the construction of the legendary
tower of Babel. Not only is this language mellow
and melodic, it is entirely made up of music and
therefore superior to any spoken language. The
Lunars, quite simply, have "music in themselves."

This aspect of the tale reflected the search among seventeenth-century scientists for a universal language. Some believed it had been located in China, where Jesuit missionaries reported the use of signs and symbols as a kind of picture language intelligible all over the country, despite the various spoken dialects. And indeed, according to Gonsales, the language of the Chinese is comparable to the one spoken by the inhabitants of the moon: the *lingua humana,* the beautiful language of Adam. Soon the traveler becomes homesick, and the gansas are due for their return migration. Gonsales harnesses the birds for the return to Earth and, after nine days, crash lands in China, narrowly escaping execution as a sorcerer.

Lunar inhabitants in the imaginative accounts are hardly ever imagined as inferior, ill-natured, or threatening; rather, they often possessed superior qualities rare on Earth. Reports of encounters with "primitive" people on real voyages of discovery on our own planet certainly helped inspire these fictional accounts.

The French satirist Cyrano de Bergerac (1619–1655), in *The Other World: The Comical History of the States and Empires of the Moon* (1657), suggests equipping the roof of a carriage with a firework in order to reach the moon. De Bergerac's moon is "a World like ours, to which this of ours serves likewise for a Moon." But the satiric portrayal calls into question the foundations of society. Old people obey the young, birds talk rather than sing, trees philosophize, and payment is made with self-written poetry rather than coins.

The noble lunarians live on the vapors of cooked food rather than food itself and communicate with simple melodies that combine to produce harmonious concerts.

Imagination had no bounds. In *Iter Lunare; or, Voyage to the Moon* (1703) David Russen employs just enough science to acknowledge the thin atmospheric layer of the Earth and the decrease of weight with distance. He dismisses all the vehicles proposed for trips to the moon—including de Bergerac's rocket carriage—and introduces a spring catapult to hurl the visitors to the moon and back again.

An illustration for Cyrano de Bergerac's tale

In Murtagh McDermot's *A Trip to the Moon* (1728), the protagonist is gripped by a whirlwind at the Pico de Tenerife on the Canary Islands and soon reaches "a Space between the Vortices of the Earth and moon, where the Attraction of neither prevail'd, but the contrary Motions of their Effluvia destroy'd one another." He simply clings to "a Cloud full of Hail" and moves "with incredible swiftness" into the sphere of attraction of the moon until he falls toward it, luckily into a fishpond. After many experiences there, he prepares himself for the return to Earth: "I design to place myself in the middle of ten wooden Vessels, placed one within another, with the Outermost strongly hooped with Iron, to prevent its breaking. This will be placed over 7000 Barrels of Powder, which I know will raise me to the top of the

Atmosphere." Before digging a hole in the moon's surface for the gunpowder to be "blown up," he fills the spaces between the vessels with water to prevent a fire. Once he is aloft, woodcocks encountered en route guide him home.

In *The flying tank* (1783), by the pseudonymous Madam la Baronne de V***, the voices of the moon's inhabitants have an angelic beauty reminiscent of flutes. These pure, sweet-smelling beings take nourishment from the water of a river. They immediately subject their malodorous guests from the Earth to an intensive cleaning. The visitors are then led through a complex cave system, where they encounter allegorical figures representing traits, positive and negative: false love is manifest as envy, jealousy, seduction, and deceit; real love finds itself side by side with confidence, security, and sweetness.

In the anonymously published *Interesting account of a new trip to the Moon* (1784), this "so much desired world" is a peaceful place without any wild animals. A single variety of fruit provides all food and drink, and table manners require a diner to laugh wholeheartedly when the host is drinking. Male inhabitants are ugly, but the females are beautiful, and they create an erotic ambience by exposing their breasts to the visitors. While the lunar women are permitted a certain degree of tenderness with their guests, the familiarity is limited to a kiss; they always stay faithful to their spouses.

Such lunar utopias were hardly limited to France or England; they can also be found in Russian literature during this time. In Vasilii Levshin's *The Latest Journey* (1784) the moon is a world of absolute equality with neither soldiers nor sovereigns, where tradition rather than progress rules and the inhabitants devote themselves to such romantic activities as farming and sheepherding. Ironically, for Levshin the "lunatics," the people in the moon, are the only sane people of the universe. In a similar vein, Mikhail Chulkov in *Dream of Kidal* (1789) portrays a moon world in which all property is held in common and creatures such as vipers, crocodiles, and tigers live in complete harmony with man. In fact, the moon is posited as being as different from the Earth as paradise from hell. While these were novels, not polemics, their utopian visions certainly reflected the authors' dissatisfaction with contemporary ideals.

The protagonist of George Fowler's *A Flight to the Moon; or, The Vision of Randalthus* (1813) casts up his eyes to behold "a milk white cloud." Examining it further, he realizes that it shrouds a beautiful, angelic woman whose "complexion was as white as the soft descending snow; her cheeks and lips were tipped with pink, her eyes were as bright as the sparkling diamonds." She then opens her sweet lips, saying, "Thou shalt see with thine own eyes that object which has so often been the subject of thy solitary meditations, and behold thou art destined to visit the Moon!" The woman then disappears in a vivid flash of light, but our hero finds himself within the very cloud in which she had originally appeared, "swiftly advancing above the confines of the Earth." The human inhabitants of the moon bear a considerable resemblance to this captivating messenger: "Their complexion is of a beautiful golden cast; their cheeks and lips are tipped with a lively red; their eyes blue; and their golden hair oft falls down their shoulders in beautiful ringlets." Fowler's protagonist observes "great symmetry and delicacy in their shape and features" and notes, "they move with inimitable grace." He remarks on their sensibilities, "They are feelingly alive to all the virtues which we possess; but angry looks never distort the beauty of their features, nor even passions, pollute the purity of their hearts. They are extremely quick in their motions, and equally quick of apprehension. They are fonder of music, painting and poetry, than philosophy and abstract studies, which they say only tend to bewilder the mind without either amusing the fancy or adding to the comforts of life."

In Jacques Boucher de Perthes's tale *Mazular* (1832), the main character falls upon a cloud that then transports him to the moon. He begins to spin so fast that he has trouble breathing and has the impression of soaring upward for fifteen days and nights until he can discern "something round and shiny"—a fairytale moon world where he is thrown down with a terrible shock. The lunar humans he encounters have only one leg, one arm, one eye, and one ear each, and no nose at all; everything on the moon is strangely cut in half. He returns home by chance: leaning out a bit too far, he simply falls back to Earth.

Some tales of lunar exploration reflect technological progress. Romantic moonflight and dreamlike devices disappear, and imaginary journeys lay
more explicit claim to authenticity. Isaac Newton
had found that "to every action there is always
opposed an equal reaction: or, the mutual actions
of two bodies upon each other are always equal,
and directed to contrary parts." The jet force, the
fundamental physical principle involved in rocket
propulsion, is an extension of Newton's law, and
Newton himself foresaw that humans of future
centuries might fly to the stars under the power
of the principle he described. Early science fiction,
still centuries shy of the knowledge needed to
make space travel a fact, nonetheless began to ex-
ploit Newtonian principles of gravity. *A Voyage to
the Moon* (1827), one of the first American novels
about an interplanetary voyage, was written by
"Joseph Atterly"—a pseudonym for the lawyer
George Tucker, who was one of Edgar Allan Poe's
university instructors. A metallic substance called
lunarium is discovered, which, "when separated
and purified, has as great a tendency to fly off
from the Earth, as a piece of gold or lead has to
approach it"; moreover, it is "in the same degree
attracted towards the Moon." A mechanism is
constructed "to penetrate the aerial void":

> The machine in which we proposed to em-
> bark, was a copper vessel, that would have
> been an exact cube of six feet, if the cor-
> ners and edges had not been rounded off.
> It had an opening large enough to receive

our bodies, which was closed by double sliding panels, with quilted cloth between them. When these were properly adjusted, the machine was perfectly air-tight, and strong enough, by means of iron bars running alternately inside and out, to resist the pressure of the atmosphere, when the machine should be exhausted of its air, as we took the precaution to prove by the aid of an air-pump. On the top of the copper chest and on the outside, we had as much of the lunar metal· (which I shall henceforth call "lunarium") as we found, by calculation and experiment, would overcome the weight of the machine, as well as its contents, and take us to the Moon on the third day. As the air which the machine contained, would not be sufficient for our respiration more than about six hours, and the chief part of the space we were to pass through was a mere void, we provided ourselves with a sufficient supply, by condensing it in a small globular vessel, made partly of iron and partly of lunarium, to take off its weight. . . . A small circular window, made of a single piece of thick clear glass, was neatly fitted on each of the six sides. Several pieces of lead were securely fastened to screws which passed through the bottom of the machine; as well as a thick plank. The screws were so contrived, that by turning them in one direction, the pieces of lead attached to them were immediately disengaged from the hooks with which they were connected. The pieces of lunarium were fastened in like manner to screws, which passed through the top of the machine; so that by turning them in one direction, those metallic pieces would fly into the air with the velocity of a rocket.

The two protagonists then undertake "a voyage, of which the history of mankind affords no example."

Tucker describes life on the "dark hemisphere":

The Sun pursues the same path in the corresponding latitudes of both hemispheres; but being without any Moon, they have a dull and dreary night, though the light from the stars is much greater than with us. The science of astronomy is much cultivated by the inhabitants of the dark hemisphere, and is indebted to them for its most important discoveries, and its present high state of improvement. . . . If there is much rivalship among the natives of the same hemisphere, who differ in the length of their shadows, they all unite in hatred and contempt for the inhabitants of the opposite side. Those who have the benefit of a Moon, that is, who are turned towards the Earth, are lively, indolent, and changeable as the face of the luminary on which they pride themselves; while those on the other side are more grave, sedate, and industrious. The first are called the Hilliboos, and the last the Moriboos—or bright nights, and dark nights.

In the novel *Trip to the Moon* (1865) by Alexandre Cathelineau, some plants are placed inside an airtight *Terrinsule* test spaceship 50 feet tall and with a capacity of 530,000 cubic feet to accommodate the necessary oxygen for the two travelers. The test craft is the basis for the later *Micromégas* spaceship, constructed with the help of the best carpenters, masons, and gardeners, aided by freed slaves. The inhabitants of the moon are attractive, amicable, and emotionally serene; they speak a single sonorous, musical language. There are no murders, wars, or sickness, and therefore no need for lawyers—"a paradise superior to the one of Adam and Eve before the fall." The lunar people build bridges and houses of wood and travel in carriages drawn by mooselike animals or through the air on the back of eaglelike birds. Although they have many common duties, they clearly reject socialism. In a religion reminiscent of a nature cult, they render homage to God with ceremonial fires. The visitors are essentially reincarnated on the moon: they remember no details from their life on Earth.

In "The Unparalleled Adventure of One Hans Pfaall" (1835), Edgar Allan Poe sends his hero to the moon in a hot-air balloon, which was the only real-life means of aerial transport available until early in the twentieth century. As the title suggests, the story revolves around a certain

Hans Pfaal's balloon trip to the moon   Hans Pfaall, who, having killed some of his creditors, goes to the moon on the first of April to escape the rest. After spending five years there, he sends one of the lunar inhabitants back to Earth in his balloon to deliver the message that Pfaall will return if the citizens of Rotterdam grant him full pardon for his crimes. The lunar inhabitant, frightened upon his arrival by the "savage appearance of the burghers of Rotterdam," simply throws Pfaall's note to the crowd and returns home without waiting for a reply. The public regards Pfaall's story as a hoax; the protagonist, for his part, "cannot conceive upon what data they have founded such an accusation."

In August 1835 the *New York Sun* reported in a series of articles under the headline "Great Astronomical Discoveries" that Sir John Herschel had discovered blue bat people on the moon with the help of a very powerful telescope. Herschel had allegedly observed "sheep, pygmy zebra, [and] unicorn roaming lunar grasslands," in addition to the "bipedal winged creatures called manbats." These creatures averaged "four feet in height,"

> were covered, except on the face, with short and glossy copper-colored hair, and had wings composed of a thin membrane, without hair, lying snugly upon their backs, from the top of their shoulders to the calves of their legs. The face, which was of a yellowish flesh color, was a slight improvement upon that of the large orangutan, being more open and intelligent in its expression, and having a much greater expansion of forehead. The mouth, however, was very prominent, though somewhat relieved by a thick beard upon the lower jaw, and by lips far more human than those of any species of simian genus.

These descriptions originated not with Herschel but from the pen of an imaginative *Sun* reporter. Sales skyrocketed, and other newspapers around the world reprinted the stories. The moon hoax

A R T I C O L O

ESTRATTO DALLA GAZZETTA DI FRANCIA

SULLE

SCOPERTE FATTE NELLA LUNA

DA HERSCHEL

A specimen of the curious race of bat people conceived by Richard A. Locke

was a landmark in the emerging sensational press of the 1830s and also a kind of litmus test. The strong reaction of the public proved that such ideas still could not simply be dismissed as nonsense; science fiction presented as fact could draw a large, fascinated audience. The publisher of the newspaper boasted, "New Yorkers read the *Sun* by day [and study] the Moon by night." Richard Adams Locke (1800–1871) soon confessed that he had fabricated the story.

For some writers, the fact that the moon is much smaller than the Earth suggested that its inhabitants would likewise be smaller than humans. In *Trip to the Moon* (1845), Jacques Bujault follows this logic: the diminutive size of his lunar *Picolins* is the only characteristic that distinguishes them from humans. In the anonymous *Very Recent Trip to the Moon* from the same year, the vegetarian inhabitants of this "other eden" reigned over by God himself are depicted as attractive beings with blond hair and blue eyes,

"men, women and children smaller than us, but of a most charming physiognomy, full of expression, of grace, of happiness." Born innocent, they await a different fate from ours, happily celebrating the arrival of an angel to take away a deceased body, as decay is unknown there.

Some authors presented a moon that would be familiar to any earthling. Georges Le Faure and Henry de Graffigny, in *Extraordinary Adventures of a Russian Savant* (1889), imagined a lunar far side with trees and forests where the

The most beautiful of the selenites in Georges Le Faure's and Henry de Graffigny's novel *Extraordinary Adventures of a Russian Savant* (1889)

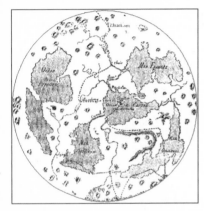

The map of the far side as provided by Le Faure and de Graffigny

giant Selenites were concentrated. The authors even provided the readers with a map of the far side, showing not only the elements typical of the visible side of the moon, such as craters, but also oceans, rivers, and cities. Other writers twisted familiar values to create their lunar fables. In Louis Desnoyer's *The Adventures of Robert Robert* (1839) the moon is an Earth where everything is topsy-turvy. Tiny elephants populate the cracks and crevices instead of ants, and sheep keep flocks of wolves together. Although gold, silver, and diamonds don't have any value, iron and pebbles are expensive. And since the rain is made of wine, water is the beverage most in demand.

The pseudonymous Pierre de Sélènes creates a vaguely communistic utopian society with a naive belief in technical progress typical of the time in his novel *An Unknown World: Two Years on the Moon* (1886), dedicated to Jules Verne. Under the surface of the moon three travelers from Earth discover a naturally created, weatherless, "supernatural" world inhabited by some twelve million people. The world has an ocean the size of the Mediterranean at its center and a delightfully beautiful city with richly ornamented houses, illuminated by cosmic and electric light. A variety of stone functions as a battery, taking its charge from the electricity-laden air. The lunar people—

who take nourishment from the air, never kill, and speak a musical language "of an extreme logical simplicity"—had withdrawn into this underground haven when the water and air on the surface began to disappear. Wages and private property have no place in this ideal society, and every citizen can claim the space he needs. Futuristic trains provide transportation through the sublunar cave world, and the inhabitants also use technology that suggests television: "transmis-

A lunar base from the book by Pierre de Sélènes

sion at distance of sensible and talking images," technology still decades away on real-life Earth. Outside, they are able to witness volcanic explosions that spew into the sky large incandescent chunks of matter, which, once they have returned to the thin and cold lunar atmosphere, simply turn into flying sparks—a natural catastrophe "nothing on Earth could have given the slightest idea of." But the visitors finally decide to return to Earth, partly because they are bored by the day-to-day monotony and the unchanging weather, partly because they find the sublime character of the lunarians too distressing.

Fantasies about lunar life certainly entertained readers of earlier centuries, but they also functioned as blueprints for speculation about life on the planets. Of course, the fantasies evolved along with the ongoing technological progress. As evidence mounted that life on the moon was unlikely, if not impossible, speculation shifted elsewhere—toward Mars, for example, where similar fantasies took root. But lunarians survived in a few exceptional works of twentieth century science fiction—both novels and movies. As we'll see in a later chapter, this generation of moonmen is not quite as sweet and lovable as earlier authors had envisioned.

In André Laurie's science-fiction novel *The Conquest of the Moon* (1875) the moon is drawn from its orbit to land in the Sahara desert. Laurie was the pseudonym of Jean-François Pascal Grousset (1844–1909).

CHAPTER SIX

Lunar Passion in Paris

The European fascination with the moon in the eighteenth and nineteenth centuries was part of a deep obsession with wild and sublime landscapes. Jagged volcanic vistas, the icy regions of the Arctic and Antarctic, and the dry vastness of the deserts attracted scientists and laymen alike, some of whom paid for their curiosity with their lives. All such environments imposed exacting physical limitations on the humans willing to investigate them. As Alain de Botton has put it in his essay *The Art of Travel,* "It is as if these landscapes allowed travellers to experience transcendent feelings that they no longer felt in cities and the cultivated countryside." The key difference from the other landscapes was that the moonscape was inaccessible to the whole spectrum of the senses, could be explored only visually, and still remained largely a landscape of the imagination.

The French writer Jules Verne (1828–1905) was gripped by this fascination. Though not a scientist himself, he had read popular works of astronomy, had a large archive, and kept current with scientific and technological developments. Until the first balloon flight, in 1783, most people saw little difference between flying above the surface of the Earth and traveling to other celes-

Jules Verne's spacecraft on its way to the moon

Jules Verne

tial bodies—both seemed utterly impossible. In contrast to many earlier tales of trips to the moon, Verne's *From the Earth to the Moon* (1865; English translation, 1867) seemed more realistic, because it took contemporary scientific knowledge into account—particularly ballistics, the scientific basis for understanding projectiles in flight, which had made many advances. The novel also reflects contemporary scientific disputes about the possibility of volcanic eruptions on the moon or whether the moon's hidden side might have air, water, clouds, plants, or even a forest. Since the development of mechanically powered aircraft and the drama of a real trip to the moon were a long way off, Verne's account was still an exercise in imagination. But the idea no longer seemed as implausible as it had a century, or even a couple of decades, before.

Like other books Verne wrote, *From the Earth to the Moon* is startlingly accurate in anticipating technological developments of the coming century. (Verne even had an agreement with his publisher to incorporate the most up-to-date scientific facts.) The book is unprecedented in its systematic treatment of scientific elements of preparations for the trip. It tells the story of three members of the Baltimore Gun Club who plan to build an aluminum cannon nine hundred feet long and six feet wide and launch themselves to the moon using gun cotton. Verne thus introduced the idea of a manned ballistic projectile escaping the Earth's gravity.

Although from our perspective Verne failed to consider some technical issues, such as the friction on the projectile during flight, he was the first to identify other salient aspects of space flight. Some details of the story even bear similarities to the National Aeronautics and Space Administration (NASA) Apollo program of the 1960s, beginning with the number of astronauts, three. The second volume of the work was serialized in 1869, exactly one hundred years before the first landing of humans on the moon. Verne correctly predicted not only that the United States would launch the first manned vehicle to orbit the moon—and would strive to be the first country to plant its flag there—but also that Florida would be the launch site. He selected Tampa, near the Gulf Coast, whereas NASA chose Merritt Island, 130 miles away on the Atlantic side. Both sites have easy access to the sea and fulfill the criterion of being at a latitude below 28 degrees north. Modern space projects favor sites relatively close to the equator because the rotational speed of the Earth is greatest there, providing a small extra boost at launch. But Verne chose this site for a different reason: to facilitate coordination of the launch with the moon's perigee—the point at which it is closest to the Earth. Verne can thus be credited with anticipating the idea of a temporal and spatial frame for the launch of a rocket or spaceship. In addition, he reasoned that a large body of water would be the safest place to land. Both Verne's imaginary moon

Brainse Fhionnglaise
Finglas Library
Tel: (01) 834 4906

mission and the Apollo flights splashed down in the Pacific Ocean and were rescued by U.S. Navy vessels. In both the novel and in reality the start of the rocket drew an immense number of people to the site. Before the invention of the automobile, Verne still expected onlookers to arrive by railroad.

Verne imagined a spacecraft constructed primarily of aluminum and weighing 19,250 pounds, whereas the predominately aluminum *Apollo 8* circumlunar spacecraft weighed 26,275 pounds when empty. The cannon used to launch Verne's spacecraft was called *Columbiad;* the command module of *Apollo 11* was the *Columbia.* The invocation of Christopher Columbus likens the trip to the moon to the explorer's ambition to find a new route to the Indies, which resulted in the discovery of the New World. That connection touches on a psychological similarity between Verne's travelers and American astronauts: both groups understood themselves as representatives of Earth, and both were preoccupied with humanitarian and even religious concerns. The journey of Verne's spacecraft took 242 hours, 31 minutes, whereas the *Apollo 8* crew spent 147 hours in space. Considering the enormous leaps in knowledge that occurred in the hundred years between the fictional and real space flights, many of Verne's predictions are stunningly accurate.

In some areas, Verne was on less secure scientific footing. It is vital, for example, to consider the effect of gravitational force on humans aboard a giant projectile accelerated instantly to a speed capable of escaping the atmosphere. Verne

was aware of this problem, but the hydraulic shock absorbers he provides for the hollow shell would not have had the desired effect. Instead, the tremendous pressure would have crushed everything—including the passengers—like a soap bubble. Verne also erroneously believed that weightlessness would occur only for a short time at the neutral point of gravity between the Earth and the moon.

In the novel a meteor sidetracks the projectile from its planned course so that instead of falling

Interior of the cannon designed to shoot a spacecraft to the moon in *From the Earth to the Moon* (1865)

on the moon, it traces an ellipse around it. Missing the moon is actually a stroke of luck for the imaginary travelers—given the logistics of their space travel, had they landed, they would not have had the slightest chance of getting back. By not actually landing on the moon, they also spared Verne the need to describe a lunar environment about which little scientific consensus existed. He didn't have to stick his neck out too far on this point. While circling the moon, his space travelers believe they recognize ruins on the far side, even canals, but they find no clear evidence of lunar life. Had his travelers reached their goal, Verne would have been forced to pin himself down to one theory. Rather than take the risk that future developments would prove him wrong, Verne simply had his characters repeat the various competing hypotheses about the moon's nature.

We may also see the "test flight" that the adventurers conducted with a cat and a squirrel as a curious precursor to later actual flights using dogs and chimps to study the effects of weightlessness. But Verne errs when he assumes that the travelers could dispose of the body of Satellite, a dog accidentally killed during the trip, "just as sailors get rid of a dead body by throwing it into the sea." In Verne's account, the travelers simply open a hatch in the capsule "with the utmost care and dispatch, so as to lose as little as possible of the internal air," and launch the poor dog's body from their luxuriously cushioned projectile into space. In fact, such a breach of the capsule's atmosphere would have killed all the inhabitants almost instantly.

Were all of Verne's accurate or nearly accurate predictions just strange coincidences, or can we credit him with an anticipatory genius, a special sense for the most likely course of scientific development? Certainly, the era during which he wrote his novel was conducive to innovation in theory and practice, and Verne's fictional account captured the popular imagination. Even though a trip to the moon long remained unlikely in real life, the possibility, however slim, had taken a firm and permanent hold on the cultural imagination. Verne's seminal work inspired most of the original pioneers in the fields of astronautics and space sciences. His novels were widely viewed condescendingly by his contemporaries, many of whom considered them unfit for adult readers; poor translations hampered the reception of his books in other countries until the mid-twentieth century. The Apollo program not only helped make Verne popular again but also led to a new appreciation of the scientific and literary merits of his work.

Although for decades Verne's work was subject to a curious combination of popular success and literary snub, he now has been accorded an undisputed place among the writers in the history of science fiction. Many of those who later made space travel a reality are known to have devoured Verne's novel, so that perhaps more than any other work of science fiction—to an extent that even a bright and imaginative mind such as Verne's would not have been able to anticipate— *From the Earth to the Moon* has had a revolutionary impact on real-world science.

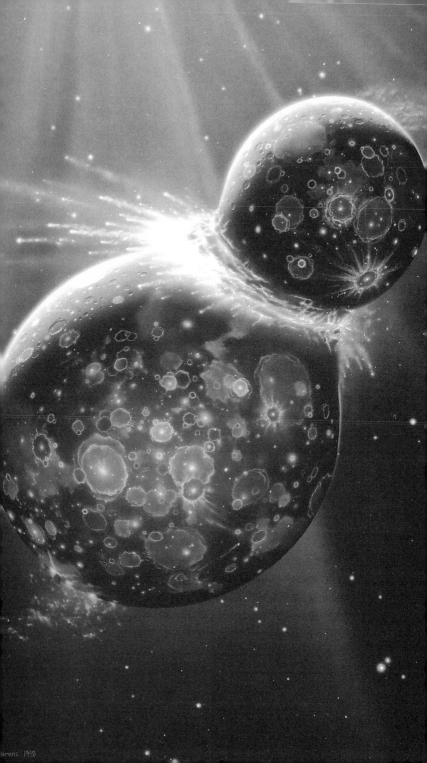
arons 1998

CHAPTER SEVEN

Accounts of Genesis

What is the origin of the moon? A simple question, perhaps, but one without a simple answer. A myth among the Native American Seneca tribes claims that a wolf sang the moon into existence. In the Western world, the study of the history of the Earth was, until the early eighteenth century, still dominated by the biblical account of the Creation, and the same account settled the history of the moon. According to Genesis, God "set two great lights in the firmament of heaven, the greater light to rule the day and the lesser light to rule the night." Just as the moon took its light from the sun, and was inferior to the sun, so secular authority, whether royal or scientific, reflected religious, thus papal, authority. But the European Enlightenment signaled that God's power was beginning to wane.

By the seventeenth century a revolution in thought was afoot, a revolution with reason, not divinity, at its center. That upheaval brought with it a new understanding of the cosmos. The French savant René Descartes (1596–1650) not only put the sun in the middle of the universe, as Copernicus and Galileo had done before him, but also imagined a universe with pieces of matter of different sizes, forms, and motions caused by different forces. This concept put the moon and its

The big splat—one theory of moon formation—as illustrated by the space artist Joe Tucciarone formation in a new framework. In fact, Descartes explained, the vortex of the Earth traps the moon and thus forces it to move around it. Descartes wrote this treatise about 1630, but its publication had to wait until 1664, well after his death, precisely because it affronted the older order. We should not forget that in 1633 Galileo's heliocentric system was condemned in a trial, and he was put under house arrest for the remainder of his life.

Toward the end of the eighteenth century, Pierre-Simon Laplace (1749–1827) proposed that the solar system had begun as a huge swirling cloud (or nebula) of gas. As it collapsed under its self-gravitation, the central mass became the sun, while rings ("Laplacian rings") of material were spun off in succession, from the outer regions first; this matter then coalesced to form the planets. Earth and moon had remained connected in terms of gravitation ever since. This model became the basis of the scientific theory referred to as co-accretion or binary accretion and was also connected to hypothesized changes in the lunar orbit—for example, that the moon is gradually receding from the Earth and must thus have been closer in the past.

George H. Darwin (1845–1912), the second son of Charles Darwin and in his time one of the world's eminent physicists, calculated that at some point the center of the moon would have been only six thousand miles from the Earth's surface. Darwin thus concluded "that if the Moon and Earth were ever molten viscous masses, then they once formed parts of a common mass." He then proposed in 1878 that the Earth had spun so rapidly in its early years that it grew increasingly elongated under the influence of the sun's tidal force until a chunk eventually broke free. This chunk, Darwin argued, became the moon. His great-grandfather Erasmus Darwin (1731–1802) had already posited the ejection of matter from the Earth to become the moon in a lengthy poem, *The Botanic Garden* (1792):

> Gnomes! how you shriek'd! when through
> the troubled air
> Roar'd the fierce din of elemental war;
> When rose the continents, and Sunk the
> main,
> And Earth's huge sphere exploding burst
> in twain.
> Gnomes! how you gazed! when from her
> wounded side
> Where now the South-Sea heaves its
> waste of tide,
> Rose on swift wheels the Moon's refulgent
> car,
> Circling the solar orb, a sister star,
> Dimpled with vales, with shining hills
> emboss'd,
> And roll'd round Earth her airless realms
> of frost.

Erasmus Darwin was not unique in choosing verse as the forum in which to express scientific thought, but his theory was not circulated widely among scientists, even though he was a member of the Lunar Society. And his mention of the South Sea notwithstanding the English geologist Osmond Fisher (1817–1914) is usually credited (or blamed) for the sensational addendum to this fantastic theory that the Pacific Ocean basin marks the "scar" left behind where our future satellite was ripped away. According to Fisher, the American continent was also separated from Europe and Africa during this process. The American geologist and chemist Richard Owen had already suggested in 1857 that the moon split off from the Mediterranean. Such ideas, though later proven to be wrong, foreshadowed the research in the early twentieth century by Alfred Wegener (1870–1930) on plate tectonics and continental drift. George Darwin didn't adopt any of these addenda, but they nonetheless became part of popular accounts of the moon's origin.

William Henry Pickering extended Darwin's lunar hypothesis, speculating that a former single continent consisting of America, Asia, Africa, and Europe had broken apart because of the moon's separation. As the so-called "fission theory" evolved, the assumption arose that the separation had occurred when the Earth was still in fluid state; the matter ejected, according to this line of thought, first formed a ring orbiting the Earth before combining into what became the moon.

George Darwin's theory was eventually replaced by a competing one. This one held that the moon had been formed in a completely foreign realm, possibly even outside our solar system, and was then "captured" by the gravitational pull of the Earth and kept within its orbit ever since. This hypothesis is generally associated with the American chemist Harold C. Urey (1893–1981), who received the Nobel Prize in 1932 and collaborated on the atomic bomb. Soon after World War II he pioneered the field of cosmochemistry. Urey saw the moon as a cold relic from the early history of the solar system. He based his argument on the different densities of sun and moon and the assumed similarity in iron of both bodies. In fact, he drew the conclusion that the moon was never a part of the Earth but had formed independently, and earlier. Urey's notion that the moon had been captured by the Earth provided a perfect scientific rationale for the Apollo program: if the moon predated the Earth, investigation of the lunar surface could produce results that could not be drawn from the geologically active Earth. He also suggested that the moon had become cold since the capture, and that the lunar maria had been formed not by lava flows but by water, possibly splashed from Earth during the moon's capture. "If indeed the surface of the moon carries a residue of the ancient oceans of the earth at about the same time that life was evolving, the Apollo program should bring back fascinating samples which will teach us much in regard to the early history of the solar system, and in particular with regard to the origin of life," Urey remarked in 1966. "I wish I could go rock-hunting with the astronauts this month," Urey said in the summer of 1969, and his addendum, "even if I knew I could never get back," proves the urgency these questions had assumed

Harold Urey

for him. His concrete influence on the space program remained minimal, though.

The three hypotheses of lunar origin available at this decisive time could be simplified in family terms: the moon as Earth's sister, born nearby out of material similar to the Earth's (co-accretion theory); the moon as Earth's daughter, born directly of our planet (fission theory); or the moon as Earth's wife, born elsewhere in the solar system and later captured into the Earth's orbit (capture theory). But by the end of the 1970s, lunar scientists agreed that none of the three theories could offer a satisfactory narrative of the moon's origin. The fission theory had been refuted, because the Earth had not been spinning fast enough to throw off part of itself, but the inescapable paradox was that the chemical compositions of the moon and the Earth held certain similarities. Key elements in the composition of the Earth—oxygen, silicon, aluminum, calcium, magnesium, iron, and sodium, to name just the most important ones—are all there on the moon, albeit in different concentrations. How could this issue be resolved?

Earlier voices had suggested a key theoretical difference, but few had wanted to listen. As early as 1911, the American geologist Howard Bigelow Baker had speculated that the Earth had once had a close encounter with a now-vanished planet, and that what is now the moon had been ripped from the Pacific Basin, thus causing a continental drift. In the 1970s the engineer James G. Baker (no relation to Howard B. Baker) suggested that Mars and Earth had narrowly avoided collision after a massive proto-Jupiter had disturbed their orbits, but his theory was widely ignored.

Before Apollo the moon had been believed to be just a rock—a relic of the solar system's formation that never formed a core. William K. Hartmann and Donald R. Davis hypothesized in the early 1970s the impact of one or more objects six hundred miles in diameter that could have ejected enough material from the Earth's mantle into orbit to form the moon. Hartmann and Davis believed that the colliding planet actually had a molten iron core, but it sank to the center of the Earth and merged with its core, thus leaving the moon relatively devoid of metallic iron. The problem with this theory is that an object six hundred miles across is relatively small—less than 1 percent of the Earth's mass, and less than 30 percent the diameter of the moon itself. Alastair Cameron and William Ward recognized that a much larger object—something with about 10 percent of Earth's mass, an object that was essentially a planet in its own right, roughly the size of Mars today—could have provided the Earth with the rapid initial spin needed to account for our twenty-four-hour day. A major conference in Hawaii in 1984 brought together about one hundred astronomers and resulted in consensus for the giant impact hypothesis.

Alastair Cameron was the first to simulate moon-forming collisions in the 1980s and 1990s. Robin M. Canup from the Southwest Research Institute, employing computer simulations, has since refined the theory with further detail. Her computer model consisted of twenty thousand components, and the collision behavior of each is analyzed. According to her research, the likeliest scenario is that the moon was formed after a glancing collision at high speed occurring four and a half billion years ago between Earth, by then nearly fully formed but smaller than at present, and a proto-planet about the size of Mars. Canup terms the early Earth Gaia and the other body Theia. The collision was the biggest blow Earth ever experienced. The impact would have erased all surface features and probably melted rocks to a depth of 600 miles—"the zero moment" for our planet and the final moment for the other one. Theia, which is believed to have been formed in an orbit similar to the Earth's, was pulverized, its matter sprayed out into a shower of orbiting debris. Within hours much of that debris regrouped to form a new body that smashed into the Earth's surface a second time and was destroyed. Most of the resulting matter was absorbed and became part of our planet, but about one-tenth, mostly from the outer parts of Theia, formed a cloud of debris, an incandescent disk around Earth. This disk was thrown into space and within a few decades developed into what is now the moon. At that distant point in time, the moon was fifteen times as close as it is now. As in Cameron and Ward's scenario, the impact set Earth aspin on its axis, eventually resulting in twenty-four-hour days. On the other hand, the presence of the moon creates a gravitational counterbalance that stabilizes Earth's slightly inclined axis of rotation, providing the congenial cycle of the seasons over a single orbit around the sun. This model also accounts for the moon's comparatively low density and its dry and refractory composition, while also including a source of energy sufficient to melt the upper layers of the early moon and produce an ocean of magma, which, over time, solidified and crystallized into the current hard crust. After the moon formed, about 3.9 billion years ago, a "lunar cataclysm" occurred—a cluster of collisions on the highlands. Rock was driven to the surface by independent eruptions from various sources deep in the hot lunar interior, filling basins with basaltic lava. Some remaining parts of this molten magma stabilized in pockets for some time before erupting to the surface.

Canup's theory includes features of each of the three mutually exclusive classical theories. Accretion plays a role, but the fragments accreted resulted from the collision. As in the double-planet theory, a second planet is involved in the scenario. And, as in the concept of fission, the moon was born of the Earth, but only after a collision with the second planet. The sequence of events as now envisioned was more complex than any of the three earlier theories had suggested.

The giant-impact hypothesis could help clarify why some lunar rocks have a similar composition to those found on Earth and some do not, but it cannot explain why some are magnetized. Perhaps, it has been speculated, the moon has an internally generated magnetic field (as does Earth), or perhaps impact events generate short-lived transient fields that could have magnetized the lunar rocks. Much recent discussion of the moon's makeup focuses on the nature of an assumed core, which may consist of iron with some sulphide. Based on various geophysical calculations, the core is thought to be less than 250 miles in radius—about a quarter of the moon's radius, compared with Earth's core, which is about half of the planet's radius. While it is possible to identify Earth's iron core by means of measurements of seismographs distributed around the planet, analogous equipment on the moon is insufficient to yield such unequivocal information. It is known that moonquakes often occur, but none of their ray paths go through the center of the moon, further complicating core measurement.

Larry Taylor, director of the Planetary Geosciences Institute at the University of Tennessee in Knoxville, has studied mare basalts, which are believed to have been created by melting in the moon's mantle and have retained signatures of that region. The analysis suggests that the lunar mantle has even lower levels than Earth's of elements that bond easily with iron—the so-called siderophile elements, such as platinum, iridium, and osmium. "What happens during the formation of any terrestrial planet is that it undergoes a melting state early in its formation," Taylor says. "In that state, you have separation of metallic iron into a core." When cores formed in Earth and other planets, these elements with an affinity for iron were, for the most part, scavenged from the mantle and transferred into the metallic core. This would explain the relative lack of these elements in both Earth's mantle and the moon's — and in return make the case for a metallic core. The depletion of these siderophile elements in the moon's basalts has been interpreted as additional proof of a metallic core in the moon.

More recently, a team at the MIT led by Ian Garrick-Bethell examined the magnetic history of samples from one of the most ancient rocks in the Apollo collection and found out that the moon in its early days may have had a liquid metal core that spun like a dynamo to produce a magnetic field. The researchers also ruled out that the rock received its magnetic properties from impacts. "This rock was heated up only twice in its history," says Garrick-Bethell. "When rocks cool, they lock in the magnetic fields around them." It remains unclear why such a magnetic field might have died off. In any case the evidence of the lunar core remains far more speculative than scientists like.

One thing is certain, though: moonstones or selenites — gems found in various locations, including Sri Lanka and India, and once believed to interact with the moon — could never have been created there. Moonstones are feldspars, and there are no feldspars on the moon.

CHAPTER EIGHT

A Riddled Surface

We have seen the various theories scientists have come up with to account for the existence of the moon, but what happened after the moon was born? With no time machine available to us, we have to rely on the measurement of radioactive elements in lunar rock samples to determine age. Predictably, the samples differ in age depending on the exact places they were taken from, but virtually none of this material is younger than 3.6 billion years, at which time a heavy bombardment of meteorites ceased. Since then the moon has remained virtually unchanged, though a relatively steady rain of mostly smaller meteorites has continued, over time, to add new soil to the moon's crust.

Aside from the well-known lower gravity relative to Earth, what does it *feel* like to stand on that crust? The lack of a familiar atmosphere means that the sky remains dark even during the long lunar day. In fact, light and dark are turned upside down from our experience on Earth: brownish gray soil contrasts sharply with the blackness of space. The glare reflected by the moon's surface constricts the pupils, weakening vision. While vibrations may be felt, total silence reigns.

The absence of atmospheric haze makes it

Landscape around a lunar pole as imagined by Ludek Pesek (ca. 1960) much more difficult to estimate distances on the moon, as well. We get a feel for how far one spot is from another only when we move, thereby shifting our perspective in parallax displacement. Whatever the actual distance, the moonscape always appears equally sharp. The color of the surface appears to change depending on the angle from which we view it: light brown when we face the sun, but gray in the opposite direction, black when we examine the soil close up. That the moon is much smaller than the Earth means that we can see only half as far into the distance, because the curvature of the surface is greater. The moon doesn't have seasons and climatic zones like those on Earth and Mars; it doesn't even have wind. With the sun creeping slowly across the lunar sky, the peaks of mountain ranges are illuminated several hours earlier than the plains and valleys, and the horizontal areas of the mountaintops are flooded with perpetual light. Temperatures can change rapidly, not only between day and night but also between sunlight and shadow. One of the most dramatic aspects of being on the moon is seeing how Earth, barely moving, dominates the sky. During the Earth's day, we could see cloud formations and watch the planet pass through phases like those we can see for the moon.

Beyond the physical aspects of the moon, we might ponder more philosophical questions about the moon's nature. Is it a *world* even though it is "cold" and lifeless? Before the invention of the telescope, philosophers were often disposed to regard the moon as a world analogous to our own. The Pythagoreans saw the moon as a second Earth. Frank Sherwood Taylor, in *The World of Science* (1937), painted the paradoxical character of the orb: "The Moon is rightly a melancholy sight as well as a beautiful one. It is a sphere of death—a picture of the state of the planet which air and water have left: it revolves, the skeleton of a world." A recurring theme in the literature is that Earth, stripped of water and its mantle of vegetation, might look to a distant observer like the moon.

Given the familiarity of our home planet, it is not surprising that observers have often focused on topographical features of the moon that seem similar to those of Earth while overlooking those that are dissimilar. Circular, craterlike outlines, for example, characterize both the Sea of Japan, between Japan and the Asiatic mainland, and the Gulf of Mexico. The island arcs of the northern Pacific could be compared to the mountain ramparts of the maria. And terrestrial plains ringed by mountains—in Thessaly, Greece, for example—resemble such areas on the moon. But circular features, exceptional on Earth, are more nearly the rule on the moon, whose surface, in fact, is not much like our planet's. The moon's Alps, Caucasus, and Apennines seem at first glance comparable to terrestrial mountain chains, but they probably resulted from faults triggered by mare-forming impacts. Oceanus Procellarum, which measures some 2.3 million square miles, has been called the Pacific of the

moon. That comparison rather shortchanges our Pacific, which is almost thirty times as big.

Similar false analogies have occurred in reverse. Various landscapes on Earth have been characterized as resembling the moonscape: for instance, the Valle de la Luna (Valley of the moon), with its distant ring of volcanoes, in the Chilean region Antofagasta; the volcanic landscape of central Lanzarote, one of the Canary Islands; and the summit of Mont Ventoux in Provence, which has only sparse vegetation and is a major challenge during the Tour de France bicycle race. The Mountains of the Moon, a mountain range straddling the border between the Congo and Uganda, supposedly owe their name to the snow-covered summits somewhat reminiscent of the bone-white lunar surface as seen from Earth. We may also remember the uneven terrain of the Icelandic volcano field of Askja, with its weird rock formations and lava fields, which NASA classified as the most moonlike landscape on Earth, sending Neil Armstrong and his colleagues there for training in 1967. And the nightmare of a "postapocalyptic moonscape" has become synonymous in the collective imagination with parts of the Earth destroyed by nuclear catastrophe.

But how, specifically, is the moon's surface different from Earth's? One key difference lies in Earth's changeability: here seismic activity, weather, and erosion have played and continue to play important roles in altering the surface over time. Craters from meteoric impacts in times long past have eroded and are no longer discernible. More than three-quarters of the surface of the Earth is less than two hundred million years old, and virtually nothing on Earth's surface is the same as it was at the time of its formation. In contrast, according to most recent estimates, 99 percent of the surface of the moon is more than three billion years old—same old, same old, you might say.

The character of the lunar surface had long puzzled scientists, starting with the particular reflective properties and composition of the material of the crust. In 1955 the American geophysicist and NASA consultant Thomas Gold (1920–2004) claimed that the astronauts (and any spacecraft before them) would sink into a yards-deep layer of dust. Gold believed that the lack of lunar wind meant that dust had accumulated on the moon for billions of years, making it virtually impossible for humans to scrutinize the surface. Gold soon corrected himself, positing that the dust would be just a few centimeters thick, and direct measurements later proved that his revised estimate was correct. Despite this vindication, other scientists referred jokingly to "Gold's dust" for many years, and in fact a theory similar to Gold's initial estimate has since been seized upon by some in support of their religious beliefs. Some who believe in literal interpretation of the book of Genesis hold that the depth of accumulated dust on the surface proves that the moon originated about 8000 B.C. If it were really as old as scientists claim, these believers say, continuous bombardment by micrometeorites would have left the layer much thicker. They usually cite an estimate of one hundred feet in controversial studies by R. A. Lyttleton in 1956 and Hans Petterson in 1960, ignoring more recent contradictory findings.

Upon investigation it became apparent that the makeup of the moon's surface was more complex than mere "dust." In 1960 the Dutch scientist Jan van Diggelen noticed what he described as "an irregular spongy character" of the lunar surface and suggested that reindeer lichen (*Cladonia rangiferina*) was the closest terrestrial analogue. Arthur C. Clarke at one point expected it to resemble "stale black bread." The Surveyor expeditions confirmed a granular aggregate, with particles from about one-twenty-fifth to four-tenths of an inch, with some larger rock fragments loosely cemented. The porous matter covering the moon's surface, debris produced by the impact of meteorites, came to be called regolith (literally, "blanket of rock"), with the finer particles usually referred to as lunar soil, even though the organic components of terrestrial soil are absent. This material includes crushed rock, glass beads, and even grains of metallic iron—another difference from Earth's soil, where this mineral would soon rust.

All in all, about 850 pounds of lunar rock was collected during the Apollo missions. Scientists in many countries have analyzed this material, which remains an object of study. Most of this rock is stored under low-humidity conditions at the Lyndon B. Johnson Space Center in Houston, where it has been repeatedly subject to theft from interns. Examination of the samples has revealed that Earth and the moon have identical oxygen isotope compositions, which vary substantially from those of Mars and most of the asteroids, but the moon rocks contain few of the volatile elements characteristic of Earth.

A meteoroid flux of mostly tiny particles constantly strikes the moon, and at times larger chunks of debris hit the surface with a great deal of energy. According to one estimate, a football-sized meteorite hits the moon every day. When a large enough meteorite strikes, a new crater results. Paul D. Spudis has described this ongoing process "as a giant sandblaster, slowly grinding the Moon's crust into dust." In contrast, the Earth's surface is protected by the atmosphere, and particles falling from space usually burn up before impact.

More than a decade after the end of the Apollo

program came a surprise that suddenly made the moon appear a little bit more like Earth. Investigations of lunar samples yielded no evidence of water, but in 1994 water on the moon had suddenly become at least a theoretical possibility. The lunar spacecraft *Clementine,* which orbited the moon for two months, provided maps of the polar regions. Some craters there are so deep that their floors are never lit by the sun; temperatures within are low enough—never above minus 270 degrees Fahrenheit—to harbor frozen water over geologic time. Light coming back from radar pulses sent into the shadowed areas seemed to confirm an icy surface. The researchers who analyzed the *Clementine* data suggested that deposits of ice might cover an area of thirty-five to fifty square miles at the moon's south pole. The possibility of ice sparked immediate interest, because the presence of water would make the establishment of a self-sustaining lunar colony plausible. But there was no clear indication where the ice, if it existed, had originated. According to one theory it didn't vaporize during the collision of the celestial body with the early Earth that produced the moon; alternatively, it may have arrived later via cosmic projectiles. In any case, there could not be much of it. Results of the probe were further put in perspective when, in 1997, the Arecibo Observatory—the world's largest single dish radio telescope, based in Puerto Rico—recorded similar effects in lunar regions at such a distance from the pole that high daytime temperatures make the existence of ice impossible. Apparently, very rough surfaces were producing radar readings similar to those produced by ice surfaces. But Donald B. Campbell, Cornell University professor of astronomy, conceded that since "neither Arecibo nor Clementine observed all the areas that are in permanent shadow, . . . there is still the possibility that there are ice deposits in the bottoms of deep craters." In 1998 scientists working with *Lunar Prospector,* a small, spin-stabilized craft designed to sample the lunar crust and atmosphere for minerals, ice, and certain gases, announced that between ten million and three hundred million tons of water-ice was scattered inside the crater of the moon's poles.

Another surprise occurred in 2008, when Erik Hauri, a geochemist at the Carnegie Institution of Washington's Department of Terrestrial Magnetism, and some of his colleagues discovered that green and orange lunar volcanic glass beads from samples collected during the Apollo missions were rich in some volatile elements and water, which high temperatures during the lunar-formation event may have evaporated. Applying a special spectrometric technique, Hauri demonstrated "that some parts of the lunar mantle may contain a few hundred or even a few thousand parts per million of water." The beads, which are coated with some volatile elements, are thought to have resulted from volcanic eruptions on the lunar surface.

This is not the end of the story, however. This long series of speculation reached a climax in November 2009 with NASA's sensational announcement that a "significant amount" of water had been located in the form of ice that had accumulated over billions of years. A month earlier, a satellite called LCROSS had been intentionally crashed into a crater near the moon's south pole. The impact carved out a hole sixty to one hundred feet wide, releasing at least twenty-five gallons of water to the surface. Does the moon bear yet more mysteries—and greater similarities to Earth—hidden beneath its monochrome surface?

The lunar craters—about 100,000 more than a half-mile in diameter, found mostly on the highlands—have long attracted attention because they present such a contrast with Earth's surface. The depressions can be round, polygonal, or quite irregular, with or without an exterior wall. Craters exist in all sizes: the smallest are submicroscopic, the largest—Aitken Basin at the South Pole—fifteen hundred miles in diameter and seven and a half miles deep. Some have jokingly compared the lunar surface to Swiss cheese, but even serious attempts to classify the various craters that pock the lunar surface have been rather confusing. The very use of the word *crater* is problematic, because the moon's craters are neither of volcanic origin nor as deep as a volcanic crater on Earth. The word's primary utility may be in reminding us of the earlier, outdated, explanation of these formations. The term *impact crater* is now sometimes used to distinguish these from volcanic ones.

Once telescopes became available, scientists quickly began discussing the origin of what

Galileo compared to "eyes upon the peacock's tail." Robert Hooke (1635–1703), secretary of the Royal Society, was one of the first to venture an opinion. *Micrographia* (1665) was devoted to Hooke's discoveries using the microscope, but he digressed to liken the lunar craters to the volcanoes on Earth and thus to attribute them to internal origin. He once dropped musket balls into wet clay, to produce pits resembling lunar craters. At the end of the eighteenth century the astronomer Franz Aepinus of Saint Petersburg also

Earlier selenologists believed the lunar mountains to have sharp pointy edges. This illustration is taken from *Recreations in Astronomy* (1879) by H. D. Warren.

concluded that lunar craters were of volcanic origin, speculating further that basins represent the first stage in the formation of a volcanic mountain. A central hillock of ejected matter forming around the vent he called molfetta. In describing the lunar surface, Aepinus wrote one of the most extreme statements about lunar volcanism:

> Anyone who, armed with this understanding of the shaping and the form of the products of volcanic action, examines the Moon, will be amazed to find its entire surface covered with examples of such products and will discern upon it all types thereof in the greatest abundance, volcanic basins with and without molfettas, as well as actual volcanoes with open and closed throats, with and without lava flows, with and without molfettas and so on.

Aepinus was aware that the moon has no atmosphere and attributed the excellent visibility of these volcanic features to the absence of erosive effects of wind, rain, and snow. He assumed the few irregularities in the dark regions to be rims of ring structures projecting about the surface of the (water-filled) lunar oceans. And the great ray craters he saw as enormous volcanoes; in fact, he held that Mount Etna on Sicily, the largest active volcano in Europe, if viewed from above, would look like Tycho. In 1778 Sir William Herschel added further detail to the theory of lunar volcanism. He perceived "three volcanoes in different places

of the dark part of the new moon. Two of them
are either already nearly extinct, or otherwise in
a state of going to break out, which perhaps may
be decided next lunation. The third shows an
actual eruption of fire, or luminous matter." The
"eruption" he described as resembling "a small
piece of burning charcoal, when it is covered by
a very thin coat of white ashes, which frequently
adhere to it when it has been some time ignited;
and it had a degree of brightness, about as strong
as that with which such a coal would be seen to
glow in faint daylight." What is more, he added,
"All the adjacent parts of the volcanic mountain
seemed to be faintly illuminated by the eruption,
and were gradually more obscure as they lay at a
greater distance from the crater."

Sir William
Herschel

From a contemporary perspective it may be
difficult to understand why people failed to question that lunar craters were of volcanic origin.
Although much larger in scale, the formations,
when viewed through early telescopes, indeed
seemed similar to the Earth's volcanic craters—
especially Vesuvius, which has a ring of hills surrounding it. Illustrations based on observations
of the moon under slanting sunlight—with shadows standing out dramatically—showed sharp
peaks where in fact there were none. Moreover,
eighteenth-century science had no clue how the
moon had come about in the first place. Finally,
no one in recorded human history has seen a meteorite fall and create a crater. From time to time
there had been more or less reliable reports on
stones coming from the sky. On March 16, 687,

for example, in China during the Chou dynasty, one observer wrote, "In the middle of the night, stars fell like rain." But in the early afternoon of July 26, 1803, at a time when even the existence of meteorites was under debate, something remarkable happened. Around the French village of L'Aigle in Normandy hundreds of people witnessed a meteorite shower of thousands of pieces, which not only eliminated all doubt about their extraterrestrial origin but also confirmed the possibility of impacts. Shortly after, on December 14, 1807, the Weston meteorite of Connecticut was the first fall recorded (by two Yale professors) in America. President Thomas Jefferson supposedly commented, "I would more easily believe that two Yankee professors would lie than that stones would fall from heaven." Still, this possibility was not immediately applied to lunar craters. Sir John Herschel, who had been the victim of the *New York Sun*'s "blue bat people" hoax, still likened lunar craters to the volcanoes known in France and Italy and stressed that "they offer, in short, in its highest perfection, the true volcanic character, as it may be seen in the crater of Vesuvius; and in some of the principal ones, decisive marks of volcanic stratification, arising from successive deposits of ejected matter, may be clearly traced with powerful telescopes."

But doubt was growing. Some scientists were not ready to recognize what Herschel discerned through his telescope. The English popularizer of astronomy Richard A. Proctor (1837–1888), in his book *The Moon,* struggled with the theory

of volcanic origin. He knew that water was necessary for the occurrence of violent eruptions, and he saw no satisfactory explanation for what he called "the structure of the great crateriform mountain-ranges on the moon." While conceding that at some point in time there may have been large amounts of water on the moon, he saw no such possibility in its current state. In addition, he asked himself how the many smaller craters surrounding the prominent crater Copernicus, for example, might be explained. In the 1870s

A nineteenth-century fantasy of the moon

Lunar landscape
with "full Earth"
in Richard A.
Proctor's *The Moon:
Her Motions,
Aspect, Scenery,
and Physical
Condition* (1886)

Proctor proposed that the lunar craters had been formed by meteors, when the moon was "in a plastic condition." "It may seem, indeed, at a first view, too wild and fanciful an idea to suggest that the multitudinous craters on the moon, and especially the smaller craters revealed in countless numbers when telescopes of high power are employed, have been caused by the plash of meteoric rain, and I should certainly not care to maintain that as the true theory of their origin; yet it must be remembered that no plausible theory has yet been urged respecting this remarkable feature of the moon's surface." Proctor imagined that "under the tremendous heat generated by the downfall, a vast circular region of the moon's surface would be rendered liquid, and that in rapidly solidifying while still traversed by the ring-waves resulting from the downfall, something like the present condition would result."

It sounds like a paradox, but in a convoluted way, more research was needed on Earth to explain the craters of the moon. The geomorphologist Grove Karl Gilbert (1843–1918), who worked for the U.S. Geological Survey, began by studying Coon Mountain (later renamed Meteor Crater), an enormous hole in the sandstone of the Arizona desert. When he compared the volume of the crater and the material on the rim, he concluded erroneously that a meteorite could not possibly have gouged out the crater. Even the meteorite fragments found near the crater could not convince him, and he declared their presence coincidental. We now know that a massive stone weighing 63,000 tons exploded there fifty thousand years ago after crashing at a speed of more than forty thousand miles per hour, about fifty times as fast as a bullet. A decade after the disastrous volcanic eruption on the Indonesian island of Krakatoa in 1883, Gilbert shifted the emphasis of his work to the moon. Ironically, he steadfastly

Barringer crater, Arizona

held that the moon's craters were indeed the result of meteor impacts.

According to a common assumption of the time, meteorites that hit the moon from various inclined angles would have caused oval or "stretched" craters. How, then, could Gilbert account for the craters' uniformly circular contours? In his laboratory he fired clay balls at clay targets under controlled conditions and discovered that the nature of the impact scars depended not only on the angle at which the clay ball hit the target but also on the viscosity of the material and the velocity of the impact. The results, he declared, would account for more or less circular impact holes on the moon's surface.

But it still took time for the new theory to gain credibility, as most scientists in the field found it too radical. The amateur astronomer Ralph B. Baldwin provided further evidence that the moon's craters had resulted from impacts. Scrutinizing photographs of the moon, he had long wondered how the straight valleys or grooves had come about that pointed toward the center of Mare Imbrium. He concluded that they could only have been carved out by massive rock formations during violent explosions. After World War II, he correlated depth-diameter values for lunar craters with holes left by bomb explosions and found similar characteristics. He assembled a list of about fifty terrestrial formations, most of which were subsequently proven to be the results of impact. Baldwin wrote in *The Face of the Moon* (1949), "all observations point to the conclusion that the great majority of the lunar craters were born in gigantic explosions, that these explosions were caused by the impact and sudden halting of great meteorites, and that the main features of the moon's crust were established in the first quarter of its life as a satellite." He also realized that the symmetric nature of the shock wave caused the crater to be circular. While working on his hypothesis, Baldwin must have been aware of the atomic bombings of Hiroshima and Nagasaki, but we can only speculate about the effect this knowledge may have had on him. His book sold poorly but got into the hands of the right scientists and is now commonly regarded as one of the most important books in the history of lunar science.

More recently, Dana Mackenzie has given us a clearer characterization of the nature of such impacts: "It isn't really accurate to think of the meteorite 'digging' a crater; instead, it compresses, fractures, pulverizes, and even melts the rock, and launches it in all directions. These secondary 'bombs' create havoc in their own right. On the moon (though not on Earth) it is easy to spot secondary craters that were created by material ejected from larger impact craters."

Meanwhile, competing theories continued to thrive. In the late 1880s S. E. Peal developed a glaciation theory of the lunar surface. He assumed the maria to be filled with frozen water and the lunar surface to be coated with ice. He imagined that the crust had emitted water vapor, which then condensed around the volcanoes and devel-

oped into icy craters. The idea of an icy moon per-
severed into the twentieth century. In 1925 E. O.
Fountain of the British Astronomical Association
theorized that insulating layers of meteoric dust
protected the moon's crust from the heat of the
sun, resulting in a permanent layer of ice just
below the surface. The German astronomer and
selenographer Philipp Fauth (1867–1941) worked
for decades on a detailed but flawed moon atlas.
He believed in the so-called glacial cosmogeny or
world ice theory—particularly popular during the
National Socialist period—which held that the
moon's crust was a hundred-mile-deep layer of
ice.

Edward G. Davis suggested in an article published in *Popular Astronomy* in 1923 that coral atolls
had built up in the lunar seas, and that some of
the craters were filled to such an extent that even
the central peaks could not be discerned. He calculated an age for the rampart of the crater Copernicus of sixty-eight thousand years. Upheavals of
the crust finally had brought an end to life on
the moon, leaving only the atolls as remnants
of a time long lost. Perhaps the strangest theory
was published by the eccentric Spanish engineer
Sixto Ocampo in the 1949 *Bulletin of the Argentine
Friends of Astronomy Association*—after it had been
rejected for publication in his homeland. Ocampo
was convinced not only that the moon had been
inhabited but that the craters and the lunar rays
were created by nuclear blasts detonated by two
warring races of beings. The different shapes of
the craters, and the existence of peaks in the cen-

ter of some, according to Ocampo, reflected the use of different kinds of bombs. Finally, Ocampo held that the last great explosions on the moon redounded to Earth, causing the biblical Flood.

That which cannot be seen, named, and classified always presents a challenge, even if its presence is clearly evident. So if the visible craters inspired waves of thought and theory, so much more a source of wonder was the far side of the moon. Until a half-century ago, it resisted both observation and naming. The Danish mathematician Peter Andreas Hansen (1795–1874), twice the recipient of the Royal Astronomical Society's Gold Medal, was convinced that the shape of the moon deviated considerably from spherical and that the center didn't exactly coincide with its center of gravity, because of differences in density. According to Hansen, all water and air had moved to the far side, creating the conditions for organic life to evolve there. He was joined by Sir John Herschel, who believed that the far side might contain an ocean of water. While the essence of Hansen's theory has long since been discarded, he was right in thinking that the center of the moon's gravity is slightly askew.

The far side of the moon is commonly called the "dark" side. But the surface is no darker than the face we're familiar with; the expression reflects only our perception of a surface so long hidden from our view. Just like the near side, the moon's far side experiences two weeks of sunshine followed by two weeks of night. We can see a little less than a fifth of the far side under certain conditions, as the result of slight oscillation or "libration"; the rest is forever invisible from Earth. Scientists and laymen have wondered what the unseen side of the moon would be like, and some writers have turned the mystery into metaphor. According to Mark Twain (1835–1910), "Everyone is a moon and has a dark side which he never shows to anyone." A century or so later, a generation of teenagers puzzled over the hidden meanings of Pink Floyd's immensely successful album *The Dark Side of the Moon* (1973), which focused on darker aspects of human existence.

Eventually the mystery of the far side of the moon fell victim to a combination of human curiosity and technological progress. The Soviet Union began a lunar program in January 1959 that eventually completed twenty missions to the moon by the end of 1970. On October 4, 1959, *Lunik 3* was launched, carrying two automatic cameras equipped with a special orientation and electronic guidance system that kept the lenses directed toward the moon. On October 7 an automated command caused the film in the cameras to be exposed as the craft was passing about forty thousand miles above the lunar surface. From the perspective of the cameras, the phase of the moon had shifted considerably. On Earth the moon appeared just a few days into its cycle, but from *Lunik*'s position in space it was nearly full. Photocells detected the sunlit side and triggered the photo sequencing. Over the course of forty minutes, a series of shots captured about

70 percent of the moon's far side. The exposed film was developed, fixed, and dried automatically on board the rocket. It took a few more days until *Lunik* was close enough to the Earth to transmit the images of the negatives to the ground stations. The photographs were enhanced by computers to produce a tentative map of the dark side.

The biggest surprise was that the surface of the far side differs significantly from the near side, with many more crater impacts. It has been calculated that maria occupy close to 30 percent of the near side but barely 2 percent of the far side. One theory focused on the gravitational attraction of the Earth, which is a bit stronger on

After the far side of the moon had been photographed by the *Lunik 3* spacecraft in October 1959, these plastic models were fabricated in the former Soviet Union.

the visible side. The force would have been even greater in the past, when the moon was closer to Earth. The average exposure to sunlight, on the other hand, is about the same on both sides, so this factor could not have been responsible for such a difference. Could the Earth have shielded the moon's visible side from meteorites? Probably not, as the Earth's disk occupies only a tiny part of the moon's sky. Ultimately, scientists ruled out external causes for the asymmetry, reasoning that analogous hemispherical differences in continents and oceans on Earth can be attributed to internal causes. The most likely explanation is that the crust on the far side is thicker, making it harder for molten material from the interior to have flowed to the surface and formed the smooth maria. It is also known that a giant impact struck the near side with such force that it created the two thousand–mile–wide Nearside Megabasin and sent vast amounts of ejecta to the far side.

Once *Lunik* allowed the Soviets to map the far side, they went on a naming orgy, dubbing every visible feature. Thus the Soviets not only put the first man in space, but trumped the United States in other endeavors as well. *Ranger 4,* the United States' first attempt to photograph the far side of the moon, crashed into the surface in 1962 before transmitting any pictures. Still, all was not lost. Later space probes, such as *Ranger 7* (launched July 1964) and especially the Lunar Orbiter program (1966–1967), succeeded in photographing the surface in much greater detail. Finally, in 1968, when *Apollo 8* circled the moon in preparation for the *Apollo 11* landing, the crew aboard were the first to see the far side with human eyes.

CHAPTER NINE

Lunar Choreography

"The higher the moon, the higher the clouds, the finer the weather." "Moon in the north brings cold; moon in the south brings warm and dry." Much traditional weather lore is based on such correlations. Poised at an uneasy intersection of astronomy and popular cosmology, astrometeorological speculations about the moon's effects on the weather are particularly persistent. The London pharmacist Luke Howard (1772–1864), who devised the system still used for classifying clouds, maintained an elaborate record of meteorological observations. He saw a correlation between patterns of barometric fluctuations and the moon's gravitational effect on the atmosphere. Some "lunarists," in turn, thought that the moon created electric or "magnetic" atmospheric disturbances. Robert FitzRoy (1805–1865), twice commander of the *Beagle* (the second time with the young Charles Darwin on board, who commented on FitzRoy's "most unfortunate" temper), later became the head of the British Meteorological Office and championed the idea that both moon and sun exert a pull on the atmosphere. His *The Weather Book: A Practical Meteorology* (1863) includes several references to the moon's impact on the weather. Such phe-

Moon + woods = moon wood?

nomena as "more than usual twinkling of the stars, indistinctness or apparent multiplication of the moon's horns, haloes, 'wind-dogs,' and the rainbow, are more or less significant of increasing wind," FitzRoy claimed, "if not approaching rain, with or without wind." Furthermore, for FitzRoy the moon at the last moment during which the obscure disk is visible "is a sign of bad weather in the temperate zones or middle latitudes (probably because the air is then exceedingly transparent)." Such were highly controversial even at the time. Skeptics of the moon's influence on terrestrial weather point out that a change in weather is always taking place *somewhere* on Earth.

An undeniable point of reference for such assumptions, however, has been the action of the tides, which result from the interplay of the gravitational and centrifugal forces of Earth, moon, and sun. This is the moon's most obvious influence on our world. The daily rise and fall of the ocean water—averaging three feet worldwide—is produced as water is drawn toward the moon and, simultaneously, the Earth's centrifugal force creates a corresponding but smaller bulge on the other side of our planet. When the sun, Earth, and moon line up—whether at the new or full moon—the spring tide occurs in which the moon's impact is augmented by the tidal force of the sun. When the moon is at a right angle to the sun, in turn, tides are at their least extreme, the neap tide.

In fact, some contemporary scientists have found a distinct correlation of earthquakes and teacher briefing a group who were about to go on a small field trip. He had issued a folder, and was explaining its function:

Each of you will have a folder.
It *gives* you a map of the walk
It *gives* you a check list of points to look out for
It *gives* you . . .

The repetitive use of the first phrase and the stress on each 'gives' with a falling mumble after each one subordinated the items he wished to highlight. For questioning, whether of individuals, small groups, or full classes, a great deal of study is required. There is too much use of questions to which we already know the answer, too much rejection of other possible answers, too many gap-filling questions, and so on. The art of questioning deserves study.

It is unlikely that experienced teachers will be willing to practise their reading aloud, yet not only is good reading aloud an asset to any teacher in a school for appetite-whetting, highlighting of key passages, and putting across vivid material which is not to be read by the class, but the aural articulation of a passage is valuable in helping pupils learn the vocal patterns necessary for inner articulation. We should all be more willing to work at this skill.

Finally, there is the more informal but even more important role of the teacher in discussion. He must adopt a way of speaking which is readily acceptable and comprehensible, and yet he needs constantly to be aware of the pupil growth which comes from their interaction with his responses, questions, suggestions. His knowledge of language theory is vital here: it is wrong to think that help is given by stopping and 'correcting'. In fact it is by involved response with the substance of what is being said that he will help the pupil towards richer and more meaningful discussion.

How to talk as a teacher can be an ongoing study for any serious whole-school learning policy.

5. Pupils' Use of Language

The place of reading

Colin Harrison and Keith Gardner

Reading is the most heavily researched single area of the whole curriculum, and yet, paradoxically, it remains a field in which a good deal of fundamental work has yet to be approached, and one of which a great many teachers would claim to be almost wholly ignorant. If this is true of teachers at the primary level, where for many an understanding of reading was founded on a single lecture in a three-year college course, how much more does it describe the position of teachers in secondary schools, who have generally had no training at all related to reading but nevertheless feel conscious that the ability to read fluently is the basis for most school learning, and one of the surest predictors of academic attainment. The Report has made it clear that only a continuously developing in-service framework can ameliorate this situation, but the report itself does provide a more than adequate introduction to current thinking on reading for someone new to the field. The report has certainly stimulated the thinking of many secondary heads of English and senior teachers, and led them to review the situation in their own schools; many are now concerned that they feel ill-equipped to interpret data on reading performance from feeder primary schools, especially when each school seems to have used a different test instrument. In many schools the part played by the remedial department is being reappraised, and senior teachers may feel vaguely uneasy about the *status quo*, but are conscious that the remedial specialists are the only staff members with any claim to an understanding of the problems of poorer readers.

The present section explains briefly something of the reading process, and then identifies some of the ways in which a secondary school wanting to develop the reading of all children, and not simply those who are poorer readers, could set about the task. The intention is not to prescribe, but to point out possibilities and alternatives in what is a relatively new area of enquiry.

The reading process

What happens when a person reads? The simplest answer might be: meanings are extracted from a set of printed symbols. This kind of definition does nothing to explain how a series of blobs of ink come to be interpreted as a message, but it serves to emphasize a crucial point in contemporary thinking about reading, which is that even from the very earliest stages reading is concerned with meaning. The implied distinction here is between two views of reading: the first for convenience we shall call the 'card-reader' model; the second is already well known by Professor Kenneth Goodman's phrase 'psycholinguistic guessing game'.

When a computer card-reader scans a punched card, the combination of holes representing each alphabetic character is read in serial order across eighty columns and the information converted into a string of binary digits. Three crucial aspects of this reading process are: every letter of every word is scanned; each letter is read in strict serial order; the reading process itself is not influenced by text content. Until fairly recently this mechanistic view was accepted or implied in most descriptions of reading. It was assumed that the fluent reader scanned each word, converted the letter-strings into sounds or sound-images, and then matched the sound images with those in memory store; in other words, meaning was something added *after* the word had been phonically decoded, or turned into a sound-image. There is now a good deal of evidence[1] to show that this is by no means a satisfactory account of what happens. A fairly dramatic experiment conducted by Graham Rawlinson[2] will serve as illustration. Rawlinson asked subjects to read aloud a number of increasingly taxing passages from a standard test, the Neale Analysis of Reading Ability. He did not tell his subjects that the final passage had been rewritten so that in every word of six or more letters all except the first two and final two letters were in random order. The following is an extract from one of the partial anagram texts:

Each April, at the re-apapnarece of the
cukcoo in its faimilar hanuts, bird-wachetrs
must mavrel at the acracute flghits
with which birds span the dinstcaes beetwen
their sesnaoal abdoes.

The unsuspecting subjects found such passages surprisingly easy to read, and hardly slowed their reading down at all. Furthermore, they grossly underestimated the number of errors in the text, sometimes reporting that they had noticed only 'a few misprints'. Rawlinson concluded that any definition of reading as a letter-by-letter scanning and matching system was ruled out by what his subjects were able to do. Similarly, it could hardly be the case that subjects were splitting

[1] Some of this evidence is found in E. A. Lunzer and C. Harrison's contribution to G. Steiner (ed.), *Psychology of the Twentieth Century*, Kindler, Zürich, 1976.
[2] G. Rawlinson, 'How Do We Recognise Words?', in *New Behaviour*, **1**, 1975, pp. 336–338.

words up into phonemes before searching for meaning, since the words in the transformed text are made up of highly irregular phonemes which often produce words which sound very unlike the original. No-one would deny that most of us sub-vocalize while reading; what seems to be the case is that in fluent readers this usually follows rather than precedes the grasping of meaning. Similarly, although one would accept that the ability to break a word down into its phonetic components can be valuable when one encounters a new or unfamiliar word, it would appear to be a strategy which a fluent reader uses comparatively rarely.

The computer card-reader, therefore, does not offer a very close parallel with a person's reading; the reading process is infinitely more flexible and can cope with missing information, misprints, anagrams, handwriting styles and type fonts never encountered before, any of which would confuse a computer. Much of this ability to grasp meaning is explained by Goodman's description[1] of reading as a psycholinguistic guessing game: reading is seen essentially as a matter of prediction of what will come next, and it works within a dual framework of rules, one set related to thought and the other to language. Goodman stresses that reading is not a precise process, but one in which intelligent guesswork based on partial information is the modal strategy, with recourse to deeper reflection and re-reading if semantic or syntactic anomalies occur. This description of the reading process is supported by a good deal of recent research, and although Goodman's own work on the analysis of oral reading 'miscues' is widely quoted, Frank Smith[2] and Gibson and Levin[3] offer a more balanced view of the contribution of individual researchers.

Goodman used the term 'miscue' to refer to errors made in oral reading. He was not happy with the implied value judgement associated with the term 'error' since his central contention was that most mistakes in oral reading demonstrate that the reader is in fact grappling with the meaning of the text and is not simply 'careless' or 'lazy'. For example, in a story written in the first person a reader might well substitute *He yelled* when reading the words *I yelled* in a story. This substitution would only marginally affect the meaning, and might simply be related to the child's greater familiarity with the third person narrative. Similarly, Goodman cites a child's saying *Might as well study what it means* instead of *Might as well study word meanings first*. The reader's version was not self-corrected, since he was quite happy with the sense. In fact, the reader has transformed the grammatical structure of the last three words of the sentence and

[1] K. S. Goodman, 'Reading: a psycholinguistic guessing game', in *Journal of the Reading Specialist*, USA, May, 1967.

[2] F. Smith, *Understanding Reading*, New York, Holt, Rinehart and Winston, 1971.

[3] E. J. Gibson and H. Levin, *The Psychology of Reading*, M.I.T. Press, Cambridge, Mass., USA, 1975.

changed the vocabulary, but he has retained the basic meaning. Goodman also noticed[1] that errors involving intelligent substitution were far more common than those related to letter reversals, which produced a nonsensical reading. He found, for example, that the substitution of *was* in the sentence *The chicken saw the cat* was rare, even though the misreading retains a verb and uses each letter of the original. More common were errors of the type *The windows were full of pets and kittens* for *The windows were full of puppies and kittens*. In this example the miscue *pets* can only be introduced by a reader who has read and understood the final part of the sentence. Further research on beginning readers has shown that provided they are offered texts which approximate to natural language, even five- and six-year-olds can use the same approach to good effect. In fact Goodman has cited examples of children who cannot cope with the pre-readers in a scheme which have sentences of the form *Ride in Sue* or *Ride in here*, but can read with confidence the more advanced stories in which there is more complex grammar and vocabulary. The difference is that the more advanced stories also have an increased load of meaning, and within this richer framework of grammatical structure, vocabulary and meaning the reader can use graphic information much more effectively. In short, it is easier to read sense than nonsense.

Reading, then, is not to be seen as a precise letter-by-letter or word-by-word scanning process. Clearly, without graphic input there would be no reading, but a model which stresses only the decoding of graphic input will fail to correspond with the experimental evidence available. The reading process is more complex, more dynamic and more flexible than this. In the past it has been normal to regard writing as an active language skill and reading as essentially passive, like listening. In fact, reading can be much more of an active process than listening. When a listener misinterprets or fails to hear part of a lecture or broadcast, it is usually impossible for him to recall the exact form of an utterance or its context after more than a few seconds, and he has to choose between reflecting on what was missed and missing more discourse, and ignoring the omission and keeping up with succeeding sentences. A reader, on the other hand, has much more control over how he chooses to process the text an author has written, and this fact is admirably crystallized in the phrase used in the Report which describes effective reading as 'the active interrogation of a text' (8.10).

What is a good reader?

The answer to this question is closely related to the concept of an active interrogation of a text, and can be linked directly to our description of the reading process. Once a child has begun to move

[1] K. S. Goodman, 'Analysis of oral reading miscues: applied psycholinguistics', in *Reading Research Quarterly*, Volume I, 1969, pp. 9–30.

away from the first reading scheme, in which there is naturally a heavy emphasis on graphic information, the meaning of a passage comes more and more from an accumulation of partial information obtained from graphic cues on the basis of which tentative decisions are confirmed or rejected. When they are rejected, for example because a phrase seems anomalous, the text is scanned again in order to find an alternative meaning which is syntactically and semantically acceptable. The good reader is therefore one who (a) has a fairly extensive working knowledge of possible language structures, and (b) is alert to anomalous meanings when they occur. It is important to note that the first requirement does not imply that all children should be taught rules of grammar from the age of five; the emphasis is on the availability of structures, not on an awareness of the rules which govern them. In fact, up to the age of nine or ten, much of what children encounter in books is no more complex linguistically than their own speech, and therefore this criterion is fulfilled.

The emphasis in the definition of a good reader therefore rests on an awareness of whether or not what is being read makes sense. This point might seem banal, and yet it is one which deserves emphasis. What is suggested is precisely that a bad reader is one who is unaware of whether or not he is understanding a text. Every reader sometimes has the experience, perhaps when fatigued, of suddenly realizing that two or three pages of a book have been read without a single fact or incident impinging on the consciousness; in one wit's words, the text has gone 'in one eyeball and out the other'. This is the most glaring example of ineffective reading, and yet there is good reason to suspect that describes much reading in schools, especially if the vocabulary and grammatical structures are so unfamiliar to the reader that the guessing-game process breaks down. The reader then is in a more hopeless position than the six-year-old child who cannot cope with a pre-reader because the language structures are unfamiliar. At least in that case the meanings of the individual words are known to the reader. With an 'O' level course book in physics, biology, or geology, for example, one would also anticipate that many of the crucial words would only be partly understood, thus introducing extra difficulty. It would of course be nonsense to suggest that a textbook should only be used when its content has already been assimilated; the reason for introducing this example is rather to throw light on some of the special difficulties associated with some textbooks as opposed to other kinds of school books.

Another important determinant of effective reading relates to a concept which will be developed and expanded later in this chapter, and this is the notion of flexibility. A crucial research finding is that poorer readers are also those whose rate of reading is least flexible.[1] In other words, those who read every passage at exactly the same pace,

[1] E. Rankin, *The Measurement of Reading Flexibility*, International Reading Association, Newark, Delaware, 1974.

regardless of whether it is complex and dense, or a relative simple story, are also the lowest scorers on tests of reading comprehension. Another aspect of this phenomenon was noted by the Schools Council *Effective Use of Reading* project team in a videotaped session in which a group of twelve-year-olds discussed the content and implications of a series of passages which together linked up to produce a story. It was noticeable that one reader was always the first to declare that he had read the instalment and was ready to discuss its implications. In fact his answers showed him to be one of the poorest readers in the group. Normally, one perhaps tends to associate fast reading with effective reading, but in this instance it was clear that while the boy's *scanning* of a passage was swift, the quality of reflection on what was read was low, which in turn meant that his understanding of incidents, and their cause and effect, was much more shallow than that of other members of the group. The good readers, however, slowed down their reading when they encountered what they took to be a section which needed closer attention, or perhaps it would be more accurate to say that they allowed more time for reflection between bursts of reading, in order to tease out meanings.

This process of re-reading and reflecting on points in the text which present difficulties or anomalies is absolutely crucial, and yet it needs to be stressed that until very recently the notion of going back and re-reading chunks of text was frowned upon in some circles and by some teachers. Experiments with tachistoscopes[1] had demonstrated that many poor readers did not read a line of print in the same way as fluent readers. Instead of having about five or six fixations per line their eyes dotted about forward and backwards searching for cues, and not surprisingly they failed to build up meanings in the way better readers could. Gradually regressive eye-movements came to be regarded as harmful, and in some reading clinics students were trained on a tachistoscope in such a way that they came to avoid regressions during reading. The term 'regressive' often has unfortunate value implications, and it has become necessary to reassert the importance of re-reading as a concomitant of effective reading. One would really want to go further and suggest that a total absence of regressive eye-movements would almost certainly imply a low level of reflection on the text being read.

Reading in school—beyond the reading scheme

During the early 1970s three major research projects were funded to consider specifically the use of reading in British Schools and the ways in which teachers encouraged the reading development of those beyond the infant stage. In Scotland, James Maxwell's wide-ranging project has given attention to the whole school age-range, and in

[1] A tachistoscope is a device that can present information to the eye for a controlled period of time.

England two Schools Council projects, 'Extending Beginning Reading'[1] and 'The Effective Use of Reading'[2] have focused on 7–9 and 10–14-year-olds respectively.

Extensive observation in the classrooms of a number of schools has revealed that reading seems to play a much smaller part in learning than one might have expected. In top junior, first-year secondary and fourth-year classes only about 10 to 12 per cent of overall lesson time in what many teachers regard as the core subjects of English, mathematics, science and social studies involves children in actually reading. In fact, the team discovered that at top junior level the salient language activity was writing, while in both secondary age-groups the language activity which accounted for most classroom time was listening to the teacher. As an example of what commonly happened one can cite an English lesson in which an eleven-year-old child spent 35 minutes working steadily on a reading comprehension exercise; the girl was fully involved with the task, and completed it successfully, finally re-writing her answers in an exercise book from an initial version in rough. The total time she spent in reading and re-reading the passage was only $1\frac{3}{4}$ minutes. She did in fact spend most of the lesson writing down and re-writing her answers. *The reading comprehension exercise, as tackled in many classrooms, is much more a writing exercise than one concerned with developing reading skills or comprehension.*

Many teachers at first-year secondary level feel that since books are so expensive they can only be purchased for groups which are approaching public examinations. Others have phased out textbooks altogether in certain subjects because they do not suit the course the teacher has planned. Many teachers also feel that since expecting a child to learn from a book is a chancy business, it is more profitable to put their faith in the spoken word and teacher exposition. When the Effective Use of Reading project began, its aims included obtaining information about what teachers expected children to gain from that reading, what in fact the children gained, and accounting for any mismatch. What was half-expected was that the team would discover that teachers tended to have unrealistically high expectations about what children were gaining from their reading. This was by no means what was found. In the event, it seems that teachers do not expect children to learn much from reading, and since, despite their high cost, books only play a relatively minor supporting role in many classrooms it is not surprising that the teacher's low expectations are confirmed.

The question which must be squarely faced is 'Is the book dead?'

[1] T. Dolan, 'The incidence and context of reading in primary and secondary schools', in D. Moyle (ed.), *Reading: What of the Future?*, UKRA conference proceedings, Ward Lock, 1974.

[2] E. A. Lunzer and W. K. Gardner (eds.), *The Effective Use of Reading*, Heinemann Educational Books (forthcoming).

The answer from the project team is emphatically negative. Their concern is rather to suggest how printed sources can be used much more effectively in classrooms, and can be given more prominence in learning, not less. One crucial reason for this response relates to a paradox which has been noted in many secondary schools. The same teachers who have tended to phase out the printed word in the lower age groups are also conscious that in many fourth-, fifth- and sixth-form courses the ability to accommodate new learning through the printed word is vital in coping with their subject. Many science courses, for example, now include options involving projects which demand wide critical reading. Some 'A' level courses have 'unseen' examination questions in which experimental data, results and tables are offered for criticism and explanation, and the answer is essentially a high-level reading comprehension exercise for the student. How adequately can a diet of workcards and blackboard notes prepare a child for the very heavy demands of textbooks in the later years of secondary schooling? Even though a teacher accepts that writing is itself a vital part of the learning process, and that discussion and exploratory talk should play a significant part in the curriculum of the first two years, this must not be taken to imply that the development of reading should be neglected.

Developing reading across the curriculum

In order to simplify some of the organizational issues related to implementing a policy for developing reading, it seems worthwhile to group the issues under two main headings: those which must be decided upon at the administrative level within a school, and those on which a headteacher or head of curriculum studies might wish to initiate discussion, but which would relate more to decisions at departmental or classroom level.

Areas for decision-making at administrative level

1. *Obtaining and using data on reading performance.* If a headteacher wishes to know the reading ability of each child in a first-year intake of two or three hundred children there are a number of ways he can tackle the problem. Currently, giving each child a screening test (discussed in Chapter 7) is a solution which many heads are considering very seriously. The advantages of the school's administering a test in the first term of secondary schooling are many: firstly, the data from feeder primary schools may be difficult to interpret, either because certain tests used are unfamiliar to the new school, or because even when the same test has been used in different primary schools results indicate that it was administered or marked in a different way; secondly, a child might have improved or regressed in reading since he or she was last tested, and the test results could well be six or nine

months out of date; thirdly, a screening test can provide valuable information about which children need fairly immediate help with reading, and in a mixed ability setting the cooperative nature of working could mask problems for crucial months.

The decision to screen a whole year is not one which can be taken lightly. Tests cost money, and although many are pirated or completed in pencil and reused, they are demanding in terms of organization and classroom time, and take many man-hours to mark. A useful booklet giving details of the most widely used reading tests in this country is available through the United Kingdom Reading Association.[1] There is considerable literature on reading tests, and the main issues are summarized in Chapter 7. A point which needs emphasis, however, is that all tests are inaccurate, and only give an estimate of true competence. At the same time the object of a screening exercise is to spot children who need help, and it is not therefore necessary to contemplate using a test with a diagnostic battery of subtests. A word-recognition test, for instance, tells you nothing directly about how well a child can read sentences. It is designed on the principle that, generally speaking, good readers will tend to obtain high scores and poor readers will tend to obtain low scores. It is then up to the test user to home in on the warning light of a low score and to establish the specific nature of a child's problems.

This leads us to another important aspect of testing: test data itself is worthless unless applied. There is little point in going to a great deal of trouble collecting data if it is not used effectively in terms of classroom action. Many primary teachers despair because their secondary colleagues take little or no account of information which has been painstakingly collected and put on to record cards for the benefit of the new teacher. This is often because the administrators in the school fail to make the information available to class teachers, but it may also relate to a desire on the part of the new teacher to avoid prejudging a child's ability. Heads and their deputies have a clear responsibility to make sure that test data from any source is put to the best possible use, and this means for example, sharing it with other departments besides English. In the classroom observation study, a researcher from the Effective Use of Reading project found more than one case of first-year pupils who had convinced a teacher in one subject that they could not read (and therefore received simpler tasks) but who read perfectly adequately for a teacher in a different subject. The problem here was one of communication between teachers.

If all teachers understood the problems associated with reading in their subject, and were sensitive to the problems of each reader, there would clearly be no need for reading tests. But while this is not the case, and where a head feels the information would be put to good use,

[1] J. Turner, *The Assessment of Reading Skills*, UKRA bibliography No. 2, UKRA, 1972.

a case can be made out for some sort of screening, though it need not necessarily be formal. A cloze test constructed and administered on a classroom basis by each English teacher gives quite an effective pointer to those children with reading problems, and it offers a less threatening situation than the 11+ atmosphere created by many norm-referenced tests.

A cloze passage is simply one in which every nth word of a normal text is omitted, and the reader is asked to choose a suitable replacement. The following is an example:

> Once upon a time, in a .... small county
> town, at a considerable .... from London,
> there lived a little .... called Nathaniel
> Pipkin, who was the .... clerk of the
> little town.

At first sight a cloze test might seem to be little different from a multiple-choice comprehension test, but in fact it has many more uses. The reading project team have used cloze tests as the basis for silent reading and group discussion activities, and in helping to measure textbook difficulty. In the present case, a cloze test can be valuable in giving the teacher some diagnostic information about how well or how poorly a child is reading. For instance, in the passage above one regularly finds that a poor reader will put *boy* in the third gap, while better readers will offer *man*. The teacher can use cloze test responses as a window into the child's mind, and returning to Goodman's model, can examine how effectively the reader is using the cues supplied for this guessing game. The word *man* can only be guessed by working backwards from the information that the person was a clerk, and therefore *boy* would not be likely. Apart from any difficulty associated with the word *clerk*, it is also the case that poorer readers tend to use what are called 'backward acting' cues less effectively than good readers. The potential value of cloze tests in the classroom is all the greater because they enable a teacher to learn something about the reading process in the reader who tackles the passage, and this is certainly not a claim one would make for multiple-choice comprehension tests.

2. *Provision for poorer readers.* Broadly speaking, there are three alternatives available as a basis for giving selective help to those who are poor at reading or who have special problems: remedial classes, selective withdrawal from lessons, and provision within the normal class grouping (the broader issue of pupil grouping is discussed in Chapter 7).

The advantages and disadvantages of having remedial classes are well known: resources in terms of both appropriate materials and specialized teacher expertise can be centralized, the feeling of constantly failing can to some extent be overcome, and learning tasks can be matched more easily to the whole group than in a mixed-ability situation. The main disadvantage is that the remedial class functions more as a long-term isolation ward than an out-patients' department, and those in such classes rarely emerge to share in the greater school learning community, or possibly emerge to join it after a three-year course of treatment totally unprepared by their workbooks and regimented lessons for the experience of a Mode 3 CSE course.

Selective withdrawal functions more as an out-patients' clinic, and given on a rotated timetable allows poorer readers to receive help while missing no more work in normal lessons than the child who goes to flute lessons twice a week or who plays football for the district team. This system has the disadvantage that the rest of the class know where their neighbour is going, but this need not be a stigma. In some schools it is recognized that this kind of help may be needed by bright children as well as slower ones. If it is decided to establish some kind of reading centre there is no reason why it should not be an attractive place with special displays and furniture, equipped with, at the very least, tape-recorder playback facilities, such that it would come to be regarded as a place it is at least interesting and at best fun to visit.

While one would imagine remedial teaching would for many children consist of one-to-one tuition in phonics or word-attack skills, it is important to mention that technical innovations such as the synchrofax and taped stories (possibly school-produced) can be used by secondary children without constant supervision. Reading while listening is currently recognized as an invaluable aid for some poorer readers, who have hitherto been held up as much by poor articulation as anything else, and the technique has brought about significant gains in reading ability.[1] Another much-discussed innovation is the use of fluent readers, usually parents or sixth-formers, as helpers of poorer readers. This help would normally simply involve listening to a reader and correcting errors, and very little is known about its potential benefits or dangers. The experience of workers in the adult literacy field, however, makes it clear that giving this kind of assistance is a delicate business, and unless the learner is highly motivated and the helper sensitive and sympathetic the result could be harmful, and destroy confidence rather than build it up. The implications of this are that despite the compelling needs of older non-readers, a successful bond is more likely to be established with a first- or second-year child, and even then one would only allow volunteers who seemed well suited to the task to give tuition. One shrinks in horror at the attitude of some schools in which this exercise is regarded as a sort of social work, begun on the basis that even if it does not work out the sixth-former will have learned something.

[1] M. H. Neville and A. K. Pugh, 'Context in reading and listening: a comparison of children's errors in cloze tests', British Journal of Educational Psychology, 44, 1974, pp. 224–232.

Making provision within normal class groupings for poor readers is in many ways the hardest problem of all to tackle, since it places on each individual teacher the responsibility for offering appropriate remedial help. The most cogent argument in favour of this strategy is that since each subject has its own special language demands, and often its own special language, it is appropriate that the problem should be tackled at departmental level. The excellent monograph *The Learning of Reading*, produced by teachers at Clissold Park School, demonstrates that teachers can inform themselves about reading development and go on to share the responsibility for achieving it in every child, rather than leaving it to the remedial department. In fact, whether remedial classes or selective withdrawal are used, it is the theme of this book that the responsibility for developing reading must be accepted at the departmental and classroom level and by every department and class teacher.

3. *The use of library and resources areas.* Before considering issues at the departmental level, some comment must be made on staffing and using library/resource areas. Norman Beswick's comprehensive books[1] deal with this complex subject very fully. The intention here is merely to raise some issues specifically related to reading and the accessing of printed sources.

The Effective Use of Reading team have studied the use of resources in topic and project work, and have made time-lapse ciné films of the use in a school library of card indexes, reference books, and special collections of materials. What they found has confirmed what many librarians suspected, which is that the crucial resource is not the indexing system but the librarian. When no librarian is present, researching a subject very rarely extends beyond the consultation of encyclopaedias, even with fourth- and fifth-year groups. In many school libraries a great deal of time is devoted to updating and cataloguing books on the Dewey system, and yet it may be consulted no more than two or three times a day. The plain fact is that from the child's point of view the system is difficult to understand, time-consuming to use, and can often lead to frustration. Since the child's need is usually immediate, any book not actually available on the shelves is of no interest to him.

Librarians have tackled this problem in a number of inventive ways. In some schools a booklet produced by the county library service is hung at strategic points; it contains an alphabetically ordered list of topics together with their Dewey number. Over the geographical section a librarian can mount a map of the world, with major countries labelled and numbered. This kind of innovation can ensure fuller use of library resources and help children to find the books they seek more readily.

[1] N. Beswick, *School Resource Centres*, School Council Working Paper 43, Evans/Methuen Educational, 1972; and *Organizing Resources*, Heinemann Educational Books, 1975.

Special collections on local topics, and 'project trolleys' containing perhaps thirty or forty books culled from the shelves or borrowed from the county library stock can also assist children with a specific task. The greater problems, however, are those associated with educating children to use the books they have chosen in a purposeful manner. The library is a complex system, and can be a valuable aid to learning in every subject, and yet in many schools children receive no more than one lesson a year on how to use it. This is comparable to a mathematician's giving one lesson a year on logarithms. Children need to be taught how to use the indexing system, dictionaries, encyclopaedias and special collections, but this is only the first step. The crucial skill of a good library user is not that of finding a book, but rather that of being able to reject a book which is unsuitable. Too often, a child sees the topic under consideration mentioned in an index and immediately assumes that the book he holds will be appropriate for his purpose. It may be that his topic is an account of the ecological benefits of the bicycle, but for him any reference to bicycles will do, even a list of Olympic Games results from 1964. As teachers we often assume that children know how to use books to obtain information, and since most teachers are reasonably fluent readers it is sometimes difficult for them to imagine the child's problems.

A video recording made by Tony Pugh of the Open University demonstrates the difficulties even a fluent reader can have in using a book effectively. His film shows a sixth-former who had been given a standard university psychology textbook and a fairly straightforward task on the lines of 'What contribution did Miller make to work on human memory?'. The girl read the question, took up the book, consulted the list of chapter headings, and began reading at page 1. Twenty minutes later she was on page thirty, and when the tape ran out she was still reading steadily, not having reached the point where Miller's name was first mentioned. Pugh's interest was in how students used books when making notes, and he had not expected to encounter a situation in which the reader did not even reach the section on which the notes would be based!

As teachers we need to be aware that for many children the only readily available strategy is to begin reading a book at the first word and to carry on to the last. By contrast the reader may care at this point to consider the route by which he or she has arrived at this line on this page; it is unlikely that it has followed word by word the path from the editor's introduction. If a different route has been taken this will be because the reader has a specific purpose in mind and is using a strategy suited to it. What Pugh's subject failed to do was to use an appropriate strategy. If children are to use books effectively, in the library or in the classroom, they must be taught two things. The first is that there are different ways of reading and using a book; the second is that they should consider what approach best suits their particular task. No doubt Pugh's sixth former could use an index perfectly well; what she failed to do was to realize that she was in a situation which required her to use that skill.

Areas for decision-making at departmental and classroom level

1. *Defining the problems at departmental level.* One clear outcome of the Report has been the need to redefine the maxim 'Every teacher is a teacher of English'. This used to be interpreted in terms of all teachers taking a share of the responsibility for improving the grammar, spelling, and punctuation of those they taught. The new interpretation rests more on the necessity of every teacher realizing that most learning takes place through the medium of language, and that each subject has its own special language demands. In some subjects such as biology and geology, there are special difficulties associated with vocabulary. In most science subjects children have to come to terms with the passive voice and highly objective writing. In history one might suggest the need to obtain information from a variety of sources and to assess bias in writing as special areas of difficulty. The implication from the Bullock Report is that teachers must ask themselves what special language problems are associated with reading in their subject, and how these should be approached.

If a head teacher or head of curriculum studies wishes to establish how well-prepared a department is to undertake an analysis of the part reading plays in its courses, he could ask the following question: 'Are books used in your subject in the same way they are used in English lessons, or do you think there are some differences?' If, despite the transparently leading nature of the question the answer given is 'No', he may assume the ground is somewhat stony. If the answer is 'Yes', there is a basis for action, since the next and most difficult task is to try to articulate the differences.

As a framework for discussion a department could be encouraged to take two year groups, perhaps first and fourth initially, and attempt to clarify what kind of written sources are used in lessons over the year's course, and the part they play in the curriculum. For example the department could note the use of worksheets, textbooks, reference books, the blackboard, posters, etc., and then examine how each was used, whether to generate activities, give instructions, provide model answers to be copied down, to convey facts, to support the teacher's factual presentation, for diagrams, or as a basis for note-taking. The group could then begin to consider whether the use of resources could be improved, and whether it emerged that specific skills were necessary for a child to cope successfully with the course.

As the next step it would be valuable for this analysis to be extended over each year group up to sixth-form level, and the teachers would be able to ask themselves to what extent the language demands of their courses were sequential, and adequately prepared for by work in preceding years. At the same time, if other departments were attempting a similar exercise it would be advantageous for cross-fertilization of ideas to take place, and for other teachers to gain some insight into their colleagues' conclusions and special problems. One would ask, for example, whether making notes in one subject does in fact involve different skills from making notes in another, how primary sources are used, and at what age most children can cope with the task successfully.

2. *The need for flexibility in reading.* It has already been emphasized that in reading the concept of flexibility is an important one, and that a child who reads every text at the same speed and in the same way is likely to be an ineffective reader. Research evidence to support this is found in the work of Laurie Thomas and Sheila Augstein of Brunel University, who designed a machine which produced a visual record of the way in which different readers read a passage for a specific purpose. They were able to show that certain approaches to a task were much more profitable than others.[1] They used their reading recorder with college students and examined which types of approach to reading an article resulted in high scores on multiple-choice comprehension tests, and which approaches led to their producing a coherent summary of a passage. They discovered that a 'smooth read' straight through the passage, followed by a sentence-by-sentence read resulted in accurate multiple-choice answers, but was not a good preparation for writing a summary. A summary based on this type of reading tended to consist of an arbitrary selection of facts, and lacked any appreciation of the overall argument and implications of the article. By contrast, they found that a successful summary writer could also score highly on multiple-choice questions, and tended to remember the facts of the passage for a longer period. The following is an example of the reading strategy of a good summary writer:

1. Smooth read straight through the article.
2. Sentence-by-sentence read of whole article.
3. Paragraph-by-paragraph read of whole article.
4. Pause for thought.
5. Search read (through whole article; much skipping forwards and backwards).
6. Pause for thought.
7. Final smooth read.

The volunteers who worked with the reading recorder gained from the insights they obtained into their own reading habits, and as well as improving their summary writing most went on to achieve higher grades under the continuous assessment scheme of their college course.

[1] L. Thomas and E. S. Augstein, 'An experimental approach to the study of reading as a learning skill', in *Research in Education*, **8**, 1972.

Since so many people have encountered references to speed reading courses, it is appropriate to make some comment about advisability of attempting to teach children to read faster. Much of the literature on speed reading has been based on armchair psychology and sold more on railway stations than in university bookshops. Two myths which have sprung up need to be firmly uprooted: the first is that one can greatly increase reading speed without decreasing comprehension; the second is that one can take in the gist of a page by attempting to fixate an imaginary line running down the middle of that page.

The first myth has been given some credence partly because many speed-reading courses only test recall of factual details such as dates and proper names. If a student learns to scan a text quickly it soon becomes possible to register such details in half the time it would take to read the whole passage, and therefore he will feel he has read it twice as quickly. In fact, what has happened is that he has really only half-read the passage, and if faced with the task of summarizing it would no doubt do it ineffectively for exactly the same reasons as Thomas and Augstein's student who read only for factual points.

The second myth has been viewed with some amusement by those with some knowledge of visual perception. When the eyes fixate a text the area clearly in focus is an ellipse not much larger than a two pence piece; certain distinctive letter-features can be perceived up to about one-and-a-half inches from the point of fixation, and the partial information from one fixation is confirmed by partial information from the next. In normal reading five or six fixations per line are necessary to ensure that a line is read completely, so one can imagine how little information would be gained by attempting to read straight down the middle of a page. The reader of this chapter may care to see how much can be understood of the following page if to begin with he or she covers all but a two-inch wide strip of print down the middle of the page. The chances are that only the occasional phrase which has been left intact will carry much meaning, and this will not be organized within any overall framework of argument. If someone were able to train themselves to achieve the considerable feat of reading down the middle of the page one would expect them to gain less from their reading rather than more.

In fairness to the authors of the less disreputable courses on speed reading one should point out that some courses do encourage flexibility of reading rate according to the task in hand, which is much more important than pure speed. Of course there are children in remedial groups whose reading is held up because they fixate every word on a line, and find it difficult to grasp the overall meaning of a sentence because its meanings remain fragmented. It may well be beneficial to encourage increased reading speed and fewer fixations per line in such a case, but this by no means implies that this emphasis should be continued once a person is reading fluently.

3. *Encouraging flexibility in reading.* It would be foolhardy to assume that all children will spontaneously develop a set of strategies for different reading tasks, although many do unconsciously in employing the 'principle of least effort'. For example, as every English teacher knows, many children approach a comprehension exercise by looking at the questions first, and if these do not demand an overall appreciation of the passage the text as a whole will remain unread. Nevertheless it is worthwhile noting three of the most widely quoted alternative strategies to a straightforward 'smooth read' of a text, not least because some experts firmly believe they should be specifically taught to children. The strategies are scanning, skimming, and SQ3R.

Scanning is the term used for the rapid searching for a specific piece of information in a text, such as a date or a name. Skimming involves a closer inspection of a text, in order to obtain a general impression of its content or argument. This would perhaps involve concentrating on a few key words or phrases, but would not go into details. For example, a reader might turn to the correspondence pages of a journal to establish whether the editors have printed his letter; this would involve scanning. He might then quickly skim all the letters in order to decide whether any of the issues raised interested him enough to merit a full reading. There is some doubt about whether readers can be taught to scan more effectively, but skimming is much more clearly a strategy which can be taught, and it is reasonable to suggest that both strategies are valuable at some stage in practically every subject. Skimming is particularly important in situations where a reader is using a number of books in order to complete some topic or reference work. It can help to avert the situation in which a child is half-way through copying out or making notes on a passage before he realizes that it is not after all dealing with the issue he wanted to raise.

SQ3R is a more complex approach to reading a passage for information than simply scanning or skimming. The mnemonic was first used by Francis P. Robinson[1] to describe a system introduced to help college students read more effectively. The characters stand for Survey, Question, Read, Recite, Review, and they can be glossed as follows: Survey—skim the passage to gain an idea of the main points (and to ensure it contains the information sought); Question— formulate questions to guide your thinking during the intensive reading (these would depend on your initial purpose); Read—read the passage intensively, pausing for thought and re-reading as necessary; Recite—say aloud the answers to your questions, the salient facts, a summary of the overall argument, or whatever (alternatively these could be recalled without speech, or written down); Review—go over what you have learned (this stage could

[1] F. P. Robinson, *Effective Study* (new edition 1961), Harper and Row, New York, 1946.

also include evaluation of the content, thinking of connections with previous learning, considering the possibility of bias, etc., as appropriate).

Some authors (see for example Niles[1]) have suggested that this technique is too difficult for school pupils to learn, but this argument can be countered by those who have simplified some of the stages where necessary and introduced SQ3R from the upper years of junior school onwards.

4. *Hierarchical models of comprehension and their relevance to classroom planning.* Much of what has been written above relates to improving what a reader gains from his or her reading, or in other words developing reading comprehension. The authors have suggested that children should be taught how to use books more effectively and to be aware of the possibility of using their reading in different ways. Good reading has been described as being related to the quality of the reader's reflection on what is read, and the strategies associated with SQ3R and the Thomas and Augstein work are valuable in that they provide a framework in which this active interrogation of a text is more likely to occur.

There remains the question of whether teachers should be seeking to enhance specific subskills of comprehension. Table I shows three examples of comprehension hierarchies, based on the work of Davis,[2] Barratt, and Dolan and Waite.[3] In what is a celebrated article Davis suggested on the basis of a complex statistical analysis of test results that reading comprehension consists of a number of separately isolable subskills. Since 1944, any number of comprehension taxonomies have been suggested by experts in the reading field, and that of Barratt is included because it is well known in this country through the Open University's Reading Development course.[4]

The great attraction of such a taxonomy is that it offers the teacher a ready-made strategy for improving children's reading. If the teacher can use a diagnostic test to establish the areas in which a child is weak on comprehension, for example in drawing inferences, it is then simply a matter of giving special attention to that subskill in order to improve overall reading performance.

It was partly with the intention of developing such a diagnostic test that Dolan and Waite carried out a major study on reading compre-

---

[1] O. S. Niles, 'Comprehension skills', in A. Melnick and J. Merritt (eds), *The Reading Curriculum*, University of London Press, 1972, pp. 228–234.

[2] F. B. Davis, 'Fundamental factors of comprehension in reading', in *Psychometrika*, **9**, 1944, pp. 185–197.

[3] T. Dolan and M. Waite, 'Does comprehension consist of a set of skills which can be separately examined?', in A. Cashdan (ed.), *The Content of Reading*, UKRA conference proceedings, Ward Lock, 1975.

[4] See A. Melnik and J. Merritt (eds), *Reading Today and Tomorrow*, ULP, 1972.

hension, using a list of putative subskills which has much in common with those of Davis and Barratt.

TABLE 1. *Three examples of comprehension skill hierarchies*

Davis (1944)	Barratt (undated)	Dolan & Waite (1974)
Word knowledge		Word meaning Word meaning in context
	Literal comprehension	Literal comprehension
Reasoning	Reorganization Inference	Inference from single strings Inference from multiple strings
Following organization		
Recognizing literary devices		Understanding metaphor
Focusing attention on explicit statements		Perception of salient points
	Evaluation Appreciation	Evaluation

Unfortunately, they discovered that despite careful controls it did not appear to be possible to isolate specific comprehension subskills; readers were broadly speaking good, average or poor at comprehension, and did not show any consistent weakness in any one area of comprehension. Furthermore, it soon became clear that the work of Davis had been the subject of the most heated debate, during the course of which the distinguished inventor of the statistical method Davis had used had denied the validity of his results, and pronounced that they showed clearly that reading comprehension was made up of one general ability.

This section, therefore, may be a slightly disappointing one for some readers, but one should not infer from it that teachers should give up attempting to improve children's comprehension. The debate between the experts is partly a theoretical one concerning what can be accurately measured. If teachers go on attempting to enhance what they feel are specific subskills it is likely that they will tend to improve overall competence, and the orientation of their teaching towards specific goals may also be an asset, simply because of the structure it provides.

Other methods of attempting to develop comprehension have been investigated by the Effective Use of Reading team. In particular, they have taken and extended some of the ideas for group reading and discussion in Christopher Walker's book *Reading Development and Extension*.[1] The work is to form the basis of a separate publication, and will elaborate a number of Silent Reading—Group Discussion (SRGD) techniques. These involve structuring reading situations in which a small number of readers (at least two, not more than ten or twelve) have to analyse the words, organization and meaning of a passage. This is done orally, although the passage has been read silently, and an initial evaluation of the techniques has proved very promising. As an example of one of the approaches used one could take Group Cloze. For this activity groups of two or three readers consider a cloze passage which has been prepared in a similar way to that on page 89. The object of the activity is for the small group to discuss which word seems to be the most appropriate for each gap, and any guess must be supported by references to the surrounding text. When two small groups have agreed on their versions, they join up and discuss points of difference, again referring to the text to support their points.

In general the team have found that while children enjoy doing this kind of exercise it is improving their reading, and also results in fairly good recall of the passage and a high level of overall evaluation of what has been read. It seems likely that SRGD techniques will become more widely used in this country.

5. *Worksheets, textbooks and vocabulary control.* In the school as a whole, printed materials tend to be used for two main purposes: for generating activities and for learning. To some extent this distinction is an artificial one, but it allows a number of points to be made which relate to one type of text rather than another.

Activity generating materials may well not be textbooks at all, but rather workcards or worksheets. The crucial need here is for teachers to monitor pupil performance continually and to rewrite or edit where necessary. In the hurly-burly of the classroom a series of questions from children about a certain workcard may simply be regarded as related to the difficulty of the task itself when in reality the fault may lie in the workcard. What is needed is accurate record-keeping on the teacher's part to help to spot poorly written or poorly organized workcards. The research of the Effective Use of Reading team into levels of prose difficulty showed that teacher-produced worksheets for one first-year mixed-ability group were comparable in difficulty with the standard 'O' level textbook in the same subject. Writing simple prose is far from easy, and it poses extra problems for a subject specialist, since he may feel unhappy about oversimplifying a complex subject, and will introduce qualifications and exceptions which will baffle the child who has yet to grasp the central point.

[1] C. Walker, *Reading Development and Extension*, Ward Lock Educational, 1974.

A further important point is that instructions need to be in simpler prose than anything else a child reads. There are two reasons for this. Firstly, if a child only partly understands an explanation or a definition, that partial understanding may nevertheless be an important interim stage in learning; if however he only partly understands an instruction he is quite likely to fail altogether to engage in the learning experience which the teacher had in mind. Secondly, one must bear in mind the fact that readers find it more difficult to understand prose which they find uninteresting, and there is very little of generic interest to most children in the instructions to set about a task, even if the task itself is a stimulating one.

A teacher must be aware however that it is much easier for a child to say 'Sir, I don't see what we've got to do' than for him or her to read a worksheet carefully, so it is important not to allow the children too much freedom to by-pass reading altogether, which a harassed teacher might be tempted to do. He could persuade the child to discuss the worksheet with a friend and then to say what they thought the task seemed to be. If the teacher gives the impression that children are not expected to work from the worksheet the group will soon conform to that expectation.

When books are used as a stimulus for learning, rather than for generating activity, the crucial factor to be borne in mind is that of vocabulary. Research into the readability of texts has shown repeatedly that vocabulary is more important than syntactic complexity in determining text difficulty, and this does not surprise many teachers. It is part of every department's planning commitment to decide what concepts are to be introduced each term, and what content should be covered. One might suggest that this kind of analysis could be extended to include specialist vocabulary. In some schools the teachers have been enterprising enough to provide a glossary for children containing words which have caused difficulties. Difficulties are not only associated with words which the teacher expects to cause problems; sometimes a simple word is incomprehensible simply because of the way it is used. Consider for example words such as *salt, relief,* or *solution*. These are used quite differently from one subject to another, and while they may be used consistently within a subject, the teacher needs to take account of the child's day consisting of perhaps six or seven different subjects, each with its own vocabulary and prose style.

6. *Motivation and readability.* A classic study on motivation in reading was carried out by Shnayer,[1] who investigated to what extent interest in a passage affected the scores children were able to achieve on comprehension tests. Shnayer first gave a standard reading comprehension test to hundreds of pupils, on the basis of which he assigned

[1] S. W. Shnayer, 'Relationships between reading interest and reading comprehension', in J. Allen Figurel (ed.), *Reading and Realism*, 1968 proceedings, Volume 13, Part 1, Newark, Delaware: I.R.A., 1969, pp. 608–702.

them to one of seven notional ability groups. Next he gave them a number of comprehension tests on other fairly difficult passages, this time asking the children to state how interesting or how boring they found the test passage. Shnayer reported two important findings: firstly, with the exception of the bottom group (which clearly had fundamental problems with reading) there were no significant differences between groups in test scores on passages which were rated as of high interest—in other words, some children's reading ability seemed to improve when they were interested in what they read; secondly, Shnayer found that his good readers found more passages interesting—the poor readers only occasionally found a difficult passage interesting.

The importance of motivation in reading is seen as the more problematical when one considers that a school textbook has been defined (by John Holt) as a book which no one would read unless they had to. In a pilot study for the Schools Council project, Harrison[1] found science and social studies texts were assessed as more difficult than sources from other subjects by both teachers and readability formulae. The special problem with science texts is that apart from often being the hardest of all to read and comprehend, they tend to be used as a support for homework and may hardly be dealt with in class at all.

The terms *supported* and *unsupported* may be used to describe the conditions under which a book is used. At the supported level, the teacher is present and either actively explains what is in the text or is at least readily available to deal with any problems. At the unsupported level, however, the child is isolated from adult support either working in a resource area for example, or doing homework. Now it is not uncommon for a reading expert to suggest that the texts for reading at the unsupported level should be about two years simpler in terms of 'reading age' than those a child encounters at the supported level, and there are various measures of prose difficulty which would help the teacher to select a suitable text. What seems to be the case, however, is that the most difficult texts in the whole curriculum are used more often at the unsupported than at the supported level.

Those interested in measures of readability may consult a monograph by J. Gilliland[2] which introduces the subject quite fully. A guide for teachers which will present data on a cross-validation study of twelve formulae, together with the results of a survey of prose difficulty levels in different subjects is being prepared by the first author of this chapter, and should be available in 1977.

For the teacher not so interested in detailed analysis, but who would wish to gain some information about how well a group can cope with a textbook, one can suggest giving the class a cloze test

[1] C. Harrison, *Readability and School*, Schools Council project discussion document, University of Nottingham School of Education, 1974.

[2] J. Gilliland, *Readability*, University of London Press, London, 1972.

based on a passage from the book. The procedure might also be considered as a revision exercise after some work has been covered. In either case the results will give the teacher some insight into how well individuals as well as the whole group have coped with the text.

7. *Additional issues for the English department to consider.* This section is brief because many of the special problems of developing reading within the English department of a school have been covered in some detail by the Bullock report. In addition, two Schools Council projects, Children's Reading Habits and Children as Readers, have examined the reading of literature in depth, and Kenyon Calthrop's excellent book *Reading Together*[1] offers many more useful ideas for tackling fiction in English lessons than there would be room for here. Most English teachers would accept the importance of fostering a habit of reading, being seen reading themselves, discussing books, encouraging library usage, developing classroom libraries, and maintaining a school bookshop.

One point which needs to be made concerns the value of accurate record-keeping in developing and encouraging reading. The Effective Use of Reading project team felt it was no accident that of all the schools they visited, the one which seemed to be making most progress in developing reading at all levels was the one which had the most comprehensive records. Each child had a book-review notebook, in which details of teacher-initiated and private reading were kept, and apart from screening test scores the Head of English had in her mark book the names of each novel read by every child, and thus could see at a glance which children were not making progress in reading, or found it difficult to find a book they would read by choice. The children were invited to share the sense of achievement they felt when they had read a book, and were encouraged by the teacher's genuine interest in what they read. This teacher was of the opinion that there are in fact very few reluctant readers; there are however many who are reluctant to read what we as teachers want them to read.

The final issue to be dealt with is one which is currently hotly debated—the use of reading 'laboratories'. The Canadian firm of SRA is the best-known one in the field, but English rivals are available, published by Ward Lock, Longman (Reading Routes), and Drake Educational (Language Centre). The Bullock Report criticized this type of approach to reading on two main counts. The content of the laboratories was felt to be arbitrary, with the emphasis on factual content at the expense of any literary merit. Also the laboratories purport to improve reading ability, and there was only evidence of a very limited sort available to suggest that they do so.

The first criticism still stands, and those who wish to attempt to improve reading within the context of specific subject areas will not use a 'laboratory' approach. The second criticism, that the evidence

[1] Heinemann Educational Books, 1974.

of improved reading ability was flimsy, has been challenged by the work of a primary head, Roy Fawcett. His Ph.D. study examined the gains in reading ability over a year of 1200 top junior, first-year secondary and fourth-year secondary pupils, half of whom began the academic year with a one term burst of SRA and half of whom acted as controls and simply took their normal English lessons. He found that in each age group the boys and girls who took the SRA course outperformed the control groups on two separate comprehension tests. Six months after the experiment was over, he gave a late post-test, and discovered that far from reverting to their previous form, the SRA groups maintained, and in some cases increased their superiority over the controls. Not surprisingly, these results have proved unpalatable to many educationalists, particularly English teachers. The Effective Use of Reading team, with whom Roy Fawcett worked, are concerned to stress that his results in no way suggest a blanket endorsement of SRA. The experiment was conducted simply because the materials are already in many schools and it was felt more information was needed about what they might achieve. E. A. Lunzer, the project co-ordinator, has stated that the results show clearly that it is beneficial to devote a specific effort to improving reading, but that there could well be equally effective methods which would allow teachers the freedom to introduce their own material, rather than being restricted to bundles of facts of somewhat doubtful educational value. His own feeling is that the SRGD techniques referred to above hold this kind of promise. What SRA does is what any reading improvement programme must do, and this is to encourage careful reflection on what is read, and to encourage flexibility in reading.

Conclusion

The main argument of this section is that reading cannot be allowed to develop wholly spontaneously, any more than learning can develop wholly spontaneously. The task of the teacher is to provide the conditions within which learning can take place, and if that learning is to be through reading it is likely to be much more effective if the teacher understands each child's strengths and weaknesses, and has designed reading assignments in such a way as to use the strengths and remedy the weaknesses. The need, therefore, is for an accurate monitoring of each child's reading, using all the information available in the school, and supplementing it during the school year. The teacher needs to have a clear idea of the reading demands within his or her own subject, otherwise it is impossible to gauge the level of a child's success or failure. Finally, having isolated the nature of the specific demands the teacher must assist the child in meeting them. This final injunction may seem banal, and yet it represents a challenge which we are only beginning to meet in any organized manner.

Subject reading strategies
Introduction

How does one assist the child in meeting the reading demands? The previous section has made it clear that we suffer from inadequate understanding of the reading process, we give pupils too few reading opportunities, we provide unsuitable material, and we do not prepare pupils to read for meaning. The majority of teaching of reading beyond the basic skills is with narrative material; the majority of reading for learning is, however, with non-narrative material.

We talk about 'the ability to read', but a moment's thought reminds us that there are very different *kinds* of reading—and pupils are not necessarily equally competent in all kinds. One person takes in most non-fiction readily, but finds prose fiction difficult to read. More commonly there are many readers who find fiction easy, but struggle with various kinds of non-fiction. There are differences between the texts in these two kinds of writing at every level; it is not merely the addition of a difficult technical vocabulary that marks off informative prose. Computer analysis[1] has shown that the frequency of even the most common words varies from genre to genre. *I, he, you, me* appear far more frequently in general fiction, whereas *than, by, which* appear far more frequently in learned and scientific writings.[2] There is also a very marked difference in sentence structures and lengths; all categories of imaginative prose have shorter mean sentence-lengths than those of informative prose. According to an American computer analysis the average sentence length for 'fiction: general' was 11 words compared with 21 words for 'learned and scientific writings'.[3]

Reading in literature, science, the humanities, and mathematics obviously shares many overlapping skills, but there are other skills which are more particular to one or the other areas of reading. It is partly a matter of vocabulary, but also of specialized ways of treating ideas, and different patterns of writing.[4] The differences are worth considering and I would suggest that different subject Departments in a school select key passages from important texts, whose reading demands can then be analysed and compared. The American reading specialist Nila Banton Smith carried out such an analysis on a large number of school texts used in science, social studies, mathematics, and literature,[5] then classified the reading response patterns

[1] H. Kucera and W. N. Francis, *Computational Analysis of Present-Day American English*, Brown University Press, 1967, p. 276.

[2] *op. cit.*, pp. 275–293.

[3] *op. cit.*, p. 376.

[4] A. Sterl Artley, 'Influence of the Field Studied on the Reading Attitude and Skills Needed', in *Improving Reading in Content Fields*, W. S. Gray (ed.), University of Chicago Press, 1947.

[5] Nila Banton Smith, 'Patterns of Writing in Different Subject Areas', *Journal of Reading*, USA, 8 October 1964.

typically required for each. For instance, we must take into account the very high importance in reading social studies texts of the need to analyse the content for cause–effect relations, comparisons, and sequence of events. There are also special problems of bringing together different viewpoints, facts mixed with opinions, quotations and exposition, narrative and comment. A special difficulty, which is almost never met in the main 'reading subject', English, is the taking in of illustration and caption so that they are read with the text. The subtle comprehension skills required for reading literature are intensively taught. From the earliest emphasis on narrative in younger classes through to analysis of character and atmosphere, the teacher is demonstrating how the reader can concentrate on certain signals and infer from these. Is there anything like this training and emphasis on either the printing conventions or the comprehension skills of non-literary writing? I am convinced that there is not. It is no wonder that pupils find reading outside English so difficult, and that they are so inept at getting the sense from the text. We simply do not teach what the Report calls 'different kinds of reading strategy'. Two of the Committee's recommendations are vital here:

68 The majority of pupils need a great deal of positive help to develop the various comprehension skills to a high level.

69 An important aspect of reading behaviour is the ability to use different kinds of reading strategy according to the reader's purpose and the nature of the material. Pupils should acquire the skills which will free them from dependence on single-speed reading.

Most of the reading done within the curriculum for learning is expository; most of the higher reading tuition is in literature. It is this mis-match between need and offer which is at the heart of the difficulties in reading to learn in the secondary school.

The strategies required need to be available to all subject teachers. Their basis is simply in the idea that we must learn to interrogate print to get the meaning out of it. Straight left-to-right, no skipping, no going back, misses the main point of print. One of the special things about it, as compared with talk, TV, or tapes, is the reader's own control over his experience. The print on the page will stay there and not change, therefore the reader can go back, or look forward; he can pause to think, make notes, or compare; he can jump and join sentences from different places. Thus reading should not be a passive occupation, pupils should be encouraged to react to a text and check back, guess what is coming, and see if the guess is confirmed by what is then read. They can think of the writer as being present and themselves as an interviewer: 'What evidence have you for that statement?' 'Will you be giving more detail about that idea?' and so on.

I am now going to suggest specific ways in which a subject teacher can encourage this interrogation by taking the pupil back into the text and showing him the way through it. Whilst I am sure that some specific teaching is required from time to time as part of a careful basic programme to make sure the pupil has a growingly elaborate foundation for his work, I suggest that the main occasions for the teaching of reading are in context and as far as possible when the teacher has helped the pupil into a sequence of work in which the need to read, to get meaning from print, is felt by the pupil. This contextual teaching requires careful preparation from the teacher, for he must call on his teaching knowledge unexpectedly, whilst in the midst of some subject matter. The teaching of reading as part of the teaching of subject learning requires a ready reservoir of teaching knowledge to be drawn on as required by the needs of the pupil and his text. I have therefore tried to divide up the kinds of help that can be given, even though in practice it may be necessary sometimes to re-synthesize them.

The sense of words

Later sections consider the teacher's task in introducing vocabulary, in building up phonic word-attack skills, and in help with spelling. All these teaching activities combine to help a pupil with the reading of words which give him difficulty. However, part of the teacher's task is to help pupils develop ways of deducing the meaning of unfamiliar words in a text. When a pupil asks the meaning of a word, there are two normal teaching responses: the first is to say 'Look it up in a dictionary' and the second is to give the meaning ourselves. I maintain that it would often be better teaching to help the pupil find ways of deducing the meaning for himself. This can be done by specifically teaching word detective work at various times. The word can also be referred to in other subject lessons, to reinforce or elaborate its meaning in context, establishing meaning by context clues. It is worth sorting out the range of kinds of context clues: syntactic and semantic:

1. *Syntactic context cues: functions of words.* The pupil stuck over a word should be encouraged first to see what function it must have. This does *not* require the teaching of the parts of speech, but simply a common-sense comment. Consider possible problems in the following:

> In the crowded hovels of the poor, lacking privacy, sanitation and a water supply, few diseases could have been more horrible, but most people were unwilling to be removed to hospital and would only go as a last resort.[1]

The reader not knowing 'hovel' would see first that it must be a thing; whereas the pupil who did not know 'ultimately' in the following example would know it could not be a thing:

> 'The food question ultimately decided the issue of this war', Lloyd George wrote after the armistice.

[1] This and other examples in this section can be found in Norman Longmate, *Alive and Well*, in *Topics in History*, Penguin, 1970.

In this more difficult case, the reader would conclude that the word had something to do with 'decided'. Other factors besides word order help pupils to guess at the function of a word. Inflections obviously help. Some inflections in the two passages are very reliable: e.g., *d, ed, ing, s, ly*. Markers (capital letters and structure words) also indicate what kind of word is being dealt with. A noun, for instance, is indicated by *the, a, an*, a cardinal number, or a possessive.

2. *Semantic context cues:* Our way of writing employs a number of devices for giving the meaning of a word or phrase the writer feels might cause difficulty to his reader. A second-year maths class were working with pin-boards, creating shapes by placing rubber bands round various combinations of the pins. The pupils were working from a maths textbook. A number of times a pupil would raise a hand. The teacher would go over:

'Please, Sir, what's "ver . . . ti"?'
'"Vertice". You remember. We learnt that last year. It's a corner, like this.'

Sir rapidly points to one on the pinboard. A look of satisfied understanding spreads across the pupil's face. Sir retires, pleased at his teaching skill. But has he been so skilful? Certainly he's explained *what the pupil needed to follow the instructions*, but:
(a) It is doubtful whether the pupil is likely to remember the word, as his eye was not taken back to it.
(b) He has directed the pupil to a meaning outside the book, without encouraging him or her to search for that meaning in its context.
(c) He has lost an opportunity to give the pupil some strategies for guessing the meaning of the word: e.g., indicating the stem 'vert' which occurs in many other words, and could serve as a memory prompter.
(d) He has in effect agreed with the pupil that reading is difficult.
   In fact the text in the book read:

The position of a shape on a pinboard can be given by stating the co-ordinates of the *vertices* (or corners).

The writer of the textbook used the device of giving the meaning of new or difficult words in brackets. The teacher should have asked the pupil to take him to the word, and then to read on, and skip back. He would thus have solved the instructional problem just as quickly *and* taught a general point for the future. Apart from brackets, other semantic cues are 'following sentences'.

Suppose your counter was at L and you threw a blue 2, taking you to N. What number must you now throw if you want to 'undo' this: that is, go back to L, where you started?

This example shows that the word to be explained ('undo') need not look difficult but still might need clarifying. Other ways of defining word meanings are by example or synonyms. In all these cases the reader has to read ahead, suspending a sense decision on the troublesome word for a while. At other times words will be defined by description, comparison, or contrast. The most important skill is to work the meaning out by the overall sense, the position of the word. The teacher should not simply give the meaning of the word, thus short-circuiting the reading process. He should help the pupil deduce it from the context clues.

If the reader were stuck at 'absorption' in the following passage, the teacher should be able to help him through the passage.

> A burning sensation in the stomach suggests that it is having difficulty in digesting food. If you have eaten unwisely, the food resists being prepared for further absorption. If you are worried, the powerful juices which break down the food into starches, proteins, and sugars, do not flow readily.

The pupil would have little difficulty pronouncing 'absorption'; he might not be much helped, but the sound might remind him of 'absorb'. Certainly the '-tion' would come out clearly. That would help him look at its parts. He might realize that '-tion' can indicate an action. If he had learnt prefixes carefully he could recognize that the prefix 'a-', can mean 'into'. Therefore 'an action into'. He is still stuck, and will need to look at the word's position. The puzzling word follows 'further'—therefore it must mean more of something already named. In this sentence the food is resisting further *something*. In the final sentence the stomach is having difficulty *digesting* the food. Then unwise eating stops further *digestion*, presumably. And this is confirmed by the final sentence. If one is worried (clearly another difficulty like eating unwisely) the juices do not flow to 'break down' the food—yet another phrase for digestion. Thus the position suggests that 'absorption' is something that happens to food as part of the digestive process.

The same process can be used to work out far more difficult words. Here is a piece from Siegfried Sassoon's memories of the First World War: *Memoirs of an Infantry Officer*. He has been describing the weary return home by moonlight:

> After this rumble of wheels came the infantry, shambling, limping, strag-gling and out of step. If anyone spoke it was only a muttered word, and the mounted officers rode as if asleep. The men had carried their emergency water in petrol-cans, against which bayonets made a hollow clink; except for the shuffling of feet, this was the only sound. Thus, with an almost spectral appearance, the lurching brown figures flitted past with slung rifles and heads bent forward under basin-helmets. Moonlight and dawn began to mingle, and I could see the barley swaying indolently against the sky. A train groaned along the riverside, sending up a cloud of whitish fiery smoke against the gloom of the trees.

Supposing a pupil doesn't know 'indolently', near the end, or has only a half-formed idea about it. There is the suffix: '-ly' always ends a word that describes an action (an adverb), such as 'smoothly' or 'happily'. This therefore describes which action? Coming between 'swaying' and 'against the sky', the word must tell us how it is swaying. Then we must guess. How can barley sway? Presumably slowly or quickly. The whole mood of this piece is slow and very quiet, and there is no mention of wind. It would be reasonable, therefore, to deduce that 'indolently' means 'slowly'. This is not far from the dictionary meaning of 'lazy' for 'indolent'. So it is possible to get very nearly the right meaning—certainly enough of it to follow the passage.

Unknown words certainly hamper reading, but the skilful reader develops ways of deducing all he needs to know. The teacher's job is to help the pupil do this.

Signal words

The key to reading argument is, as I have stressed, seeing relationships. The key to relationships between ideas is a rapid and unconscious understanding of connectives, which the Americans rather nicely call 'signal' words. It is curious that virtually no teaching time is given to these words. Pupils who have devoted classroom hours to discussing the atmospheric associations of various adjectives, may never once have discussed 'accordingly' or 'nevertheless'. As the bulk of the pupils' reading will have been in fiction, which has a very low use of such words, the pupils will have had neither experience nor tuition when they come to tackle such connectives in their subject reading. And yet, as one writer shrewdly pointed out, there is less difficulty in reading if the reader comes across an unfamiliar noun, such as 'elephant' or 'helmsman', than if he comes across a connective, such as 'since' or 'although', without fully grasping its function. A sentence in an historical text in which 'despite' is not understood is rendered meaningless. Even an apparently easy small word can be difficult, and yet vitally important to the sense. How often do any of us teach the sense of the word 'if'? Yet consider these three uses from a school textbook:

(a) If the range was close, the guns fired grape-shot or cannister.[1]
(b) If the Americans were going to fight against a great power like Britain, the Congress had to set about raising an army.[2]
(c) If British soldiers had suffered much, the Loyalists had suffered more.[3]

None of them are simple conditional uses ('If it rains, I'll stay indoors'). (a) has the meaning of when. The sense in (b) is really very different. By this stage in the book it is in fact clear that the Americans were going to fight. The point of the 'if' clause is the nature of 'a great power'. 'If', here, really means 'as' or 'Because of the fact that'.

[1] R. E. Evans, *The War of American Independence*, *Cambridge Introduction to the History of Mankind*, CUP, 1976, p. 17.
[2] *ibid.*, p. 12.
[3] *ibid.*, p. 41.

Paraphrased, the construction could be rendered: 'As Britain, against whom the Americans were to fight, was a great power (with all that that means in terms of a professional army), Congress had to . . .'. In other words, this sentence is simply not a straight conditional. 'If' is used in the sense which the Shorter Oxford English Dictionary gives as 'given or granted that', a sense with which our pupils are less familiar. Example (c) is similar, except that the conditional is even weaker. The reader knows by the time he has reached this sentence that the British soldiers had indeed suffered much. The sentence means something like: 'Certainly the British had suffered much, but in fact the Loyalists had suffered more.'

It is clearly insufficient to expect word recognition to be adequate for connectives. Equally, pupils have inadequate experience from fiction, and must be helped to see the *function* of these words. I find the following analysis a helpful one. It classifies connectives by the kind of signal they give the reader about the sense relationship.[1] Obviously these are not precise or exclusive categories, but a consideration helps the reader understand the structure of ideas and to respond to the writer's intentions:

1. *Go Signals.* These indicate a continuing idea, an equivalent example, and the same line of thought:
and
first, second, third
next
furthermore
likewise
in addition
similarly
moreover
at the same time
also

2. *Caution Signals.* These warn the reader that the writer considers the point following requires concentration; it is likely to be a conclusion or summary:
thus
therefore
consequently
accordingly
in retrospect
hence
in conclusion
in brief
as a result

[1] Adapted from H. Alan Robinson, *Teaching Reading and Study Strategies,* Allyn and Bacon, 1975.

3. *Stop Signals.* These are stronger than the former, and clearly indicate that a statement of special significance is to follow:
without question
significantly
without doubt
hereafter
unquestionably
absolutely

4. *Turn Signals.* The reader is warned that he is approaching an opposing idea, or a change in the direction of the discussion. They are all varieties of 'but', and I consider are worth special emphasis:
yet
on the contrary
nevertheless
notwithstanding
on the other hand
otherwise
in spite of
although
despite
conversely
however
I find these are frequently not recognized for their real meaning. In the following sentence, for instance, very many of the pupils arrived at precisely the opposite meaning to that intended:

> Although Hinduism is the main religion of India, most of the Indians in Britain are in fact Sikhs.

5. *Relationship Signals.* These are the most important. They articulate the structure of ideas, and above all they clarify the argument. The most important are:

cause and effect	because
	since
	so that
	accordingly
condition	if
	unless
	though
illustration	for instance
	for example
apparent contradiction	despite the fact that
	although

I do not suggest that the teaching of the analytic groupings is itself a great deal of help. However, I believe that they should be taught specifically to provide an underlying understanding. This is best done both by taking the words out of a passage, and by working to a passage from the words. One device is to find a long form for the compressed meaning of the word, e.g. 'although' in example 4 means: 'We have just learned one thing which will make the next thing I am going to say surprising to you.'

This specific teaching is useless, however, without being allied to contextual teaching whenever print is used in the content subjects.

Paragraph sense

Our emphasis on reading texts through without re-reading, our withdrawal from the teaching of reading after the initial stages, and our avoidance of non-narrative material leaves the pupil to cope unaided with more complex argumentative writing—the stuff of school textbooks and virtually all information sources! In such writing, paragraph structure offers a problem. Narrative is basically linear, but argument rarely is. The central problem of reading non-narrative paragraphs seems to me to be summed up by this comment by a linguist:

> Problems of clarifying meaning are constantly with us, too pressing to be evaded. If we find it hard to understand well, the fault is not altogether that of the writer or speaker. Even at its most lucid, discourse is inescapably *linear*, doling out scraps of meaning in a fragile thread. But significant thought is seldom linear: cross references and overlapping relationships must be left for the good reader to tease out by himself. Much, also, must be 'read between the lines'.[1]

Thus the problem for the reader is re-organizing the linear thread of discourse into the three-dimensional structure of thought. It is assumptive teaching to presume that this is understood, even unconsciously, by the pupil. It is over-hopeful to presume that the thirst for knowledge, even in a well-motivated project, will itself lead pupils to develop the necessary skills. These cannot be truly isolated; they cannot be fully taught. However, considerable help can be given by analysis of sense structures as specific teaching and by adequate preparation before reading assignments. The subject specialist should have studied the structure of the paragraphs of the material he is asking the pupils to read. His prime task is to help them see the main idea of each paragraph and the ways other sentences support it.

Most paragraphs have a main idea, and the experienced reader readily picks this out, usually without thinking about it very consciously. When a reader has found the main idea, the supporting ideas seem to fall into place in one's understanding quite naturally. The key sentence may give an answer to the general question of the paragraph; it may define a term which is essential to the argument; or it may sum up the point of a series of examples or details. Obviously

[1] M. Black, *The Labyrinth of Language*, Pall Mall Press, 1970, Penguin, 1972, p. 21.

the main idea sentence may come at any point in the paragraph. Sometimes it is not placed obviously. Here for instance are two paragraphs, from an essay by an art critic, Eric Newton, called *Looking at Pictures*:

> Perhaps one of the most valuable and unexpected results of what is roughly known as 'modern' art is that it has given us all a shock and compelled us to ask a question. We may admire Picasso's pictures or we may detest them (almost everyone seems to pick on Picasso as the most extreme example of modernism, so one may as well quote him as a type rather than as an individual); but all of us, with the exception of the arrogant philistine—to whom this article is not addressed—agree that he needs explaining. And the explanation we all hope for is in the answer to a question.
>
> Let me put the question in its simplest form—the form in which the intelligent but uninstructed layman, anxious to learn rather than to scoff, would phrase it. 'I like pictures to represent something I can recognise: I like them to be more or less realistic—if that's the right word. I also like them to be beautiful, though I'm not such a fool as to think that the beauty of a picture is the same thing as the beauty of the object or objects it represents. My difficulty is that modern art seems to me to fail too often in both respects: for me, it lacks both realism and beauty. Yet responsible critics praise it. I assume, therefore, that there must be some quality they find in it that I don't. Please tell me what that quality is. I'm willing to believe that I've missed something but I want to know what that "something" is.'

The first paragraph has the main idea at the very start: '. . . one of the most valuable . . . results of . . . modern art is that it has . . . compelled us to ask a question.' The rest of the paragraph defends and explains that point. The paragraph ends with a link, and in the next paragraph we think we are going to get an immediate statement of that question, and probably at the start of the paragraph again. In fact, the question is not stated until the penultimate sentence: 'Please tell me what that quality is.'

Inexperienced, nervous, or superficial readers frequently miss the main idea, and pick as the most important something that catches their eye because it is well known, or startling.

For instance, in this section of an article by Gillian Tyndale in *The Guardian*, titled 'Gillian Tyndale talking about easy marriage',[1] I found many sixth-formers thought that the writer was writing mainly against difficult mortgages. Housing was an easy subject to understand, and it leapt to the eye. The main idea had become subordinated in the pupils' minds to what was only an example.

> Why, when divorce is protracted, often expensive, and occasionally (even today) unobtainable, when to get a job you need an employment card and to have a baby in hospital you have to book months in advance, when a passport requires two reputable referees, to obtain a mortgage on a house is a complex manoeuvre, and to go to university requires a sustained effort of form-filling and form-studying apart from the examinations themselves
> —why, in this intricate, inevitably docketed existence that is civilised

[1] Reprinted in *Relations Between the Sexes*, The Humanities Curriculum Project, Heinemann Educational, 1970.

life, do we treat one of the most important steps most people ever take with an airiness bordering on levity?'

There is also often confusion between the main idea and a topic. In some textbooks, for instance, as in some magazine articles, a word or phrase is printed about a paragraph or group of paragraphs, telling us what this paragraph is going to be about. However, it is often not the main idea, because it does not convey the gist of the content. For instance, an article might have 'The Balanced Diet and Calories' as a topic heading. The main idea of the first paragraph could prove to be an explanation of what a calorie is, and the second could define a balanced diet. Thus the top heading would be a guide towards the main idea, without being it: it would be the topic of the paragraph.

Although a writer does not usually plan a paragraph in any particular way, it is possible to see that each paragraph of factual and argumentative writing especially tends to have one of a fairly limited number of patterns. The pupil's ability to grasp the logical structure of the ideas is helped if he is shown what these patterns are. It is also a considerable help for such procedures as note-taking, summarizing, abstracting, and, of course, the comprehension questions of examinations. The underlying patterns show the relationships between the main paragraph ideas, their supporting points, and the examples or details that illustrate those points.

1. *Lists.* After a general statement the writer can give a number of examples, which are of broadly equal importance and could be rearranged in any order. You could draw a diagram of such a paragraph like this:

In these cases not only is the precise order of the listed examples not important, but also the number of listed ideas does not matter: an editor could take a couple out or put another in without spoiling the logic of the paragraph. It is important for pupils to realize that in note-taking or summarizing they must find a general description for all the examples, and not pick out the one or two which especially appeal or are rather better known.

2. *Chronological order.* In historical writing the ideas and details are obviously arranged sometimes in the order in which they happened. You could show the simple time chain in a diagram like this:

3. *Comparison and contrast.* It is sometimes harder to follow the thread of a writer's ideas if he uses neither 1 nor 2, but is showing the truth of an idea which has a comparison built into it, such as: 'In many ways ours is the most successful civilization that has existed'. Such a paragraph is likely to have many comparative words like *more, less, never, so, rather than, so many,* etc. Pupils sometimes lose the thread of the writer's point of view in such examples.

4. *Cause and effect.* Often a writer is building his argument round the relationships between cause and effect. He may be giving a list of effects and deducing their cause, or describing a cause and then listing its effects. Frequently the thread of the argument is complicated by alternative causes being compared.

5. *Expansion.* Perhaps the most common pattern is one of the loosest, in which a main idea is supported and expanded. Sometimes the writer builds up to the main idea, and at other times he states the main idea first, and goes into great detail to expand it. It is important to watch out particularly for the function of 'signpost' phrases like 'Nevertheless . . .', 'However . . .', 'On the other hand. . .'. Sometimes these phrases mark the turning point in the argument.

Obviously such a catalogue of organizational patterns is artificial. However, if a reader is to grasp the argument of a piece of writing, it is important that he senses the logical structure of the passage. This usually means seeing how secondary ideas support or explain the main idea, and how examples or details back up the secondary ideas. An important skill is to be able to sift the supporting detail from the main ideas and within the supporting detail to see that which is essential and that which is unimportant. Of course, nobody draws diagrams when they read, but a diagrammatic analysis may help pupils see how our unconscious mind should see the *relationship* of facts to each other and of the facts to the ideas. It is well worth getting the feel of this supporting structure, for much mis-reading results from not sensing this relationship.

Consider these two paragraphs from a widely used second-year humanities textbook. They are on the basic effects of climate on man's life. Each has a logical structure which the reader must sense if he is to follow the argument:

Man does not like cold climates. Polar and sub-polar regions have the lowest human population densities of all. In these areas people are very likely to catch respiratory diseases. Also the shortness of the growing season cuts down the amount of food which can be produced, and it may be that the long periods of darkness and lack of sunlight have other physical and psychological ill-effects on man, about which we do not yet know much.

High temperatures, when combined with high humidity, allow plants to grow very rapidly so that more than one crop can be harvested per year. The warmth also cuts down the need for food and shelter, so that high densities of people are possible in tropical and subtropical countries. However, in these warmer climates, insects, fungi and bacteria can reproduce in large numbers so that there is an increased chance of the spread of disease in man, crops and domestic animals. In colder regions the winter serves as an annual check on reproduction, but there is no such seasonal control in the tropics.

The first paragraph makes a statement: *Man does not like cold climates.* Then it gives evidence to prove this: *Polar and sub-polar regions have the lowest human population densities.* It then gives three reasons for this: *increase in respiratory diseases; reduced amount of food produced; possible psychological effects.* And for two of those reasons it gives causes: *shortness of growing season; long periods of darkness and lack of sunlight.*

The reader must sense that there are four levels of logic here, which can be shown diagrammatically, the lowest explaining the one above it (except for the increase in respiratory diseases, which the writer leaves you to infer is caused by the cold):

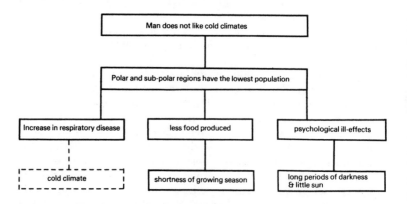

The three reasons explain the basic fact, and that justifies the main idea. In this kind of pattern you can cut off the diagram at any low level and still have the basic sense quite firmly. If you wanted to shorten intelligibly, however, you would really have to remove *all of one level.* Similarly, when summarizing, it would be wrong to include, say, the reduced food and leave out the psychological ill-effects, each of these being of the same level of significance.

The second paragraph is far more complicated. Although it is mainly about warm areas, it is not a simple catalogue of advantages, but contrasts the good with the bad, an example of pattern 3. The writer explores four results of 'high temperatures . . . combined with high humidity', and contrasts the good with the bad. Further in the last sentence he is contrasting one of the results with the colder regions.

This argument can also be represented in a diagram. The dotted lines lead to a point which is not *stated*, but which the reader is expected to *infer*. The *x* indicates a contrast point.

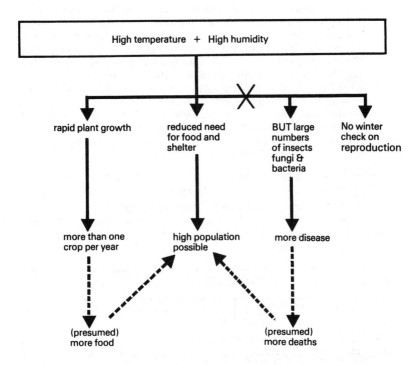

It is noticeable how many key signpost words there are in this paragraph. This is in itself an indication of the greater complexity of the argument. Simple listings, as in the previous paragraph, require fewer signpost words. We find: *when* (with the sense 'if'); *so that*; *so that*; *However*; *so that*; *but*. In each case, *so* signals an effect. *However* warns the reader that a contrast is to be expected, and that the paragraph is swinging round to look at the bad effects of the heat.

It is interesting that the last sentence looks at first as if it is changing the subject back to colder climates, the polar and sub-polar of paragraph one. But it is not. 'In colder regions . . .' presents a contrasting condition not experienced in warmer climates, and the final sentence is actually giving a reason for the fact in the sentence *before* it.

Here is a passage from a more advanced book. It is likely to deter the reader because of the rapid introduction of difficult words, but, as in many mathematical or scientific arguments, the *point* is actually very simple, virtually common sense. Many readers get lost, however. One picked out the point about mis-spellings and claimed the author was saying that it is a bad thing that careless people let mis-spellings get into books:

> What is commonly called perception is a mixture of perception, illusion, and hallucination. Thus, in reading, some of the words which we feel ourselves to see are not seen at all, and others are seen as quite different from their actual printed forms. There are mis-spellings in almost every book, but they pass unnoticed, unseen by the mental eye while reading. Parts of words, even whole words, are often not present as sensory stimuli at all, the mind making them up out of whole cloth. So also in listening to spoken language we hear words which the ear does not hear at all. If one says rapidly, in the proper context, 'What time tis it?' or 'Please pass me de butter,' the error will often be undetected. The letter *t* is often pronounced as *d* in such words as *ability, certainty, falsity,* but only experts in phonetics notice the fact. Again and again in rapid speech, words are totally omitted without anyone being the wiser.

It would be possible to show how the argument goes in a diagram. There is one major idea, and it is explained by two facts. Each of these is supported by some lesser details. It is important to see that the parts taken down into the third line in the diagram for the paragraph are not of the same order of importance as those in the second line.

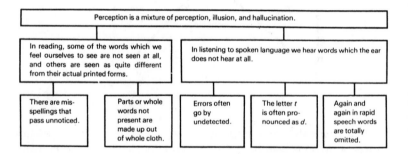

Probably the most important reading skill after the initial stages is the ability to re-create the argument the writer converted into the linear sequence of a paragraph. The pupil can be helped to see possible paragraph structures, sometimes as an isolated exercise (when the paragraphs can be chosen specially for their structural clarity), but more often in context (when the specific teaching can be drawn upon by pupil and teacher alike). Usually the method will be simply to highlight by question or pointing the structural articulation of the paragraph. Sometimes, however, visual techniques, such as the flow-chart diagrams I have demonstrated, listings, underlinings, etc. may be used to make the sense structure clearer. Each of us must give the necessary help in our own subject and with our special material.

Overall structure

We have a long, vigorous, and on the whole very successful tradition of helping pupils see the structure of narrative, drama, and poetry. We almost never do the same for the factual. Yet a chapter from most textbooks *has* a structure, and the pupil needs to be able to follow it if he is to grasp the argument. Consider, for example, a typical piece of biology reading, a chapter from the widely used Nuffield CSE Biology textbook;[1] called 'Smoking and Health', it is a 4-page essay with diagrams. It is also appropriate to consider such an example as it is very similar to articles that adults are likely to meet in a range of papers and magazines. If the reader is really to understand the writer's meaning, he must involve himself in the structure of the argument. I have found that fifth-year pupils reading this essay lose the steps in the progression, partly being pulled away by the illustrations and partly giving too much attention to details. I suspect that the pupils whose reading of this passage I observed had never had argumentative structure explained to them. They could not see the wood for the trees. Attention to paragraph structure would have helped pupils grasp the overall structure of continuous pieces of writing, whole chapters, papers, essays or articles. Clearly as much help as is given to using a test tube should be given to *using* the overall argument of such a chapter.

The first task is to help pupils realize what is meant by *argument*, which is the pattern of the ideas. Arguments can be built up diagrammatically in flow-diagram form, using very few words, in an expansion of the method used to interpret a paragraph.

To understand overall structure it is also worth teaching that paragraphs usually *have a purpose*; they do a job in building up the sequence of an argument. Each paragraph is then a step of one sort or another. Thus paragraphs *do something* for the argument, they don't just 'go on'. Even the youngest pupils in the secondary school can have this idea made explicit, but by fourteen there can be some classifying of paragraph functions:

1. *Explanatory.* These are the most common paragraphs outside narrative; they outline facts by informing or explaining.

2. *Definitional.* Many chapters have one or more paragraphs devoted to clarifying the meaning of a word, phrase, clause, or concept. This may be necessary, for instance, before a sequence of explanatory paragraphs is embarked upon. The reader needs to study these especially closely, and mentally keep his finger on such a paragraph for back reference. Frequently such a definitional paragraph needs to be read *before* the explanations, but then again *after* them, for only then will the full force of the definition be clear. It is worth asking oneself *what* is being defined and *what* it means.

3. *Introductory.* These are often broad, focusing the nature of the

[1] *The Maintenance of Life*, Penguin/Longman.

argument that is to follow. Sometimes they are signalled by obvious phrases: 'This chapter is about . . .', 'Now we shall see . . .', 'It is important to outline. . . .' However, not all are so baldly signalled, and, indeed, many introductory paragraphs deliberately start dramatically but obliquely, a combination which throws most pupil readers. Here are two introductory paragraphs from different chapters (4 and 2) of a book on the pyramids for younger secondary pupils.[1]

> How did the ancient Egyptians move the stone from the quarries to the pyramid being built? As we know, the larger pyramids used up millions of tons of stone, most of which was in two-and-a-half-ton blocks. This was no easy task, yet the Egyptians managed to find a way.
>
> The pyramids still stand today and some of them look much as they must have done when they were built thousands of years ago. Most of the damage suffered by the pyramids has been at the hands of men who were looking for treasure, or more often for stone to use in modern buildings. The dry climate of Egypt has helped to preserve the pyramids, and their very shape has made them less likely to fall into ruin. These are good reasons why the pyramids can still be seen today, but perhaps the most important reason is that they were planned to last 'forever'.[1]

Both paragraphs are introductory but they work in different ways. Whereas the first announces the subject of the chapter in its opening question, the second deliberately postpones it to the final 'but . . .'. It is worth pointing out these techniques in general terms. If the pupils are expected to read the chapters, it is worth, as part of the readiness preparation, establishing the announcement function of the paragraphs and sorting out 'which words tell you what the chapter is going to be about?', and 'why are the other facts put in?'

4. *Summarizing and concluding.* Pupils tend to switch off towards the end of a piece of reading as their stamina wilts. In fact the summarizing and concluding paragraphs are vital for review and check. (Indeed, they are often worth looking ahead for as part of a previewing technique.) It is important to note that not all conclusions are summaries—often the conclusion gives the writer's retrospective judgement or future speculation. Review signal words should be thought of: 'Thus . . .', 'We can therefore see . . .', 'It is therefore clear . . .', 'The evidence can be seen . . .'.

6. *Transitional.* I consider the transitional function especially important but easily missed by the pupil:

> Although Hinduism is the main religion of India, most of the Indians in Britain are in fact Sikhs. The word 'Sikh' means 'disciple', and Sikhs follow the teaching of Guru Nanak, who lived in India from A.D. 1469 to 1539. The title 'Guru' means 'Teacher', and although Nanak was brought up as a Hindu he later combined what he thought was the best in Hinduism with what he thought was the best in Islam. He taught that 'There is no Hindu, there is no Muslim, but only one human being, the Sikh'.[2]

[1] J. Weeks, *The Pyramids*, Cambridge University Press, 1971.
[2] Roger Street, *Focus on Faiths*, Nelson, 1974, p. 11.

The 'turning word', 'Although', needs pointing out. This is *not* going to be about Hinduism. Like most transitional paragraphs, this one refers backwards first. It is worth discussing the structural function of transitional paragraphs, and finding actual examples. They are important for the structure of the piece, and for the pupil's understanding of the argument.

7. *Narrative.* In expository writing, narrative paragraphs are often used to illustrate points being made, although they can also be used simply as introductions or to add atmosphere. The reader needs to be used to the idea that the story is there for a purpose, and to ask himself what that purpose is. The narrative thus requires a closer and rather different kind of detailed scrutiny to that given to prose fiction.

It needs adding that a diet totally of worksheet and gobbet will ill-prepare pupils for continuous reading of argument. It has been shown that of the reading done in class very little is continuous. Indeed, about half of all reading in class is in bursts of less than a quarter of a minute in any one minute. The table shows how little continuous reading there is. Obviously we are not training our pupils to understand structure in continuous reading!

Patterns of continuity in reading for each age and subject group, given as a percentage of the overall reading time recorded for reading in the group

	Type 1	Type 2	Type 3	Type 4
ENGLISH				
Top Juniors	46	29	11	13
First year Secondary	53	27	10	10
Fourth year Secondary	34	25	14	27
MATHEMATICS				
Top Juniors	41	47	11	0
First year Secondary	70	22	8	0
Fourth year Secondary	72	24	4	0
SCIENCE				
Top Juniors	69	26	5	0
First year Secondary	75	21	3	1
Fourth year Secondary	57	36	4	3
SOCIAL STUDIES				
Top Juniors	60	28	8	3
First year Secondary	60	28	8	3
Fourth year Secondary	44	42	9	5

Type 1 = Total reading during time-sampled minute was 1–15 seconds.
Type 2 = Total reading during time-sampled minute was 16–30 seconds.
Type 3 = Total reading during time-sampled minute was 31–45 seconds.
Type 4 = Total reading during time-sampled minute was 46–60 seconds.
Thus Type 1 represents only intermittent reading, and Type 4 continuous reading.
Source: Schools Council Project, *The Effective Use of Reading*, Second Interim Report, 1975, Table 4.

It is interesting to take a piece of argument from one of the most popular papers. How would school leavers cope with it? How should teachers help? As an example, we have reproduced an editorial from the *Daily Mirror*. Similar pieces of writing can be taken from the popular daily papers almost any day. It has three difficulties for pupils:

(a) range of reference and background knowledge required;
(b) breadth of vocabulary;
(c) complexity of argument.

The courage of Jack Jones

JACK JONES is a man of the Left.

There's no mistaking that. He is against Britain remaining in the Common Market. He favours more state participation in industry.

He is leader of Britain's biggest union — the Transport and General Workers — and a dedicated supporter of the Labour Party.

But Jack Jones is a constructive radical. Not a ritual extremist.

Builder

Where many Left wingers are self-confessed wreckers and apparently proud of it, Jack Jones is a builder of bridges.

● He has striven to hold all sides of the Labour movement together.
● He has striven to heal the self-inflicted wounds in Britain's industrial relations by successfully advocating worker participation.
● Above all, he has striven to uphold the social contract as a means of creating a more equal society.

More often than any other TUC leader he has spelled out the importance of unions sticking to their side of the contract.

His appeal at the Scottish TUC annual congress was the most powerful yet.

Where is the chorus of TUC voices that should be supporting him?

For the most part, they are as silent as the grave they are digging for the social contract.

Blunt

Today it is sometimes hard to remember that the social contract is the official policy of the **whole** TUC.

Some of those who voted in favour of the contract last year are now undermining it at every opportunity.

The good sense of Jack Jones has gone hand in hand with the blunt warnings of Denis Healey.

If more trade union leaders had taken their advice in the last twelve months, today's Budget could have been a lot less severe.

The far Left might even have got the reflationary Budget it wants but obviously now cannot have.

Such an example, reproduced in facsimile, could be used either *in context* (e.g. current affairs, social studies, history, newspapers, the media) or *specifically*, as an example to illustrate written argument.

In either case the teacher's approach would be broadly the same, though the emphasis would be somewhat different.

The teacher's approach, I should suggest, would be to tackle the difficulties in the order in which I have just listed them. First, through discussion, establish the background facts: 'Left', 'TWU', 'worker participation', 'social contract', 'Denis Healey'—whichever of the facts need filling in. Secondly, the teacher would pick out, or have the pupils pick out, the difficult words, and explain them: 'participation', 'ritual', 'striven', 'advocating', 'congress'.

Then, the real point: here are some fairly simple sentences, placed, however, not in a simple series, like a string of beads, but arranged by the author to advance some opinions. Each opinion is justified by one or more reasons, and sometimes the reasons are supported further by examples. Interestingly, there is also a relationship or progression between the main opinions—the third opinion, which is in the form of a question ('Where is the chorus of TUC voices that should be supporting him?') being the central one. Interestingly it is cast as a question to give it more force, but it *means*: 'We consider the TUC should be speaking up for him.'

Pupils can annotate the originals to analyse the argument, or can produce their own list, as shown. By one means or another the teacher can show the structure of the argument. This benefits the current lesson objective, and the wider aim of teaching reading.

Opinion 1: Jack Jones is a man of the Left.
Evidence: 1 Against Common Market
          2 For more state participation in industry
          3 Leader of biggest union
          4 Dedicated support of Labour Party.
Opinion 2: He is constructive.
Evidence: 1 Builds bridges
          Examples: (i) works to hold all sides of Labour together
                   (ii) tried to heal industry by recommending worker participation
                   (iii) tried to uphold social contract.
          2 Spells out importance of unions keeping contract.
          Example: (i) recent appeal to Scottish TUC.
Opinion 3: TUC are not supporting him.
Evidence: 1 They keep quiet over contract
          2 Some even undermining it.
Opinion 4: **(conclusion)** If they had, Budget would have been less severe and nearer to what the Left want.

Pupils' work on overall structure should be done in three ways:
1. specifically by theoretical study on limited occasions, of the idea, with illustrations;
2. as part of the study of an argumentative piece of writing for its own sake, but chosen to allow a study of structure;

3. by readiness techniques, help during reading, and subsequent review by all subject teachers who use print.

Graphic conventions

A publisher uses a number of graphic conventions to help convey meaning. Leaving aside those best thought of as punctuation, these range from type fonts to the use of illustrations.

1. *Typographical aids.* I do not wish to exaggerate reading difficulties, but it is important to realize that the teacher of English takes trouble to explain the typographical conventions of prose fiction (for instance, the layout of story dialogue), of verse (for instance, the problem of line endings), and of drama (for instance, the italicized stage directions in speeches, which are not to be read aloud). Any experienced teacher knows that most pupils do not just pick up the ability to read these typographical conventions. The teacher returns to them year after year—often spurred on by the failure of the pupil to use the tuition in his own *writing*. The result is a fairly methodical approach to the teaching of the typographical conventions of the presentation of literature forms, even with the least able. Nothing similar is done for conventions used outside literature. I suggest that there should be specialized teaching of the possible use of bold type, italics, sans serif (e.g. to differentiate captions from text), etc., and the subject teacher can and should then draw on this knowledge in guiding the pupil through texts.

2. *Illustrations.* Consider the complexity of layout of many of the most carefully and ingeniously designed modern information books. It is generally felt that illustrations help the poor reader. Actually this seems to me an arguable and unproven point. For flipping through, an illustrated book may be more attractive visually, but it is not necessarily easier to read its text; indeed, it may even be harder. This is because the continuous text is broken by a variety of visual devices, chosen and reproduced to be eye-catching, and laid out by a graphic designer to make a pleasing *visual pattern* of the double-page spread, and a variety of such patterns from spread to spread.

Modern technology—particularly offset litho—has opened up new design possibilities. Of these the most important is the simple one of making it possible to print half-tone illustrations on ordinary paper. Previously a coated, glossy paper was required. This led to a choice either of line drawings, or of grouping the half-tone on 'plates' gathered in one or more sections of the book. The new freedom, encouraged by other visual trends in our culture, has led to a 'graphics explosion' in information books, including school books. Picture researchers use their ingenuity to extend the range of kinds of illustration available—a diversity which, whilst considerably enriching the evidence available also makes it harder for the pupil to grasp the total effect.

The pupil needs help in the art of synthesizing graphics and text. The continuity of reading is severely fractured by captions, drawings, and photographs. The ability to step forward and backtrack is made more difficult. The mental organization of caption, illustration, and main text is extremely difficult.

Pupils need help in coping with this kind of book. Such help involves discussing what elements a book has; how they are arranged; how to differentiate captions; the possible orders of reading; and how to integrate the elements. Time must be taken to explain to the pupils the various ways in which graphics are used:

(a) Immediate reference within the text, with the sentences directly referring to and or broken by the graphics.

(b) Reference from the text to a nearby graphics, probably with its own caption.

(c) Added graphic, to be studied separately and synthesized by the student.

Within the curriculum specific attention should be given to study of visual sources: graphs, maps, photographs. When books are used, the interaction of graphics and text should be prepared for and discussed. To reintegrate the static breaks of the illustrations with the impetus of the text is difficult. This is a reading skill—paradoxically, one which modern attempts to make books easier for the 'less academic' have often made more acute.

Preparing for reading assignments

The subject teacher who has planned that his pupils should read as part of a sequence of work should normally, I would suggest, devote class or group time to pre-reading preparation. This is especially necessary for subjects such as science and mathematics in which not a great deal of reading is normally done in class.

The first aim is to generate interest in the passage. Too often we say: 'read pages 22 to 34', leaving the author the entire task of raising the interest. Excerpts, non-print material, key questions, anecdote—there are a variety of possible approaches.

It is considerably easier for the pupil if reading purposes are defined. The ideal reader does this for himself, but when all the pupils are to read, the teacher and group can do it together. Too often the reading is purposeless, or too unfocused ('find out all you can about ...'), or the purpose is discovered only retrospectively in the following lesson when the questions are asked too late. Rather, a question (or variety of questions) should be posed, and the reason for reading the passage made clear.

It is frequently necessary for the pupils to be led into the text. I see six possible techniques that will need to be used on occasion.

1. *Background facts.* The text may presume and depend upon knowledge that the pupils have not got. Whilst avoiding giving a

subordinate rival lesson, it is worth planting this information. I have seen pupils unable to make sense of a passage, which is well within their competence, merely because of some unfamiliar references in it.

2. *Previewing*. Attention to headings, sub-headings, and typographical devices ('Why is that sentence in bold type?') will help the pupil see where the reading is going.

3. *'Stopper words'*. This American phrase nicely describes those words which 'stop' reading. Again, a brief pre-reading session should not be converted into a laborious vocabulary session, but key vocabulary should be picked out. In my observation it is not mainly the true technical vocabulary of the subject being studied which confuses, but use of a less common non-technical word. This passage, for instance, is from a 'Geography for the Young School Leaver' pack:

> Some of the most exciting new parks and open spaces in heavily built-up urban areas have been created from derelict industrial areas. The city of Stoke-on-Trent, for example, is creating a chain of parklands out of colliery slag heaps, pottery claypits and disused railway lines.

The style is very simple, but 'derelict' is likely to stop some readers. Admittedly the word is virtually defined by illustration in the following sentence, but I suggest that most pupils should be given some help—either by pre-definition, or by reminding them that whatever 'colliery slag-heaps, pottery claypits, and disused railway lines' have in common must be what is meant by 'derelict'.

4. *Structure words*. In many cases the structure signal words that I talked about on page 109 are worth highlighting in advance.

5. *The structure*. On other occasions the structure of the argument is worth outlining in advance: e.g. 'The first three paragraphs are about this point of view. Compare them with the next five, which put the other point of view.

6. *Graphics*. If a graphic representation is a crucial part of the argument, it may be worth reminding pupils of how it works and its relationship to the text. Graphs often give difficulty, as do certain numerical tables. It is sometimes helpful simply to point out in advance: 'When you get to the last paragraph, the graph opposite needs studying. It tells you the same thing as the words in the paragraph'.

The aim of this preparation, therefore, is to give the necessary background, purpose, and way in. It helps to guide the pupil reader by showing him where the reading is going and how it is getting there. It is, after all, no more than a teacher would do in preparation for any other study activity, such as experiment, visit, or simulation. If helped in this kind of way a pupil not only gains more from *this* reading: he gains more reading experience to help other reading.

Reading aloud

Reading aloud is not favoured by teachers, a generation of whom are still reacting against plodding through a novel from pupil to pupil week after week. Mixed-ability groups have now made the use of the technique less easy, and where it is used, it has become associated solely with English lessons.

I would suggest that the technique has value in helping pupils grasp the intonation patterns that even in unvocalized silent reading are necessary to make sense.

The value of reading aloud is illustrated in the following example from mathematics. Merely *looked* at, the passage seems nonsensical: it has to be vocalized to be understood, with intonation patterns clarifying the sense groups. It is also a good example of the extreme importance of understanding the place of the comma in clarifying sense groups: a pupil who has not considered the comma would be less likely to understand this, and a pupil who has been helped to understand this in his maths lesson would be helped to gain a feel for the comma:

> If we start at 0 each time, the shift plus 1 takes us to the point plus 1, the shift plus 2 to the point plus 2, the shift plus 3 to the point plus 3, and so on. In the same way, each time starting at 0, the shift minus 1 takes us to the point minus 1, the shift minus 2 to the point minus 2, and so on.

(It seems to me that quotation marks round the phrases 'plus 1', 'plus 2', etc. would have made the sense clearer.) The point is that there is nothing *mathematically* difficult about the idea in this passage which refers to shifts on a number line:

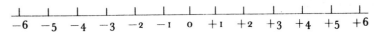

There is, what's more, nothing difficult in the words, or the syntax. There are merely two factors that create difficulty:
(a) The need to understand the sense of 'takes us', a phrase which is omitted from the clauses following its first appearance.
(b) The repetitions marked by commas.
If the teacher, or a good pupil reader, reads it aloud, not only is the sense immediately clear, but help has been given to the pupil to similarly reconstruct sense by his intonation patterns for other examples. Obviously the pupil should be encouraged to vocalize such a passage.

Subject teachers will need to choose carefully when they ask pupils to read aloud. It is at its best for highlighting key passages, or supporting pupils' opinions. Obviously less able readers should not be abruptly exposed to the whole class, but small groups on prepared occasions can get round these difficulties.

Conclusion

It is too readily assumed that all that is said about the teaching of reading in the content area applies only to the less able, or maybe at most to the average and below average. This is not so. Fluent reading by competent pupils can mask the fact that they read the text merely as a string of statements, and that they do not read the reasoning process. These are the pupils of whom one American declared: 'There are many students who read too well for their own good in most schools'.

I have suggested teachable clues to help grasp meaning. Mere teaching of these clues as an analysed series of skills does not itself improve reading, but equally leaving the pupil to develop ways of working out sense entirely on his own is to leave him with insufficient help. We must keep providing opportunities, and support those opportunities with pointed help.

The question of time will naturally be raised. If one is not careful the idea of teaching reading is seen as an added task laid on top of a subject teacher's present load. But the teaching of reading outside English is not *something* to teach, but a *way* of carrying out part of the teaching, one which not only helps the pupil gain the ideas, knowledge, or skill currently being explored, but also increases his ability to learn other aspects on other occasions, independently and at will. It will be rare that a lesson sequence or the work of an individual or group of pupils will be stopped for specific reading tuition, but whenever print matter is used as a source, the teacher can integrate his reading help with the lesson. Thus time is saved—for the time of the pupil devoted to the text will be more valuably spent, the knowledge will be more thoroughly grasped, and the pupil will have been helped for the future.

A subject-teacher's reading policy

A school might consider these points for its whole-school policy:
1. Make sure that print is a necessary source.
2. Judge the suitability of each item.
3. Provide a variety of kinds and sources, including continuous passages of some length to teach structure and short passages for intensive study.
4. Structure the moments of reading, to provide both group guided reading and individual varied reading.
5. Devise functions to focus the purpose of the particular reading occasion.
6. Prepare for most reading occasions by introduction and pointer.
7. Allow adequate uninterrupted time for reading.
8. Help individuals through the text where necessary (not by offering alternative explanations).

The use of talk

'I can't even speak English proper because you never talked about anything important.'[1]

The pupil's limited range of language may be a serious barrier, preventing his access to learning. This is, as we have seen, especially true of reading problems. Conversely, however, the school's limited uses of language may restrict the pupil from deploying his language in learning situations. A language policy is concerned with helping him get *at* the understanding and knowledge on offer, and it must also help him get *into* the understanding and knowledge, since without comprehension, access itself is of little value. The way into ideas, the way of making ideas truly one's own, is to be able to think them through, and the best way to do this for most people is to talk them through. Thus talking is not merely a way of conveying existing ideas to others; it is also a way by which we explore ideas, clarify them, and make them our own. Talking things over allows the sorting of ideas, and gives rapid and extensive practice towards the handling of ideas. It is also, as Nancy Martin's contribution to this chapter makes clear, a valuable step towards writing.

John Dixon has asked: 'How did classroom dialogue, a central method—*the* central mode?—of teaching and learning come to be so neglected? Why is there still so little published discussion among teachers and research workers on dialogue between pupils as a mode of learning? Why is linguistic theory, so original in other respects, still so weak on the most elementary form of discourse?'[2] And he might have asked why a method so long recommended has been so little used. After all, the Report's predecessor, the Newbolt Committee, fifty years earlier had spoken of 'the conversational method'. What that committee had to say is now agreed to by all linguists: 'So-called knowledge is not knowledge if the thinking powers are not applied to it, and . . . the only way to get a child to think about it . . . is to get him to talk about it.'[3] This is what the Bullock Committee had in mind when it stressed the value of talk as an educative activity in itself. From infant conversation to university seminar, the student needs the opportunity to talk about his learning, indeed to learn through his talking.

The failure can be seen only too clearly when the pupil gets older. It is, I think, well put by a speaker from an Agricultural College talking about the Suffolk boys who come on from school:

They make statements—big, flat statements. They never explore ideas. Nobody has taught them how to use words to convey theories, so now that

[1] Beatie Bryant in *Roots* by Arnold Wesker (Penguin edition, p. 55).

[2] John Dixon, 'Talk and Collaborative Learning', in H. Rosen (ed.), *Some Appraisals of the Bullock Report*, University of London Institute of Education, 1975, p. 28.

[3] *The Teaching of English in England*, HMSO, 1921, para. 74.

all the old village story-telling has died out their talk is very poor dull stuff. Although sometimes you'll get a flood of the old richness when somebody has to relate a bit of gossip. There is terrific animation. Eyes light up. The man isn't really gossiping in the trivial sense of the word, he is story-telling.[1]

The teacher's task is to help the pupil extend from the possibility of story-telling (and to nourish that tradition too) into the 'conveying of theories', and not merely into the dead-end of 'big, flat statements'.

The first problem of such aims is the need to establish an atmosphere in which talk is possible and to build up relationships in which talk is wanted. We know from commonsense and linguistic research that genuine talk is not possible in certain atmospheres, nor is it possible on one side of certain relationships. We have all experienced occasions when we or others have been made virtually speechless by an intimidating atmosphere—whether in interview, or because of a particular individual. Labov, the American linguist, has demonstrated[1] how fluent black children can be reduced to mono-syllables in teacher-dominated situations. We know that from our own memories or reading. This exchange, although from a novel, records the kind of relationship which makes a pupil literally speechless. Saville is unable to answer the teacher's questions, not because of any deficiency on his part, but because of the nature of the verbal exchange:

He examined the register once again. 'Father's occupation.' He wrote something down on a piece of paper. 'Now "colliery worker" means that he works at a coal-mine. Is that correct?'

'Yes', he said.

'Yes, *what?*'

'Yes, sir', he said.

'Now there are any number of people who might legitimately say they work at a *coal*-mine. The *manager* of a *coal*-mine might say he works at a *coal*-mine.'

He waited for an answer.

'Yes', he said, then added, 'Sir.'

'I take it, of course, he's no such thing.'

Colin waited, unsure of what to add.

'He's not the manager, Saville?'

'No.' He shook his head.

'No, what?'

'No, sir', he said.

'He's not the deputy-manager, either, I imagine.'

'No', he said.

'Does he work on top or, as they have it, underneath?'

'Underneath.'

'Does he superintend the men down there, or does he actually hew the coal itself?'

---

[1] Ronald Blyth, *Akenfield*, Allen Lane, 1969, p. 169.

'He hews the coal', he said.
'E'ews the coal.'
'Yes, sir.'
The class had laughed.
'In other words, Saville, he's a miner?'
'Yes, sir.'
'Then why couldn't he put that on the form?'
He bowed his head and wrote in the register for several seconds.
'Well, Saville double l, have you got an answer?'
'No, sir.' He shook his head.[1]

Saville (who was grilled earlier about his father's mis-spelling of his surname) is made to look a fool, indeed to be a fool, by the teacher's use of language as a weapon to maintain his superiority. There can be few of us who, however we feel schools have changed since that kind of exchange, do not recognize the linguistic pattern as one which we use from time to time, thus driving a pupil into inarticulateness. However, we try more often, and frequently succeed, in creating an atmosphere and setting up an exchange in which we can build on the pupil's natural way of putting things.

Secondly, and related to that, pupil talk is most likely to flourish when there is a genuine interest in what is being communicated. Although speech education still has a place in schools, for some pupils and on some occasions, its tradition can be very unhelpful indeed. There will be times when concentration on clarity, modulation, expression, and projection is relevant and meaningful for the pupil. Indeed, the art of public or semi-public use of the voice is one which will always be worth working on, whether for use in reading aloud to children round the fire, at the rugby club dinner, or in quasi-professional roles. However, that is virtually a separate concern, and its methods have little to offer the encouragement of communicative talk in a school. Indeed the vestigial remains of elocution and performing can cramp talk for the many by emphasizing the manner rather than the content.

The teacher's response must be to the totality of the statement, not to the triviality of part of the expression. An important statement can accommodate a vague confusion of phrase, just as a printed statement can carry a blurred word, and still be fully explicit. The teacher has to choose whether to respond to the overall sense, which he has probably taken quite clearly, or to pull out the weak phrase for examination. If he does the latter, he will lose the former. In general, a school policy will want to place emphasis on a genuine interest in and response to the pupil's ideas and intention.

But this is to describe pupil talk as if it were merely talk to and with the teacher. Certainly the teacher as sympathetic adult will be important, as will any other adult audience that can be arranged, such as visitors, other professionals, parents, local people being

[1] David Storey, *Saville*, Jonathan Cape, 1976.

interviewed, and so on. However, the greatest speech opportunities, and potentially the most valuable, are those between pupils. On these occasions the pupils can think on their feet. They can explore ideas in a way which can sharpen, intensify, and even speed up the process of thinking. During and through the speech the pupils can create new perceptions, new ideas, and new knowledge.

The humanities are too often seen as the main talking subjects. Possibly, though this is no more than a personal hunch, there is too much reliance on the exchange of views in subjects like social studies. More certainly there is a gross underestimation of the value of pupil discussion in other areas of the curriculum. Music, for instance, is rarely mentioned in this context, yet the exchange of opinions and reactions in small groups is a real possibility. Group music-making offers an even more promising opportunity, in which small improvised compositions are planned and rehearsed. These demand cooperative talk, as the suitability of musical material is offered, doubted, considered, and decided. (Do we get louder or softer at the end? Should we repeat that note pattern? When should we bring in the metalophone?) Similarly, in three-dimensional crafts, design problems lead to small-group talk, bringing memories and experience of life to bear on the challenge of materials and techniques. Similar group planning, analysis, and thinking through can be a part of science, mathematics—indeed most subject activities.

A school considering the place of talk in learning is likely to want to study classroom approaches in some detail, for increasing the value of pupil talk is probably the hardest actual pedagogical problem within a language policy. Crudely translated into: 'They've got to do a lot more talking', the ideas of the Report could be at best a waste of time, and at worst could produce disruptive chaos. Some schools might not yet feel ready to go far in this direction. But a school planning to do so might well consider in seminar and staff meeting how pupil talk is encouraged, the kind of occasions, and the kind of teaching required. The following headings could be looked at:

Contexts

Talk flourishes best when it grows in a context. The kind of purposeful developmental talk that Douglas Barnes has described in his two books[1] usually comes in a carefully arranged situation, with some form of suitable learning material to use in discussion, and with a suitable setting. What kind of contexts can a teacher set up to encourage such talk? The staff of a school can compare contexts that seem to have worked and ones that do not. They can work towards developing their own guidelines that will help others create promising talk contexts.

[1] *Language, Learner, and the School* and *From Communication to Curriculum.*

Disciplinary patterns

The traditional school injunction of 'silence' has the practical virtue of clarity. No doubt there will always be occasions on which it will be the most appropriate disciplinary injunction. When it is discarded, however, and talk is expected and encouraged, with what is silence to be replaced? There can be a genuine disciplinary problem to which too few educationalists give any thought. Talking situations can get out of hand, and the noise level with a full class in an ordinary room rises so high that the aim of productive talking for learning is broken by the sheer hubbub and the teacher's eventual need to shout above the noise. The injunction of 'reasonable noise' is too vague, and offers no guidance to the participants. Teachers therefore need to develop in a school alternative acceptable disciplinary guides, working out alternative injunctions such as 'No talking outside your group' or 'Each chairman will decide who is going to talk'.

Time constraints

Are there any generalizations about time and talk that would be helpful to a school? Obviously the question 'How long is a good discussion?' is a fatuous one. Yet a staff can help itself develop good classroom talk if it does give attention to the effects of increasing or decreasing time for talk: too little and the session barely gets going; too long a session for the material and the talk flounders.

Discussion techniques

Most valuable talk will be virtually unstructured. Nevertheless there is advice which can helpfully be given to pupils about talking together. We do, after all, give them advice on most learning techniques, and it would be strange if this were omitted. Some consciousness of the problems and methods of group discussion can be of great help. This worksheet below is from the Abraham Moss Centre in Manchester, a school which takes pupil discussion very seriously and has developed ways of discussing discussion.

When you are taking part in a discussion remember the following points. They are *very* important.

1. Remember you are taking part in a discussion
   (a) to learn
   (b) to help other people to learn.

2. Always have a paper and pencil handy during discussions.

3. Ask for anything you do not understand to be explained.

4. If two people have misunderstood each other, help them out.

5. If someone has said something which you think is important, remind the group about it.

6. Always let other people 'have their say', even if you are bursting to say something.

7. Be prepared to change your mind if you have been proved wrong.

8. Do not shout people down just because you don't agree with them.

9. Do not show off in a discussion.

10. You must take part in a discussion even if you feel shy.

11. Listen carefully to what everyone else in the group says.

12. Always have one member of the group writing down the most important points discussed and any decisions that the group has made.

End product

Talk is frequently, though by no means always, sharpened by a function—that is if the talk is leading towards some form of end product: establishing a hypothesis, choosing a design, planning a strategy, drafting a report, making a decision. It will often help if the teacher focuses the talk by choosing such an end product in advance. A school staff can consider what kinds of end products have proved the most profitable.

Perhaps the hardest problem in talk is that of the place of the teacher in situations somewhere in that rich but uncharted area between teacher-dominated 'discussion' (which certainly has a place) and pupil-to-pupil discussion without a teacher. If the pupils are always to be dependent on the teacher for the sense of direction in talk, how will they learn to cope without him? Yet his ability to alter course, to focus, to weave in the relevant reference is valuable not only for the discussion of the moment, but also for the pupils' growing awareness of the opportunities and methods of purposeful talk. It is possible for teachers to develop strategies for participating without dominating, but these techniques are at their hardest in subjects with few periods and much explaining—such as science. They are at their easiest in the more personal parts of the humanities, and for this reason those teachers whose experience has been wholly in literature, for instance, often have great difficulty in even understanding the difficulties of a teacher trying to find participatory ways in pupil discussion in other areas. The core of the problem seems to be the knowledge which needs to be deployed in much of this discussion, and the teacher's fear that the conversation will confirm wrong ideas if he does not oversee, intervene, and correct. Preparation seems doubly valuable in these situations, and it is vital to ensure that the background information to the talk has been learnt earlier. Sometimes information cards can be a useful source of facts, basic ideas, and key vocabulary, available to all the participants.

In the next section we shall see the uses of such exploratory talk, between teacher and pupils and between pupils and pupils, in the context of a science practical lesson. Following that we shall look at the uses of writing, seen initially as a form of recorded inner talk. A final point about talk, however, needs making. It is not a question of how to find time for talk, as if this were to be an *added* activity competing with the main learning and therefore requiring spare time. The opposite is true: the speed of real learning in most subjects is likely to be *increased* by talk—so that the task is therefore not to find time, but to choose the best time.

Talking and learning in science, an example
Irene Robertson

As a starting point we could look at a common-sense distinction that most of us work with: the distinction between a science ('biology' or 'physics' etc.) and the language of science or scientific language (I use language here to denote all signing systems, e.g. algebraic, pictorial, that a scientist may use). We can consider that science *is* the language in which it is reported, taught, received. This could apply equally to a doctoral thesis, an 'O'-level textbook, the talk that goes on when a teacher gives a lesson on electricity, or the talk that goes on among pupils as they conduct their experiments in school labs.

We can view the teacher then as teacher of his subject-cum-language. In learning that subject-cum-language, the teacher has come to see the world in particular ways. He has organized it scientifically. I may see a jug as an everyday object to hold water; in a pottery class I may see it in terms of the structure of its clay and the movements of its making; in a science lab I may see it in terms of a concern with the physics of liquid.

The science teacher can share particular ways of seeing with his pupils. I use *share* here, because I want to get away from a simple notion of the direct transmission of knowledge from teacher to pupil. For if we agree that coming-to-know-in-language is the issue, then we must recognize that this is a process within each individual. All of us, teachers and pupils, are responsible for our own enquiry (here scientific enquiry) and we can truly know only what we have come to experience and express for ourselves. Nobody else can do it for us. As teachers, we probably recognize this most clearly when we distrust work that a pupil has simply copied from a book. We feel that he has not understood or made it on his own. We should be suspicious for the same reasons when we find ourselves filling blackboards and getting the pupils to copy our own formulations.

How to share these ways of seeing with our pupils? We probably feel, perhaps unfairly, that when pupils initially come to us they are ignorant of science's ways of seeing the world. After all, that is what they come to us to be educated in; that is why we are teachers. We should equally remember that those pupils are already competent in working with language. We all live by explaining the world and ourselves in it. Our pupils' explanations may in detail be different from ours, just as yours may be different from mine. The detail is not the issue; the possibility of enquiry is.

If we can respect the pupil as his own enquirer, and remember that becoming educated in science is doing your own scientific enquiry, then we should bear this in mind when we make provision for him in schools, in lessons. When we as adults do our own enquiry we can turn to books, and are probably willing to admit that this may be a particularly difficult or uncomfortable thing to do; we turn and talk to others; we talk with ourselves. We are then in a whole network of dialogues with ourselves and books, ourselves and others—and it is the interplay of these dialogues that are our learning and coming to know. For instance, whatever we may gain from this paper, whatever we may count as knowledge for us, will be what emerges in the talk we engage in with the paper and with others sharing it. In turning to this account of lessons, we could consider how enquiry was enabled; where it fell down; and we could listen to what these boys say of what doing physics was all about.

In the physics lessons which I shall describe, boys were learning about speed and internal combustion engines. Speed was already a common-sense notion which the boys had experienced and talked about on countless occasions previous to the lessons in which they measured speed in terms of the formula distance/time. The boys weren't being handed new knowledge; rather they were coming to focus differently.

There are, of course, ways of knowing to which we need never give utterance; an example of this can be found in the account which follows when pupils at work do not need to express the fact that they know how to start a two-stroke motor, since their knowledge is displayed in their activities. But in schools we are largely concerned with public, communicable ways of knowing and understanding; we talk with pupils, we read what they have written—listening and looking for their understanding.)

These activities of shaping and sharing understanding are what I was listening for in the lab which is the setting for this account. I try to place them in the context of the lessons as a sequence; the organization of space and time; the teacher's concern with his purposes; the pupils' views of what they are doing when they practise science. Other issues could have been brought into focus, and may well be by those who read the paper. The account is in no way intended as a blueprint for a set of practices to guarantee good lessons, but rather is offered as an occasion for us to continue our own enquiry into our endeavours as teachers.

Working in the lab

What is described here is a sequence of lessons in physics experienced by a group of mixed-ability fourth-year boys at Sir William Collins School. The lessons dealt with speed, the measurement of speed, two-stroke and four-stroke engines, the kinetic theory of gases. The teacher of the class, Tony Hoskyns, usually plans a piece of work to last for about five weeks. The class has two one-hour physics lessons per week. This points to what I see as a major concern of the way this teacher works. As the boys are preparing for examinations, the teacher makes an initial decision about lesson topics with this in mind, but also draws upon his sense of the boys' interests. The boys have enough time to work on a project in a way that enables them to feel a responsibility for their own learning. To put it another way—for five weeks the boys know that they will be working in a particular area; the twice-weekly lesson is not a surprise given to the boys by the teacher—the rabbit out of the hat.

In the lessons on the measurement of speed, Tony Hoskyns had suggested to the class that they made trolleys. For this he provided wood, wheels, tools and a design for a trolley which he and another teacher had produced. In the eventual making of the trolleys the boys used the design simply as a guide, and adapted it considerably. How the materials were used depended on the discussions and decisions made by the boys themselves.

Initially the problem presented to the boys was—If we wish to measure speed, what kind of movement do we want our trolleys to make? Answer—straight movement. For two weeks boys worked together in groups of two or three, constructing trolleys, and afterwards fitting them with small diesel two-stroke motors (thus anticipating the topics to be looked at in the later stages of this lesson sequence). They were asked to measure the speed of their trolleys—first when pushed by hand and secondly when driven by motor.

During these lessons very little of what we usually take to be formal teaching or formal learning was taking place. To the layman the physics room probably looked more like a workshop than a science laboratory. Boys were grouped round the benches working on their constructions, or making an enormous row and smell in trying to start-up their two-stroke motors.

Overtly the boys were paying little attention to the problems of physics as such. Problems arose and were worked on in terms of getting the motors to start and keep going; making the trolleys strong enough to take the weight of a motor; adjusting weights to get the balance to make the trolleys run straight.

It was at this stage that I experienced my first dilemma as a researcher for the Oracy Project, committed to a notion that children, like adults, need the opportunity to talk out, to enquire. Many of the activities I've described took place without very much talk, or at least without the kind of talk that researchers usually offer as evidence of learning.

Yet when asked (in very much the way of their teacher who was visiting groups and asking), one group of boys told me readily that their motor would start once it had heated up; that once hot, it would keep going; that cutting down the supply of fuel would slow down the motor; and that the up-and-down movement of the piston turned the propeller.

How did the boys know all this when they had been given no formal teaching, but had simply been asked to make a trolley, measure its speed, and get their motors to work? One answer, I feel, lies in the teacher's trust of the boys' own common-sense. They were left to read the maker's instructions that came with their diesel motors; to spend time pushing the propellers to create the heat to get the motors going; to play with varying the fuel supply. At this stage, as indeed at any stage in their work, the boys could also consult their textbooks. At the beginning of the year, each boy is given a copy of the book *CSE Physics* by Nelkon, which he is asked to keep always at home, and which he is invited to use as and when he wishes. It becomes another resource, rather than a programme of learning.

Using resources in the course of problem-solving is, I think, well illustrated from the actual measuring of speed. Tony Hoskyns simply handed out clocks (second-timers) and a metre measure.

One group measured out a length of five metres on the floor and marked it in chalk. One boy propelled the trolley down this; a second boy operated the stop clock. Their problem was to get the trolley to go straight and it was one they did not solve until they saw a second group at work. The other group, on their own initiative, threaded through the trolley a piece of fine string, tied to a stool at one end and held taut by a boy at the other end. When they pushed their trolley it went along the straight line of the string, making the measuring of speed easy.

The blackboard note for the lessons on speed indicated: *speed = mt/sec*. The teacher assumed that the boys already knew about speed, and could make explicit this knowledge in measuring the speed of their trolleys, converting feet per second or miles per hour (their already familiar notion) into metres per second. The formula for speed was on the board then, but was not taught as such.

The boys would push their trolleys over the measured distance. A typical exchange between a boy and the teacher was:

B Just over a second for five metres (pause) five metres, 1.1 second.
T So what's the speed?
B Five metres in 1.1 seconds.

This I see as the teacher simply checking and affirming the boy's understanding of speed.

The second part of this sequence of lessons concerned heat, engines, and the kinetic theory of gases.

Pinned on the notice board was a paper, a guide for work for the boys, which read:

> *Unit:*
> *Heat. Engines.*
> *The Two-stroke Motor.*
>
> Diagram with parts
> How does it work?
>
> Advantages over four-stroke
> Disadvantages
> Kinetic theory of gases
> Petrol/diesel

Throughout the lessons concerned with this topic, an overhead projector was set up, showing a mobile transparency of a two-stroke engine (mobile so that the action of compression and combustion could be shown). This was then available for the boys to come and look at and work with as they needed to, either alone, or with a friend, or with the help of the teacher. The transparency showed in diagram form the essential parts and piston movements of the engine.

In the following exchange the teacher was working with a boy (B1) and was joined by a second boy (B2).

T Now if you look up on the board! How does it work?
  Just give me a minute on how it works.
  (The teacher then leaves the boy time to think.)
  (The boy calls over his friend, B2.)
B2 Four-stroke. I know it.
B1 It's a two-stroke.
B2 Two-stroke, it's just the same.
B1 How does it work then?
B2 Right. First of all the piston comes down, opens the inlet valve. In comes the fuel. They're both shut. The piston comes back up, compresses it. As the piston comes down the outlet valve opens and pushes it all out of the exhaust.
T That's the four-stroke, though.
B2 Aah—but they're the same principle.
T Yes, what actually happens? How does it work?
B2 On the two-stroke you haven't got a spark plug. A two-stroke ignites itself under pressure and you get a spark off the thing.
T Now just a second (turns to B1) can you give me a minute on how it works?
B1 (The boy uses the transparency as he talks.) Well, you see, it lets the air in and lets the fuel in and this comes up like this and ignites it.

T   Why does it heat up?
B1   Because of the pressure?
T   Because of the pressure inside here? (indicating)
B1   Yes
T   Fine.

What interests me in this exchange is the question of the rightness or wrongness of B2's knowledge—the question of his understanding. In looking at the model of the two-stroke engine he in fact talks about the four-stroke engine (only the four-stroke has valves, not the two-stroke).

What the boy insists upon is the principle behind the two kinds of engine. It seems to me that the teacher, by not saying to the boy 'you are wrong' accedes to this. The wrongness of the boy's explanation is in the detail. He does know well what an explanation of an engine requires: an account of piston movement and fuel ignition. Instead of having his explanation discredited, he is able instead to listen to B1, who does indicate clearly the two strokes of the two-stroke engine and the way in which fuel ignites. B1 in turn is not sure of this second point. The teacher, I feel, senses this, and so indicates the pressure chamber to remind the boy. It's an important bit of teaching smoothly incorporated into the boy's own explanation.

B1, along with many of the boys in the group, is also unsure about the petrol/diesel distinction, as the following exchange shows. The teacher and B1 are talking about the disadvantages of two-stroke engines. They are looking together at a duplicated sheet of information about engines, a further resource available in the physics rooms.

T   There's a tendency to overheat. Why do you think this is?
B1   It ain't sparking. It's heating up from the pressure inside there (indicating the pressure chamber).
T   This could have a spark-plug here just as easily.
B1   I know but that ain't made for a spark-plug is it?
T   Well this would be a two-stroke petrol engine, if it had a spark-plug up here, instead of a diesel. So we're comparing a two-stroke with a four-stroke. Why do you think it would overheat compared with a four-stroke? You do of course get this pressure cycle every two-strokes instead of every four strokes. So you're getting heating. . . .
B1   Twice as fast.
T   Right.

The boy then is not clear of the petrol/diesel distinction. But he has learnt that on the model in front of him there is no spark-plug and so the firing of the fuel and air is caused by pressure. His suggestion that this is what causes the engine to overheat is a plausible one, and in fact not entirely wrong. It needs only a bit of prompting by the teacher to point to the cause of the overheating: the *frequency* of the pressure cycle, and he can finish the explanation for himself.

The boys of course shared enquiry with each other as well as with the teacher. Here two boys are at work on a diagram of the parts of an engine:

B3 Bill, what's that called?
B4 Piston crank.
B3 Piston Rod.
B4 Piston Rod.
B3 What's that one?
B4 That's connected on to the crank shaft (pause) that is, yes that's the
     crank shaft. That's what's meant to be connected on to there really,
     because that's what pushes it. That's just there to measure it see—
     that's just a diagram. The piston rod is connected on to the crank
     shaft which makes the piston go up and down. See.

B3 is the one who starts the conversation by asking for clarification,
but I think we can argue that B4, in starting to answer B3, making
his own thinking clearer; is coming to name the parts properly; is
seeing that a diagram has to be read as such and isn't the real thing;
and is giving a very clear explanation of the connection between
crank and piston. The boy doesn't know what he knows, until he
says it. And this remains, perhaps, the single most powerful argument
for all kinds of classroom talk.

Writing in these lessons is linked to the talking that goes on, not
least because Tony Hoskyns is constantly talking through work that
the boys have written. In the task assigned, a boy would come to the
teacher when he had completed a stage in his writing. The teacher
then would read aloud to himself and to the boy, that particular
piece of work: 'At compression the molecules of air and fuel crash
into the sides of the cylinder and bounce back at the same speed.
This means that the molecules are elastic; when they bounce back,
they collide with other molecules, thus producing heat from the
collision.' Then followed the teacher's remark: 'Beautiful.' And later
from the same piece of writing:

T   'When the air and fuel is let in, the exhaust opens and some mixture
     goes out, so there isn't as much mixture being compressed as there
     would be in a four stroke'. Yes that's good, that's well put. (teacher
     remark)
     'This would mean that it wouldn't go as fast as a four-stroke engine.'
     Why wouldn't it go as fast as a four-stroke engine? (teacher question)
B5 Because there's not so much pressure inside.
T   Beautiful. Thank you very much indeed.

Here it seems writing takes its place not as a burden for teacher and
pupil, but as another occasion to share understanding. The boy here
could go away and add his point about pressure, arrived at in that
exchange. The writing, too, is treated not as teacher's way of testing
pupil knowledge, with all the misery for teacher and taught that that
involves, but as another resource among the many in the room that
two people can meet over. Each boy has an open-top clip-file in a
filing cabinet where completed work is banked, available for reference.

A group of the boys talked of the importance of writing for them as a way of checking their own understanding. In coming to write, they felt that they had to work out their ideas more clearly. Sometimes, they would go to the teacher with a problem, but sometimes, too, they could get a sense of what they didn't understand from trying to write.

Some of the boys, too, had a sense of the public nature of writing, and of science as a public discipline. As one boy said: 'If you didn't know what a thing is called—you wrote down that blue thing with the red knob on, they (whoever was referring to your work) wouldn't know. And you've got to know what you're talking about'.

Many of the boys did express a concern to get through exams, to get a job in which their physics would be useful—and a useful physics is one that is communicable with others. The teacher shares this point of view with the boys. Written work is brought along as a communication for people to share.

Still interested in seeing the work from the boys' point of view, I asked a group of them what they reckoned an experiment was. One boy said: 'An experiment to us is reality to the teacher because with us at first we're just trying it out. We don't know what's going to happen but the teacher knows what's going to happen and how it's going to happen. We've got to find how, what, when.'

If the teacher already knew, and if the boys knew that he knew, I wondered if they would not simply prefer to be told. 'No, we want to find out for ourselves,' one boy said. Another added: 'You've gone wrong somewhere, go and check it. He'll never tell you exactly.'

One boy talked of building a circuit: 'Building helps you understand. If you're just given it on a piece of paper you wouldn't know because you wouldn't know what went into building it.' Another boy explained: 'If you've done the experiment yourself and you know in your head that if the piston goes up, and it says well what happens when it goes up, well you think, well when it went up, it sparked and made it come back down, so it must compress when it goes up, and you'd work it out that way.'

What I want to argue from this is that the boys feel that being told something does not necessarily enable learning; that the knowledge in a teacher's head cannot simply be transmitted to the pupils' heads; that to understand, one has to be responsible for one's own enquiry; that science is about tackling problems.

From the boy's point of view, where does talk feature in all this? I asked: 'Does it help if you talk?'

'If everybody talks about what they're doing and how they're figuring it out, everybody takes an interest in everybody else.' And: 'Discuss between you, you get more and more to do it properly first time. There's an incentive to do it properly first time.'

Writing[1]   Nancy Martin
Why write?

Every day in schools children write—in exercise books, in rough work books, on file paper, on worksheets. They write reports of experiments, essays about climate or causes or persons, stories, recipes, poems, occasionally journals; they write exercises and answers to questions on what they have just been told or read; they copy a great deal off the blackboard or from books. Why do they? What is all this writing for and what does it achieve?

When the Project team (Writing Across the Curriculum) asked pupils and students about their recent experience of writing in school we found, not surprisingly, that the teacher figures prominently in their recollections. He structures situations, he makes demands, he influences both implicitly and explicitly by his responses to what is offered. The teacher has his own ideas about what his subject is and what learning is and these inform his practice. He may be more or less aware of the criteria on which he bases his teaching but in either case his pupils soon know what pleases him. But what pleases the teacher, what he considers to be the appropriate language of his subject and the appropriate way of using it, may not be helpful—indeed may actually impede—the understanding of his pupils.

> Our history teacher used to make us put down—she gave us a load of facts to make into an essay. Well, I couldn't do that. When I was confronted with a whole list of facts I just couldn't do it and I failed my history exam and she told me I'd fail 'cos I couldn't do it. (Technical College Student)

> I knew what he was on about but I only knew what he was on about in my words. I didn't know his words. (Technical College Student)

> In my exams I had to change the way I learnt, you know. In all my exercise books I put it down the way I understood, but I had to remember what I'd written there and then translate it into what I think *they* will understand, you know. (Technical College Student)

The teacher as the only audience for the pupils' writing is a point which will be discussed later. Meanwhile, it is clear that if the teacher sees writing mainly as a means of recording and testing this will inevitably influence the expectations and attitudes of his pupils.

> . . . our lessons consisted entirely of bending pieces of glass over bunsen burners and copying down endless notes of dictation. Our involvement in the learning procedure can be measured by the accuracy of my notes. Every time he dictated punctuation I wrote down in long-hand, taking 'cover' instead of 'comma' so that a sentence might read: 'Common salt cover to be found in many kitchens cover is chemically made of sodium cover represented Na cover and chlorine cover represented Ch full stop.'

---

[1] This section refers to work undertaken by the Schools Council Writing Across the Curriculum project, and includes quotations from the book based on the project, *Language Policies in Schools*, Ward Lock Educational 1977. For a full list of the books based on the project, see page 303.

> Dictation was similarly rife in other subjects. (College of Education Student)

> In RS right up to the fifth year we were not allowed to make our own notes on the Apostles, everything was copied from the blackboard. This was a weekly exercise in neat writing and nothing else; we never discussed the work, nor was there any homework set. What I do not understand is why everybody passed the O level. Maybe there is some loose connection. (College of Education Student)

A major concern of some teachers was technical accuracy which took precedence—or so it seemed to the pupils—over content. Handwriting, spelling and punctuation were frequently referred to:

> Girl 1: Well, mostly they just mark spellings and they're so involved in getting all the spelling right that they forget the story and there's so many papers to mark they can't go over it twice.
> Girl 2: And they just make sure you've got the punctuation right and everything. (Third-year pupils)

The power of the teacher was dramatically illustrated by the recollections of some of the students who could remember clearly how a single remark by a particular teacher influenced their feelings about writing for months—even years—afterwards. Sometimes the effect seemed beneficial, sometimes not, but either way it seems that teachers may often underestimate the effect that their opinions can have on their pupils.

> My first notion of the change in emphasis between junior and grammar school came when I had to write an essay on Neolithic man for my first piece of history homework. I started 'My name is Wanda and I am the son of the head man in our village.' The history master read it out to the rest of the class in a sarcastic voice—everybody laughed and I felt deeply humiliated. I got 3/20 for covering the page with writing. I hated history after that until the third year. (College of Education Student)

We also asked many teachers what they set writing for and what they thought it achieved. We received very varied answers which I think are representative both of the different ways teachers see the functions of the writing they set, and also of the varying purposes for which writing is used both in our society and in schools. Some idea of this span is illustrated in the following examples.

A science teacher said he used written work to provide a record of work done which could be learned. He thought the act of writing itself helped pupils to learn what had been presented to them. He said it had not occurred to him that writing might aid independent thinking and the making of connections by pupils. He reckoned this was done more appropriately in talk. A history teacher said: 'you can't arrive at judgements about evidence—which I take to be the aim of history teaching—without envisaging of all sorts. So I set a lot of writing about hypothetic situations and events and ask the pupils to place themselves in these. But they must have something to work on. So I do a lot of input teaching before they do this kind of writing. I am pleased with the results—as a history teacher.' Another history teacher said that independent writing (not notes) helped memory, clarified unfamiliar information and ideas and allowed the writer to discover what he thought and felt about the information.

Teachers of most subjects said they were concerned for their pupils to pass their examinations and they therefore set their pupils a good many written exercises involving written answers to comprehension questions and copying accurate information from the blackboard. Other teachers said, 'you have to survive. It's no good discussing what would be best. That won't help you on Monday morning when it's how to survive that matters. So I make them copy, or write answers.'

An English teacher said: 'they write to explore—themselves, their experiences—growing up—their reading—literature—and what they learn from their own writing of stories and poems. I ask them to write to help them learn—about themselves and about the world.'

A history teacher said about his first-year tutor-group 'What they think English is depends on what junior school they came from. Some of them thought it just consisted of handwriting and punctuation and spelling. "Are we going to do spellings everyday, Sir? Are we going to have a test everyday?" The other response was, well, it was creative writing. Already by 11, they've got it drummed into them what a subject involves. And it's very difficult to break them out of it.'

It is all there, in these documentations, if one has time and patience to unravel it. Teachers set written work to test, this is universally recognized and practised. After this the intentions vary, not so much by subject as by individual pedagogical philosophies. Writing is to record, to aid memory, to induce specific imaginings, to clarify information, to discover what one's knowledge and opinions are, to learn specific information, to explore the self and the world, to learn what the different subjects are by the kinds of writing the teachers demand—and as a means of class control.

Within these varied uses for writing one can detect different functions. There is an inner role which is concerned with learning and thinking, and an outer role which is to do with external features, the skills of presentation, paragraphing, punctuation, spelling, titles, spaces, underlining, etc. These skills carry powerful social prestige but would seem to have little to do with the inner role of using language for thinking. History may account in part for these differing aspects. When the majority of children had only elementary education the focus was on basic literacy—to read and to write with some measure of educated spelling and punctuation. With the coming of universal secondary education the focus began to change. Many more children began to move into further and higher education where good hand-writing and spelling were not enough. The student needed in addition to master the kinds of writing shaped by the thinking problems which he encountered. The Bullock Report is concerned with both these complementary aspects of writing. How is it that they are so often seen as polar opposites? The reason may perhaps be found in increased numbers of pupils who, since 1944, stay on till 16 to take public examinations. In a more affluent and socially mobile society most school students want the ticket of attainment offered by examinations, and the types of examinations that have been most widely developed in this country have put a premium on the giving back of information presented to the students by the teacher in accordance with a prescribed syllabus; furthermore, the fact that the students are examined by written answers has put a premium on writing of a certain kind, impersonal and classificatory. Thus, the shaping of writing by the thinking problems of the student has not been much in sight before the sixth form because the public examinations at 16 have not generally been concerned to test the pupils' capacity to think. There are, of course, some exceptions. Some of the CSE examinations have attempted to break away from this pattern of academic tennis (returning the ball of information served by the teacher), and most notably some science teachers and mathematics teachers have been aware of the long-term danger, and have developed programmes of work leading to specific examinations which have set out to test the student's capacity to use his knowledge in new situations, i.e. to think for himself. It is perhaps not insignificant that the Project team (Writing Across the Curriculum) has found more interest from science teachers in the ideas we were briefed to present and more cooperation in attempting to widen the range of writing available to school pupils than from any other subject group apart from those teaching English.

Writing, and Contemporary ideas about language

On the whole people tend to regard language as they regard breathing —something you take for granted until you notice something wrong with it. Thus language becomes something to be put right, cured, corrected; its active role in learning is not understood. Written work suffers particularly in this context. It is a visible product, there to be corrected, and while talk is, today, often seen as a useful part of the process of learning, writing is seldom seen as having this function. The written language of school pupils cannot be dealt with at anything but a superficial level without taking account of language more generally, when one is in a position to single out the features and problems peculiar to writing. At the risk of giving altogether too simplistic an account I want to try to identify those ideas about language which need to be understood if writing is to make a more effective contribution to learning.

First, there is the fact that language is learnt by use—by encountering it and using it in a variety of situations. It is therefore, important to note the difference between the use and the study of language. Essays, science reports, notes, stories, written accounts of all sorts are examples of use—in these, language is the means not the end. Study of word meanings, parts of speech, clause analysis, literary devices and style, analysis of texts and other matters of this kind undertaken by English teachers, are aspects of the study of language. Language is here the end as well as the means. There is no research that I know of at school level which shows that the *study* of language has any effect on how pupils *use* language, and much that shows it does not, and since these researches are fairly widely known many English teachers have given priority to experience of *using* language, whether it be in the form of purposive use or practice exercises.

Second, because language is closely linked with thought, the utterances which each individual constructs are outward expressions of his thinking and are therefore a crucial part of his learning. To assimilate new information and incorporate it into our existing knowledge we have to restructure it for ourselves in our own formulations. We have seen how children have expressed to the Project Officers their difficulties in learning in the words of their teachers. This is particularly true of younger children, and it is interesting to note that some of the special science programmes referred to above, advise that younger children write up the reports of their science work according to their own lights and not in a predetermined form. Thus, while notes copied or dictated may be the most effective way of getting children to learn specific material, or of maintaining control of the class, they rob children of opportunities to use the language as an aid to their own thinking. If teachers understand this dilemma they can solve it in terms of their own priorities.

Third, conversation is the most immediate and effective way for most people to penetrate and assimilate new knowledge. The feedback from others checks and expands one's own understanding. Consider the following example.

Keith Harvey is a probationary physics teacher. One of the Project team tape-recorded a group of children working together during one of his lessons. The pupils had previously done some worksheets on ways of measuring time. For this lesson they had been asked to bring a candle or jars and tins (which could be fixed together to make sand or water timers). Some of the pupils had had difficulty in deciding how to measure the amount that their candles had burned in a given time but it was clear that different candles had produced different results. This is a short extract from the end of the transcript.

Julie   Sir, Mr H. You wouldn't be able to make a reliable candle clock.
Teacher Why?

Julie	'Cos both our candles burnt down different times. It depends on the candle.
Girl	Unless you buy the same candle.
Julie	Because her candle burnt slower than our candle.
Teacher	There's all sorts of things (indecipherable).
Julie	And the wind.
Teacher	And the wind, yeh.
Girl	Like the candles you put in candelabra last longer than these ones.
Teacher	Well, they're different thicknesses, aren't they?
Boy	No they're not, they're posher.
Teacher	What, they're made out of something else?
Girl	And then twisted.
	That's what I mean.
Voices	Yeh, those twisted ones.
	Burning slower
Girl	Yeh, because when we had a power cut two weeks ago we had a candle like that and one of them fancy ones in a candelabra. And the candelabra lasted longer than that one.
Julie	So it depends on the candle.
Boy	Those twisted ones do burn slower.
Teacher	I can't see why.
Boy	'Cos it has to go round the bend (Laughter).
Girl	It's a different wax, I suppose.

During the lesson there was a great deal of talk about the details of measuring and the variables that influenced their results. Now, with the teacher, these pupils are here reflecting on the implications of what they have been doing—and are bringing experiences from outside the lesson to bear too.

Keith Harvey commenting on the transcript said: 'I like practicals. I think you can get a lot more—not what you expect—but you get a lot more from the children. It seems to be a good way to teach children to think for themselves. I think the learning itself is far more diverse than simply having an objective. The objective here is connected with the idea of time and yet I'm sure they learned far more about candles—and general principles of exactly what they're doing. They're out on a limb—they haven't been told what to do.'

Fourth, in the absence of this kind of talk, expressive writing is a good substitute. Practical work and field studies by their very nature allow for exploratory talk as part of the operation, which might make exploratory written work redundant. But as this kind of talk is more difficult to organize in classrooms, personal, near-to-speech reflective writing will fulfil some of the functions of exploratory talk, and in addition provide opportunities for sustained reporting and reflection which talk does not. Here is another illustration concerning Sue Watts, a chemistry teacher. She asked her first year mixed-ability class to keep a diary of their work. These diaries are in addition to the record of experiments which the children keep. I quote from two of them.

Carol

Today we played another sort of guessing game on one of the benches we had three trays half full of water one was hot one was cold and the middle one was warm. first of all we put our hands into the cold and hot trays then after a minute we put our hands into the middle tray and they felt very different my right hand felt cold and my left hand felt warm but this was wrong because both of them should have felt the same because the water was all the same temperature. After we had done this we were given a thermometer and we found out the temperature of my fingers and my temperature is 35 °C.

Ann

In the science laboratory we have been looking at many different things, I was particularly interested in mercury and why it is so heavy. I once learned that you get mercury in a thermometre and that if you drop the thermometre and it smashes all the mercury will spread over the floor and turn into silver balls. You mustn't touch it because it is poisenous. My mother once dropped a thermometre when she was a child and that is how I know.

These diaries give the children the chance to reflect on what they are doing. Thus Carol thinks that it is 'wrong' that her two hands felt different in the same bowl of water. She is initiating a question about the relativeness of subjective judgements and her teacher will respond to this when she reads the diary. Ann selects the things she is 'particularly interested in' and brings together school knowledge and knowledge from other sources. Neither of these children could have initiated this reflective awareness of the work they have been doing in their formal reports. Both kinds of writing are important and have different functions within the subject area—in this case, science. The diary entries are like talk in some ways, though there is not the instant feedback that comes from discussion. On the other hand they give an opportunity for coherent and extended reflection which talk does not.

Many teachers have said to us—and pupils too—about this kind of writing, 'But that's English'. This is both sense and nonsense. It is sense in that everyone interprets the present in the light of expectations built up from past experience. In as far as school experience has taught both teachers and children that different kinds of writing are, in school, associated with different subjects, they are right, but in the world, language alters according to the situations of which it is a part, i.e. what it is intended to do, and who it is for—so the diaries being primarily for the writer with the teacher as a subsidiary interested reader, are different in purpose and audience from the formal reports. Neither is functionally peculiar to science or to English.

A school-wide view of writing activities

One of the features of the subject organization of secondary schools is that teachers of one subject have only the haziest ideas of what goes on in other subjects. A valuable result of the Committee's recommendation with regard to language policies is that teachers have begun to look at the overall picture of language activities in their schools. A recent sample offered by a school to the Project team for discussion was a surprise to the staff. It consisted of all the writing done in one week by three school pupils of varying abilities from three different classes. Each pupil was found to have done: (1) a piece of independent, extended writing for English, set after class discussion but free to be treated in any way; (2) notes or extended pieces copied from the blackboard; (3) answers to comprehension questions on the blackboard or on worksheets. (2) and (3) covered history, geography, European studies, religious education, and science. So a bird's eye view of written language activities offered to these three classes in one week showed a very narrow range, with five-sixths of the writing (coming from five subjects) giving the pupil little or no chance to use his own language, or to write extended prose as distinct from short answers. *No teacher pursuing his own teaching conscientiously was aware of the heavy weighting in one direction which the total picture revealed.*

Given that teachers are beginning to ask how much and what kinds of language there should be, it is clear that they need some tool of analysis. Teachers need to look at the part played by language in their lessons and communicate this to each other, then the first steps towards a language policy will have been taken. To develop a useful policy further, teachers of all subjects must then be prepared to incorporate in their thinking the sort of knowledge about language and its import for learning which has been indicated in this book. What are the implications for writing?

The writing research models of discourse

The writing research team, funded by the Schools Council to investigate the development of writing abilities at the secondary level was faced with essentially the same problem as teachers examining language use across the curriculum; they asked themselves the same questions—what kinds of language should pupils have opportunities to use? Are some kinds more important than others to the pupils' learning? Are some kinds more difficult to acquire? What are the circumstances most favourable to acquisition?

A first step in attempting to answer such questions was to survey a wide sample of school writing (2,000 scripts). To do this and to describe them we developed a model of mature written discourse. We assumed that the socializing effect of school would cause school writings to approximate—more or less—to adult models, though we did not discount the possibility that we might find some writing which existed only in schools. This model allowed us to place each piece of student writing at some point on one of two scales; not scales of quality, but scales of (1) function and (2) a sense of audience—what the writing was for and who it was for. Though other dimensions might be examined, these we felt, were of primary importance in affecting what and how the writer wrote. Models are developed for particular purposes and we were seeking a model which would allow us to classify school writings by other categories than those of subject and thereby enable us to study the comparable intellectual problems which would be, we predicted, reflected in the writing from various subjects. These would be masked, as they are in practice today, if we viewed various kinds of writing as being proper only to various subjects. We thought that intellectual growth transcended subject boundaries and we should be able to trace this more generally by looking at kinds of writing across the curriculum since verbal–mental operations such as report, explanation, argument, classification, comment, narrative, etc. are not peculiar to any one subject.

1. *The Function model.* First, then, the Function model was designed to help us look at what writing is for. This is not exactly the same as a writer's purpose because our culture has itself developed distinct language forms which are typically associated with certain situations. We know quite well, for instance, when we are in a story, or a sermon, or being persuaded to buy something—because we have internalized these kinds of language from our day-to-day encounters with them. When children come to write in similar situations they draw on their pool of language experience which helps them to know what kind of language to use in certain types of situation. For instance, no-one teaches children in an infant class to begin their stories 'Once upon a time', or 'Once there was a little pig . . .' but most of them do this because this beginning is so obviously a 'marker' of stories which they have listened to or read.

The boundaries are by no means hard and fast but it has been found possible and useful to distinguish three broad categories of function to which recognizably distinct kinds of writing belong. What distinguishes them is that writer and reader mutually recognize the conventions that distinguish one 'job' from another. It is true that there are often linguistic differences but these are the markers of the different functions whose essential difference lies in the sort of things the writer takes for granted about his reader's responses. To take a simple example, if we read 'Once upon a time there was a flying horse' we know the writer is taking it for granted that we recognize a story and shall not quarrel about whether horses can fly.

Our model distinguishes three main 'recognized and allowed for' functions of writing, EXPRESSIVE, TRANSACTIONAL, and POETIC.

THE EXPRESSIVE FUNCTION

The central term of the model is taken from Edward Sapir,[1] who pointed out that ordinary face-to-face speech is directly expressive

[1] E. Sapir, *Culture, Language and Personality*, University of California Press, 1961.

and carries out its referential function in close and complex relationship with that expressive function. Since much writing by young children is very like written-down speech we described the expressive function as that *in which it is taken for granted* that the writer himself is of interest to the reader; he feels free to jump from facts to speculations to personal anecdote to emotional outburst to reflective comment as he wishes, feeling that none of it will be used against him. It is all part of being a person *vis-à-vis* another person. It is the means by which the new is tentatively explored, thoughts may be half-uttered, attitudes half-expressed, the rest being left to be picked up by a listener or reader who is willing to take what is expressed in whatever form it comes, and the unexpressed on trust.

The Writing Research Team said: 'The more we worked on this idea of the expressive function, the more important we felt it to be. Not only is it the mode in which we approach and relate to each other in speech, but it is also the mode in which, generally speaking, we frame tentative first drafts of new ideas. By analogy with these roles in speech it seemed likely to us that expressive writing might play a key role in a child's learning. It must surely be the most accessible form in which to write, since conversation will have provided the writer with a familiar model. Furthermore, a writer who envisages his reader as someone with whom he is on *intimate* terms must surely have very favourable conditions for using the process of writing as a means of exploration and discovery.'

We quote in illustration three examples taken from metal-work, middle school science and social studies which contain many of the features of expressive writing. The first is an excerpt from a long account of a project in metal-work by a boy of twelve.

The becoming of a gnat killer
the name Gnat Killer came to me after the project for maths I had been doing, you see I was going to do a project to see how many people spelt Gnat rong they ither spell it Kat, Gnate, Knate, Knatte, and even Knatt. well when I started to do my project, I thought well I will call it the Gnat then Killer came into my mind because I am mad on Horror. So I called it the Gnat Killer, it may sound violent but it isn't well at least I don't think, it is only another name been created to the world for the perpose of a project, I dont know if my turnin two the dictionary, it will read

> Gnat Killer
> name for
> a dagger used
> by Ricky Baker
> in 1972 8th dec:
> (not spelt nat killer)
> in italics:

that is only my idear thoe not that it will work.

In the second case, the children watched an egg hatching in the science lab. They were asked to write about it in any way they liked.

We quote two examples to show the variation in perceptions and focus.

The Chick Hatching (In the science lab.)
The egg was begginning to crack, the chick inside was using all the strength it had to get out from being trapped inside the shell. I thought to myself, if just I could get a knife and crack the shell gently, then I could see what the chick look like. I kept seeing bits of the chicks wing as it was slowly cracking. The chick was gradually pushing it's way out, but could not go no further, because the other eggs were in the way, so our science teacher (Mr bowman) gently lifted the lid of the incubator and the egg that was cracking was moved away from the other eggs, Now the chick was not having such a struggle. you could see it pushing up against the top of the shell, and then going down for a rest. I said to Mr bowman will the chick be hatched by the time the bell goes, and mr bowman took it as a joke, put his head down to the incubator and said will you be hatched by the time the bell goes." don't worry the chick did not reply.
The chick was just about hatched now, all it had to do was struggle out of the top of the shell, it was just about out, as it got out all the class shouted, "it's out!
It was just standing up properly, with all it's feather's sticking up on end, it looked around in wonder, when the bell went and mr bowman but the lid back on the incubator.

'The Egg . . .'
The egg begins to hatch, the crowd of children cluster around the humid inqubator. The egg cracks about 3cms, the egg tooth on the chick's beak can be spotted, the chick pants as he catches his first breath of fresh air then he goes back to work the egg opens about another centermeter then the chick lays back gasping for air it rests for about a minite then on it goes to crack the egg it pants and gasps catchies abreath then back to work in about ten minutes the egg is hatched the chick's feathers are wet and flat and the chick lays panting in about two minits time he is stood up then the feathers begin to dry off and fluff up then it becomes steaddy and soon enough it will be walking around.

The third is an excerpt from a fourth-year boy's account of his work in social studies. Of his project on 'The British political system' he writes:

This was the first big project I did and now when I look at it I marvel at how pathetic it is . . . I have placed emphasis on the wrong things and made all the mistakes I was warned about . . . so it ended up in my eyes as a great failure. But I have learnt more from that project than any other piece of work I have ever done at school so it was only a failure in terms of what it was, not what was learnt from it. I learnt not only how to improve my work from this project but believe it or not I also learnt a great deal more than I already know about the British political system although most of it is not in the project.

These writings have many of the features that characterize speech— the utterances stay close to the speakers, reflecting the ebb and flow of thought and feeling, and like speech this expressive language is always on the move. It moves according to the demands of what it is for—all are towards the transactional in that they are giving information—what the reader wants to hear, and how the writer's language resources allow him to meet these demands—his own and other people's. All three would seem to have the self as audience and also their teacher whom they believed when he told them to write as they wanted. In addition Ricky Baker and the fourth-year boy had some vague sense of a wider audience in that they knew their writing was for someone at London University to read.

THE TRANSACTIONAL FUNCTION

The transactional function is that *in which it is taken for granted* that the writer means what he says and can be challenged for his writing's truthfulness to public knowledge, and its logicality; that it is sufficiently explicit and organized to stand on its own and does not derive its validity from coming from a particular person. So it is the typical language in which scientific and intellectual inquiry are *presented* to others (not necessarily the kind of language in which these activities are carried out); the typical language of technology, trade, planning, reporting, instructing, informing, explaining, advising, arguing and theorizing—and, of course, the language most used in school writing. It can be more or less expressive too, but the more familiar the writer is with the information, or the arguments or whatever, the easier it is to respond to the demand of explicitness and the less is the need for expressive features.

Ian was in the same metal-work class as Ricky. His project was a brass dish and he wrote a long piece about brass and how you work it and how he made his dish; but, although there are expressive features in his writing, his purpose is firmly transactional. He feels himself to be knowledgeable about brass and sets out to *inform* his readers. The differences between his piece and Ricky's (quoted above) illustrate the differences between writing that is focused on a transaction and writing following the free flow of the writer's thoughts and feelings.

About Brass
Brass is an yellow alloy of copper and zinc in varying proportions with quantities of lead tin and iron brass is very malleable, fusible ductile and readilly cast and machined, Muntz or yellow metal is a variety of brass containing 60% copper, As it resists corrosion well, it is largely used for shop propellors, bows and fittings, corrosion means rusting away.
Some examples of what brass can make:
1. electrical fixtures
2. unexpensive jewelry
3. metal decorations
4. table lighter
5. trombone
6. screws

7. marine hardware
A percentage table

copper	60%
brass	37%
white alloy	3%
	100%

In the famous Praha football stadium, nearly everything in side the walls are made out of copper including the goalpost, the team that play here are Sparta Prague. Mallable means easy to work with, and very dicible.

THE POETIC FUNCTION

This function is that *in which it is taken for granted* that 'true or false?' is not a relevant question at the literal level. What is presented may or may not in fact be a representation of actual reality but the writer takes it for granted that his reader will *experience* what is presented rather in the way he experiences his own memories, and not *use* it like a guidebook or map in his dealings with the world—that is to say, the language is not being used instrumentally as a means of achieving something, but as an end in itself. When Huck Finn said all Tom Sawyer's great stories were lies he was mistaking the function of the stories (the poetic function) and operating the 'rules' of the other 'game'—the transactional. So a reader *does* different things with the transactional and poetic writings: he uses transactional writing, or any bit of it, for any purpose, but who can say what we 'do' with a story or a poem that we read, or a play we watch? Perhaps we just share it with the writer; and not having to 'do' anything with it leaves us free to attend to its formal features—which are not explicit —the pattern of events in a narrative, the configuration of an idea and, above all, the pattern of feelings evoked: in attending in this non-instrumental way we experience feelings and values as part of what we are sharing. The term 'poetic' is used here in a narrower and more specific sense than in its general usage. We use it here in the sense of the Greek from which it was derived to mean 'created'. Its examples in school use are the stories, poems and plays which children write.

In the context of these function categories it is perhaps worth drawing attention to the ways in which knowledge—and, by implication, learning—in secondary and higher education is organized; systematically by subject disciplines and presented in the explicit, often specialized and generally formal language of those who know to those who are beginning to know. In contra-distinction the kind of knowledge (and way of learning) that goes on outside school is random, unsystematic and may or may not be linked by the learner to school knowledge. The bridge between systematic and unsystematic knowledge would seem to be the learner's own language, his own formulations which are the means by which he incorporates the

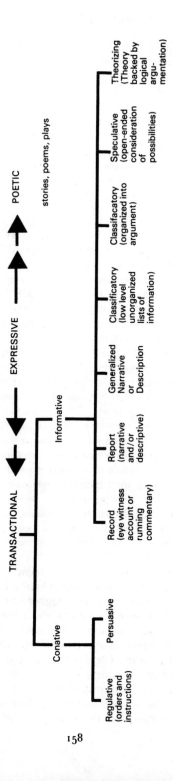

TRANSACTIONAL → EXPRESSIVE → POETIC

stories, poems, plays

Conative

Regulative (orders and instructions)

Persuasive

Informative

Record (eye witness account or running commentary)

Report (narrative and/or descriptive)

Generalized Narrative or Description

Classifactory (low level unorganized lists of information)

Classifacatory (organized into argument)

Speculative (open-ended consideration of possibilities)

Theorizing (Theory backed by logical argumentation)

new into his existing corpus of knowledge. Most people do this by talk, but in school the expressive and poetic functions of writing give children opportunities to recognize and evaluate new knowledge and new experiences and thereby gain self-knowledge.

Our transactional category clearly needed to be broken down in terms of the kind of transaction undertaken, since the great majority of school writings are in this category. If teachers want to gain some idea of the range of writing opportunities that their pupils are getting *across the curriculum* the diagram we used is one way of enabling them to do this.

To quote from *The Development of Writing Abilities*,[1] 'This, then, is the broad picture, and taking the view currently being put forward that linguistic competence embraces rules of two very different kinds, grammatical rules and "rules of use", we suggest that the expressive function may mark out an area in which rules of use are at their least demanding, an area of comparative freedom in this respect. Thus, it may be that, as a writer moves from the expressive into the transactional, he increasingly takes over responsibility for rules of use that, in sum total, constitute one kind of order, one mode of organisation by which we encode experience—the cognitive order described by psychologists and codified into laws of logic by the philosophers. And it may be that, as a writer moves from the expressive to the poetic, he takes on responsibility for rules of use that represent, in sum, a different kind of organisation: one that we can recognise in a work of art, while at the same time we know little about the general principles that govern it.

'All this has obvious bearings upon our developmental hypothesis. From the area of least demand as far as rules of use are concerned, the learner-writer progresses by increasingly recognising and attempting to meet the demands of both transactional and poetic tasks, and by increasingly internalising forms and strategies appropriate to these tasks from what he reads and incorporating them into the pool of his resources: thus in the course of time he may acquire mastery of both varieties of rules of use.

'There exists always, of course, an alternative hypothesis: if you limp about long enough in somebody else's language, you may eventually learn to walk in it.'

2. *A sense of audience: the child and his reader.* The notion of audience, like that of function, has powerful implications for school writing in spite of the fact that almost all of it is produced for one sort of reader —the teacher. What makes for differences between the pieces of writing is not, objectively, who the reader is, but how the writer *sees* his reader; and children see their teachers in very different ways. Not only may different children's views of the same teacher vary, but a child may see his teacher as a different sort of reader on different

[1] Britton *et al.*, *The Development of Writing Abilities (11–18)*, Macmillan's Research Series, 1975.

occasions. Furthermore, the process of writing is more complicated even than this suggests. The monitoring self is always a part of a writer's sense of audience, as well as those others—the class, friends, examiners sometimes, parents perhaps, and a shadowy 'public'— who lurk behind his shoulder. In speech the actual or potential feedback from the listener continuously modifies what is said because the audience is physically present and sooner or later will speak too and thereby modify what you say next. In writing there is no such feedback, but somewhere, out of the direct focus of attention, a sense of an audience remains, ready to take shape when summoned, or stepping forward uncalled. We think it likely that one reason for the great amount of inert, inept writing produced by school students is that the natural process of internalizing the sense of audience, learned through speech, has been perverted by the use of writing as a testing or reproductive procedure at the expense of all other kinds of writing. When a writer's focus is on returning as exactly as possible what he has been given, the sense of any other audience, including the monitoring, reflective, independent self, disappears, leaving incomprehension, resentment or despair, or alternatively the satisfaction of producing something to satisfy someone else's demands. We think the following conversation throws light on some of these matters.

Andrew (a third-year boy) is talking to one of the project officers who knows his school and his teachers well. They are talking about Andrew's humanities project on 'The industries and populations of towns' and are referring to an anecdote he had included about slavery:

BN   There's an anecdote about a grave . . . it was something about slaves —it was something about, er, in a cemetery near Harlow there's a gravestone which says on it 'Here lies so-and-so, the property of somebody-or-other . . .' How did you find that out?

A   Well, we just moved—'cos we don't live far from that end, we'll go to church fairly often, and, we went round to see that church 'cos we saw it from the road and I like looking at gravestones and that and I saw it sort of fitted in with that.

BN   And who are you asking the question of 'Rather unusual to put on a grave, don't you think?'

A   Well, that's to anyone who reads it.

BN   Do you think of your reader?

A   Yea

BN   Who do you think of?

A   Anyone who reads it. Mainly Sir, 'cos he reads it. And anyone else.

BN   But I mean, when you put a thing like that, or your jokes, do you think to yourself 'How will Mr S . . . react' or do you think in more general terms, how will anybody react—you know?

A   More general, I think, how would anyone.

BN   You do? So you're not just thinking of him.

A   Oh, no.

BN Would you, I mean, it's kind of a bit odd because you say that mostly these just get stockpiled, you know, so they're not read by anyone else normally.

A Not really, no. But we're always told to write it as though we're trying to explain to someone else of our own age—so we just get used to it.

BN So you do it, I mean you do actually think of other people reading it even though you don't expect them ever to do it?

A Yea, that's how we have to. They always write out 'as if to someone of your same age', and you've got to tell them about it. . . .

BN In some of these things you give your opinions, don't you?

A Oh yea, most of it.

BN You like giving your opinions, do you?

A Yea

BN Do you very often get asked for them?

A Mostly, yea. Especially in this subject. (Some talk about the humanities projects of the class.)

BN It looked to be as though you were looking for the chance to give your view.

A Well, perhaps, yea . . .

BN This is a bit you wrote about *To Sir, With Love*, you said 'On reading this section from the book . . . I felt thoroughly annoyed to think that those businessmen wanted a man for the job but when they found that he was coloured, they rejected him and what was more annoying was that those men probably gave the position to a less intelligent white man.' That was just how you began. So already you're expressing an attitude—you're saying what you think about it.

A Yea. I don't think we were asked to put our opinion on that but I was annoyed. When I usually do something, I always position myself as the person like now, I mean I was treating myself as the coon who got turned down.

It seems that Andrew, in his writing, takes account both of his actual audience—his teacher—and some potential wider audience. Earlier in this conversation he had said that he often puts jokes in his writing because his teacher enjoys them and he himself likes them. This seems near to a trusted adult audience, which we think the most fruitful in developing writing ability because it allows a writer to experiment with safety. Andrew also understands the convention which requires him to write *as if for* some particular audience—'they always write out "as if to someone of your own age", and you've got to tell them about it'; and he understands that this convention allows him to express his own opinions and feelings within the framework of the 'as if' situation.

We have found an increasing number of teachers who have tried to create situations involving an audience other than themselves— real or imagined—as a context for the writing they have asked their pupils to do. On the whole this tends to come from teachers concerned with the humanities. In the sciences and among older students there is perhaps an unspoken sense that this is a device for younger children or for less serious students, yet the American psychologist George Kelly in a paper called 'The language of hypothesis' describes an attitude and a corresponding use of language which would help to bridge the gap between the known and the unknown. Kelly calls this the use of the invitational mood. He says that at moments of risk we would be greatly helped if we deliberately abandoned the indicative mood and operated in the invitational mood with its language form, 'let us suppose . . .'. He says that this procedure suggests that things are open to a wide range of constructions and 'there is something in stating a new outlook in the form of a hypothesis that leaves the person himself intact and whole'. When students reflect on their work or even report what they know, they put themselves at risk in terms of their teacher audience, or any audience for that matter, and we would stress the value of the invitational mood both with regard to a hypothetical audience as well as hypothetical ideas. We quote an example here because the 'as if' situation is directly one of audience, 'Suppose that you wanted to explain what diffusion is to a younger child who has not done any science and does not know what molecules are':

Explaining diffusion (Frances, 14)
We all know that gases are invisible, in other words we can't see it there but it is present. Gases are made up of molecules, for the moment let us call molecules 'marbles'. These marbles are present in solids as well, but here they are squeezed together and are unable to move. But in a gas the marbles (which cannot be seen with the human eye, only with an electronic microscope) move about freely and do not move in one specific direction, they move at random. . . .

This, of course, was an 'exercise', but it was also a game played by the children and the teacher—a combined knowledge and language game. The pupils had to manipulate their knowledge and their language to serve an audience other than themselves and their teacher, and, as their teacher observed, you only really understand ideas like these when you can play with them—supposing this, and supposing that, in Kelly's terms.

Our next examples come from situations where the writing was directed to real audiences other than the teacher. First, the *Cleveland Evening Gazette*, where the English adviser arranged for the paper to carry a weekly feature page, 'Report from the Cleveland classrooms'. A different school would be responsible for the page each week. This feature would run for six or seven weeks, and then be picked up again later in the year if it interested the readers. One issue was contributed by pupils of a sixth-form college. The page carried reports, a story—'What a north-east town might be like when the iron and oil had run out'—opinions and a poem).

Continuously, monotonously,
the tall concrete towers

> Churn out clouds of never ending smog.
> As time goes by,
> Slowly the whole landscape
> will become engulfed by a
> network of structural steel,
> While beneath this mechanical mass
> Life still goes on.
> Life without aim,
> Life without meaning,
> Just one repetitive day after day.
> Each morning, men of all ages,
> Hurry to their industrial employment.
>
> *Karen Robson*

We include below the categories of a sense of audience which the Writing Research team devised to help them in their descriptions of school writing:

child (or adolescent) to self
child (or adolescent) to trusted adult
pupil to teacher as partner in dialogue
pupil to teacher as examiner or assessor
pupil to other pupils, or peer groups
writer to his readers, or public audience.

The two models of 'Function' and 'A Sense of Audience' make it possible for us to look at the range of writing which children are doing in all their school work; and it allows us to be quite precise about the development of their ability to write appropriately for different purposes and different audiences which was how we defined progress.

'Development is to a large extent a process of specialisation and differentiation: we start by writing the way we talk to each other—expressively, and envisaging as our reader a single, real, known person; gradually we may acquire, *in addition*, the capacity to use the transactional and the poetic, and to write for an unspecified, generalised or unknown audience. Further, this differentiation ideally occurs as a result of coming to hold those adult purposes for which the mature forms of writing have been evolved, and of learning to anticipate the needs of different and remoter audiences. So the development of writing abilities is partly conditional on the more general development of the child out of egocentrism; but that general development may itself be aided by the practice of writing.

Yet, the findings of our survey showed that the range of kinds of writing that children have opportunities to use in the secondary school narrowed drastically as they moved from year one to year seven. The dominant kind of writing was classificatory for an examiner audience. James Britton (*et al.*, 1975) commenting in the Research Report says:

If, as we had predicted, the development of writing abilities showed itself as a growing range of kinds of writing shaped by thinking processes, we should have expected to find in the sample a great deal of expressive writing in the early years, in all subjects (instead of the 6% we actually found), and an increase in the later years of classificatory, speculative and theoretical writing, as well as persuasive and poetic—all these compensating for a reduction in the expressive; and at the same time, a proportion of expressive writing maintained and developing into its maturer forms and purposes.

It would appear, however, that the pressures to write at a classificatory level of the informative—and in the main for an audience of the teacher as examiner—were great enough to inhibit early expressive writing and to prevent all but minimal development into the more abstract levels of the informative; strong enough at the same time to cut down drastically in the seventh year the output of the poetic. We believe an explanation for this unexpected narrowing of the range must be sought *in the whole curriculum and its objectives* [author's italics]. The small amount of speculative writing suggests that, for whatever reason, curriculum aims did not include the fostering of writing that reflects independent thinking; rather, attention was directed towards classificatory writing which reflects information in the form in which both teacher and textbook traditionally present it.

The Writing Across the Curriculum project team—and many of the teachers they worked with—have this to say about these findings:

Writing (like talk) organises our picture of reality, and, at the same time and by the same process communicates it to someone else. It is true that for educational purposes the first is more relevant than the second: we ask children to write so that they will organise their world picture, not so that we can learn things from them. But it is a fact of life that we can't have one without the other. Language *has* two faces and we are well advised to take account of both. The moral we draw from this is that, if we want writing to be a means of thinking and of active organising, we must make sure that the writer feels he has a genuine communication to make and is not merely performing an exercise—which was how David, aged thirteen, saw it: he said, 'You write it down just to show the teacher that you've done it, but it doesn't bring out any more knowledge in you.'

'Genuine communication' for a child, in whatever context—writing for history, English, science, a letter, a diary—is very often going to mean an inseparable blend of giving an account of the topic and expressing a response to it. If this is so, we should accept the mixture; if we discourage the personal element in it, we risk making writing an unwieldy and alien instrument instead of a natural extension of the child's own mental processes. And accepting it means more than simply allowing it to happen: it means agreeing to be communicated with in that way and making ourselves a real audience to a child by giving an authentic response to the communication as a communication rather than by giving back an evaluation of how well he has accomplished the task.

What emerged from the Project, 'Writing Across the Curriculum'

The Project team reported what seemed to us the most significant things that emerged from four years' work with teachers in schools. We identified also some deep dilemmas which teachers will need to resolve for themselves in the light of both their individual priorities, and the school priorities arrived at by collective discussion and decisions.

'We observed that the examples of children's written and spoken language which seemed to reveal language operating as an agent of thinking came from individuals or small groups of teachers who were interested in our research and made use of it. They did not generally represent changes in the work of a whole class, let alone a whole school. We therefore began to suspect that the changes we found were 'exceptions' or 'isolated cases'. We think these children were probably already 'committed' to learning and therefore responded most positively to enlarged opportunities. We think the majority of the children remain generally uncommitted.

If this is right it means that special programmes for this or that will have little effect. This hypothesis suggests that every piece of writing (and the circumstances that gave rise to it) represents a network of past experience, relationships and expectations linked to a continuum of other such networks. We have attempted to identify what have emerged for us as the most significant features in these webs of learning and language. But, as might be expected, the items which we see as significant are themselves networks. This interrelatedness throws light on why efforts by individual teachers to make a change here or a change there have only resulted in what we have referred to as 'isolated cases' of progress. We think changes have to be more wide-spreading to be effective.' We summarized the chief enabling or disabling features in the circumstances from which school writings arise as follows:

1. *How a learner sees himself.* His view of himself as a school student depends a great deal on how his teachers see him. How his teachers respond to his efforts, or lack of effort, depends in turn upon how the teacher sees *himself*, and his view of his role as a teacher is closely related to his view of learning. When we look at language in these contexts we find the kinds of experience and of language reckoned appropriate to the classroom is closely related to the teacher's view of learning and his role relationships to his pupils.

2. *A writer's sense of audience.* We think it likely that one reason for the great amount of inert, inept writing produced by school students is that the natural process of internalizing the sense of audience, learned through speech, has been perverted by the use of writing as a testing or reproductive procedure *at the expense of all other functions of writing.*

3. *Assessment and criticism.* The most dramatic changes in writing that we observed came when teachers moved out of their role as examiner into the role of an adult consultant. The unresolved problem remains —at what points in children's engagement in the process of writing does advice and criticism assist rather than divert? (See the analysis of phases made by Medway and Goodson in the pamphlet *Language and Learning in the Humanities*)'

4. *The role of everyday language in learning.* The things we choose to put into words, and the words which come to hand to express these meanings reflect our unique interpretations of experience. We need to begin by accepting and encouraging children's own language whatever it may be. Particularly in the written language this view collides with some notions of 'correctness'.

5. *Conditions for moving into good transactional writing.* (a) These occur where circumstances encourage the interplay of first-hand and secondary experience. The validity of first-hand experience and/or reflection in writing about history or science for example is generally ignored. (b) There should be opportunities for pupils to encounter good and varied 'models' of transactional language—to be encountered rather than imitated. Generally such 'models' are limited to textbooks.

Changes in notions of progress in learning and language

With a group of science teachers we kept returning to creative thinking and creativity. With a comparable group of humanities teachers the central issue was co-operative learning. We see these as related, and also related to the constraints on intention and choice which arise from general school procedures. We think that these are important in getting more children to commit themselves to learning. We also think for such commitment to happen on a wider scale than our observed 'isolated cases', whole environments for learning need to be changed, and for this to happen there need to be enough teachers in a school who share a view about learning and language to create a different set of possibilities in the school.

How these observations have been reflected in our most recent work with teachers

1. We have moved towards case studies
(a) of individual children, their writing, transcripts of their talk and conversations with them about their writing, reading and other work.

(b) of particular teachers—transcripts of their lessons, and of conversations with them.

(c) of exam work in a particular school.

2. In work with teachers we have arranged more workshops run by teachers of different subjects which put us all in learning situations. Transcripts and discussions of what was going on for each of us, and subsequent extrapolation to classroom situations have all been fruitful.

3. Our focus has been shifted from spending the available time on the research findings and their implications to getting teachers to plan and subsequently report back on specific steps they can take to get discussions going in their particular schools about learning and language, and to plan joint action where channels for collaboration exist. Essentially we needed to shift the initiative from us to them.

4. Teachers who had worked with us for some time were able to take this lead in meetings. Reports and suggestions from them were generally more effective than presentations from the Project team. Research findings became less important and the outcomes of teachers reflecting on their work more important.

How teachers can cope

Let us look briefly at the sharpest of the dilemmas which these findings present for teachers. There are three that I should like to draw special attention to.

1. *Intention.* Teachers have intentions, aims, objectives etc. for their pupils; their pupils have them too. Where they coincide conditions for learning are probably at their best. In *Writing and Learning* we wrote:

'Intention is said to be the motivating force for thinking as well as action, so creativity (in a very general sense) would seem to be related to intention. This seems a long way from creative writing, dance-drama and work in the art room which is, perhaps, where many people would locate creativity in school. But it is clearly absurd to believe that creative thinking can be attached in this way to a limited number of activities. We needed to look at creativeness more generally and our 'across the curriculum' brief forced us to do this. In discussing problems of commitment to learning we kept returning to intention and the kind of environments which would permit it and sustain it. Young children are curious and persistent and eager. One could call intention robust at this stage. In the secondary school it is, for whatever reason, a delicate thing. If it exists in relation to school work at all, it is easily snuffed out, or it drifts away, or is superseded by another feeble interest, but since so much of schooling pays such limited attention to children's intentions it is not surprising that creative thinking or creativity don't flourish, or at best merely survive in those activities that are least susceptible to other people's programming—improvised drama or writing poems for instance.

It is unfortunate that paying attention to children's intentions has become equated in many people's thinking with giving children sole responsibility for what they do. Clearly before most of us can exercise any choice of what we do, we need to know what the possibilities are, so it is not a simple matter of lessons structured by the teachers versus unstructured lessons in which pupils do what they like, but a much more complex operation of a teacher building opportunities for pupils' reflection, choice and self-direction into the overall intention of his own scheme for a series of lessons. The following piece of writing by a first year girl for her science teacher seemed to us to illustrate both the way in which children need to be made aware of the possibilities in the work they are learning and also the value of opportunities in the work they have done.

> *The most interesting things I have done in science at school*
> The best piece of Science I have do so far is the experiment with water the first experiment we did with water we had different kinds of water like pond, Canal, Destiled, tap water. With the water we added a soap flake to each sample of water. And if it bubbled a lot we had to tick it on the chart. After we did that piece of work we had to find out were the water in our taps came from. The next experiment we had to do with water was called water and floating. With this we had things like a cork, Glass, block of wood, foam, copper, stone, polystyrene and we had to float them and on our chart we had to tick off what it did. The thing I liked about this work was that we were experimenting and finding things out for selfs by doing this work.

Of course the problems become more acute as children move up the secondary school where the bulk of school knowledge is transmitted through the written language in its transactional forms. Reading books and writing attract increasing attention and most students find both difficult. We think it important to say that we found that project work, where it was well done, gave by far the most opportunities for students to make their own sense out of new information, to re-structure it to fit into their own expanding world picture, and to remain at the centre of their own learning. In short, pupils need more opportunities for their own choices, self-direction and self-criticism, within a framework of guidance and criticism by their teachers.

2. *The conflict between children's own language and the pressures for them to use 'correct' English.* We suggest that this topic needs much discussion between members of school staffs. One solution might be to distinguish between things that children write which are essentially their own learning operations (using their own formulations and expressions) on the one hand, and on the other hand the things that they write which are presentations of information for other people (these being as 'correct' as possible). A teacher will function both as a reader of their work for themselves, and, for other people, real or imagined. The Report suggests we may conclude that written 'standard English'

is only desirable in transactional public writing and for all other forms of writing the most appropriate and acceptable will be individual, more or less expressive writing. What seems important is that pupils themselves should understand that the writing they do should vary according to who it is for and what it is for.

3. *Assessment.* Closely tied up with this matter of standard and non-standard English is assessment, and I have referred earlier to the problems and ambiguities inherent in the situation. Teachers must be concerned with their pupils' progress; children like to know how they are doing. Yet the risks and anxieties of being tested constantly operate against the children's taking risks in their thinking or their language. Again we suggest that teachers might vary their roles. They will get very different results in writing when there is no assessment attached, but only interested comment.

Listening

Of the four modes of language use, listening is the least researched and the least specifically taught. Ironically in secondary schooling as a whole, and despite notable exceptions, it is in fact the language 'activity' to which by far the most pupil time is devoted—past research has indicated well over half the time! However teaching methods change, listening will still remain important in a range of adult occupations from professional jobs to telephone sales clerks.

The Report considered whether listening abilities should be included in the suggested national language monitoring scheme. This was rejected for the twin reasons that we do not really have the knowledge of how to measure listening skills, and anyway the practical problems of testing on this scale are daunting. However, the Committee did feel that attention needs to be given within schools to the use of listening and the encouragement of improved listening ability: 'The difficulty of listening is commonly under-estimated' (10.19). Research reviewed by Andrew Wilkinson[1] showed that students listening to lectures comprehended half or less than half of the basic matter. The two questions are of what does listening consist, and how can the resultant skills be best fostered?

Attempts to analyse separate sub-skills, in the manner of analyses of reading, have not proved successful. However, the work of the Schools Council's Oracy project[2] does, it seems to me, make it clear

[1] Andrew Wilkinson and Leslie Stratta, *Listening and the Study of Spoken Language*, University of Birmingham Educational Review, Volume 25, No. 1, p. 3.
[2] Andrew Wilkinson, Leslie Stratta and Peter Dudley, *The Quality of Listening*, Evans/Methuen, 1974.

that, as in so many other aspects of education, by stopping and focusing on an activity teacher and pupil can note certain aspects. In listening, for instance, the pupil can be helped to consider how communication is conveyed. He can also be made aware of the activity required in the apparently passive act. The Report puts emphasis on what I should call a *conscious* contextual approach: 'In our view the ability can best be developed as part of the normal work of the classroom and in association with other learning experiences. But deliberate strategies may be required, for it cannot be assumed that the improvement will take place automatically.' (10.21) The 'association' involves the consideration of the listening demands of subject activities, and the 'deliberate strategies' involves both the planting of listening activities and, I should suggest, some specific practice and tuition. A whole-school policy might consider the following recommendations:

Examine the subject

Each subject should consider the listening expectations of the classroom activity. They should ensure:
1. Listening plays a part.
2. There is a variety of listening sources, e.g. not all teachers' exposition. This variety should be of speaker(s), speech situations (talks, reminiscence, discussion, announcement, etc.), technical presentation (live, tape, visitor, regular teacher).
3. The sources are of suitable length, frequency, and difficulty.

Involve pupils

Whenever possible pupils should be required to take some action. The immediate function of the listening should be clear. This may mean:
1. Simple note-taking from news bulletin, talk, or discussion.
2. Carrying out an activity after listening to instructions.
3. Analysis—by extracting one person's point of view, or one set of arguments from a debate.
4. Making a decision after hearing a discussion.
5. Simulations—such as making an order out from a phone call, taking the train times from a phone call.

Specific tuition

This should be given on occasions. That is, whereas the activities above are inserted primarily to provide material for the project, activity, or study, I am suggesting here the occasional use of listening material to consider the problems involved and to give practice. This could involve:

1. Listening comprehension tests, which have the advantage of helping pupils become conscious of the skills.[1]
2. An analysis of speech from the point of view of the listener.
3. Practice in skills, e.g. aural recognition.
4. Note-taking based on sound sources.

Monitoring communication for learning

Douglas Barnes

Teachers have always wanted to know how well their teaching was going, and have gained reassurance from signs of pupils' interest, and from the quality of work done, especially in tests and examinations. Thus their sense of success is based upon the products or outcomes of their teaching. What has been missing has been the means of monitoring what actually happens in the lessons—the events which have given rise to pupils' satisfaction or dissatisfaction, and have led to written work of better or worse quality. Teachers' attention has tended to be directed towards products rather than towards the processes which led to these products, because of the difficulty inherent in both participating as a teacher in a lesson, and at the same time observing what happens in it. The tape recorder, however, now allows us to be both participant and observer: we are the first generation to be able to monitor our own speech-behaviour

[1] *Learning Through Listening* (listening comprehension tests for 10+, 13+ and 17+), tapes, teachers' manual, pupils' answer booklets, Macmillan Education, 1976.

The tests comprise three tapes, for the 10+, 13+ and 17+ age groups respectively, plus accompanying test materials for pupils and teachers. The first two tapes play for about an hour, the third for about one hour twenty minutes (playing speed $3\frac{3}{4}$). The tapes include the following tests:

tests of content—designed to measure the ability to follow and understand a piece of informal exposition;
tests of contextual constraint—which measure the ability to infer missing parts of a conversation from what is actually heard;
tests of phonology—designed to measure the ability to understand differences in meaning brought about by different emphases;
tests of register—which measure the ability to detect changes in the appropriateness of the spoken language used;
tests of relationship—which measure the ability to detect the kinds of relationships existing between people from the language they employ.

The tests use unscripted speech for assessing listening ability, rather than readings from the written language.

and to understand its effects on others. It is now possible for teachers to extend their insight into their teaching beyond the evidence of pupils' interest and achievements to the monitoring of communication, which constitutes a central part of teaching and learning.

Looked at in this context, the Committee's recommendation that schools should discuss a 'policy for language across the curriculum' seems to strike a different note. Certainly my emphasis would not fall upon grammatical forms, or upon appropriate usage, or on spelling and punctuation; it would not be concerned with the forms of language but with the part played by talking and writing in learning. Nor would emphasis fall upon individual pupils' language skills, but rather upon how the contexts which teachers set up in lessons encourage this or that kind of talk, this or that kind of writing. But this must have been the Committee's intention too.

When teachers, working alone or with colleagues, plan a course they are about to teach, they may have at the back of their minds a picture of how it will go in practice. This picture includes the kinds of presentation, discussion, demonstration sessions they will have with the class, the range of individual tasks they will set, the strategies they will adopt when going round to talk to individual pupils or groups while they are working, and so on. Part of the Committee's suggestion is that teachers should be more explicit with themselves and others about the teaching and learning processes that they are envisaging Especially when teachers are working as a team or a department, it should not be taken for granted that they all mean the same by such expressions as 'discussion', 'project', 'enquiry', 'making notes', and so on. The readiest way of finding out what one means by familiar educational terms such as these is by looking closely at what pupils and teachers actually do, and this is what I am recommending.

We can approach 'what teachers and pupils actually do' through what they say and what they write: in this case the writing would be considered not as a final product to be evaluated but as evidence of learning activities. (Teachers would also wish to look at non-verbal work done by pupils—models, art and craft, and so on—but that takes me beyond my present topic.) It is clearly possible to tape-record communication between the teacher and a whole class, and it should also be possible for a teacher to carry a lightweight cassette recorder and record his communication with individual pupils and small groups. Almost all teachers engage at times in this latter kind of teaching but we know very little about the range of strategies which they commonly use, though this is probably an important part of their work. A teacher who frequently organized small group work —for example in science practicals—might wish also to have information about the kinds of discussion which his pupils were able to engage in without his leadership. Thus teachers can have four sources of information about what they and their pupils do:

1. Spoken communication: teacher–class
2. Spoken communication: teacher–individual/group
3. Spoken communication: group without teacher
4. Written communication.

Let us suppose you are a middle-school teacher with a class of ten-year olds, and you are interested in finding out more about what happens in that part of your pupils' time which is devoted to social and environmental studies. Some of the time you are presenting new ideas and possibilities to the class and encouraging them to talk about them, but most of the time your pupils are working alone and in pairs on topics which they have chosen in the light of your presentation and suggestions. During this time you go round giving help and advice, and, when you have time, asking particular pupils to explain to you clearly what they are trying to do. Every now and then you plan a session in which pupils show the class what they are doing, and you endeavour to encourage general discussion of this so that they may learn from one another. Let us further suppose that you have tape recorded two presentation sessions, and have recorded yourself on several occasions going round the class giving individual advice. You have the written work done by pupils, and perhaps a recording of a 'report-back' session, and with all this bundle of material you turn back with the question: What should I be looking for when I listen to the recordings or read through the written work? There are some suggestions below, but in the end it is you who should decide what you are looking for—or you and your colleagues together, if you are working in a team.

What you should be looking for when you listen to tapes—preferably with one or two other teachers with similar interests—depends on the patterns of learning which you are aiming for. Thus one of the first benefits of discussing recordings with colleagues is that this is the ideal way of focusing discussion on the details of how children are learning and how best to support them in doing so. I am therefore not imagining that many schools will have a clearly-spelt-out policy document which describes the patterns of spoken and written communication to be adopted, but rather that deeper understanding of where you agree, and disagree with colleagues will come from looking closely at real examples. It is not a matter of achieving consensus—which is unlikely—but that after studying classroom communication you and your colleagues will be clearer about what kinds of talk and writing each of you values.

The best support I can offer for such debates is to suggest some questions which might start a discussion. These suggestions are far from exhaustive; indeed they may well prove unsuited to your values or to your situation. They are offered here merely as starting points.

Spoken communication: teacher–class

The central question to be asked is: Are you carrying out the teaching strategies which you intend? In answering this you might consider questions such as these:

1. Are you requiring your pupils to think for themselves or mainly asking them to feed back information from a book or from an earlier lesson?
2. Are you eliciting the pupils' existing experience and understanding and working from that, or does the way you are planning lessons tend to make their present knowledge irrelevant?
3. Do your responses to pupils' contributions include replies which use and develop what they have said, or are you predominantly evaluating replies as right/wrong, or good/bad?
4. When you present information, or give a demonstration, or read a poem, or discuss a visit, are you requiring your pupils to explain and hypothesize, or are you telling them what it means?
5. Do you ask pupils to expand what they have said, to respond to one another, to ask questions, to offer evidence, to consider alternative explanations, to plan lines of action?
6. Are your pupils in fact contributing at some length to the lessons, or answering merely in brief phrases?
7. Are they raising questions of their own, offering experiences and opinions, joining in the formulating of knowledge?

The way I have framed these questions tends to indicate what I consider to be good teaching and to contrast it rather crudely with an alternative view. If your own view of good teaching is different you will need to frame different questions.

Spoken communication: teacher–individual/group

The seven questions suggested above would be relevant here also. Indeed they are even more relevant, since in talking with one or two pupils a teacher is able to pursue such a teaching technique without her normal concern to hold the class' attention by keeping the discussion moving. The following may however be added.

8. What range of matters are you discussing with them? Are you mainly concerned with methods or are you asking them to talk about the meaning of what they are doing?
9. Are you succeeding in eliciting your pupils' thinking, so that you understand their idiosyncratic ways of interpreting things, their problems and misconceptions but also their strengths?
10. Can you sit in on a group discussion and gather how the work is

progressing without all the group addressing their remarks to you? Do you join in pupils' discussion as a participant rather than an evaluator?

Spoken communication: pupil–pupil

If a teacher asks a group of pupils to tape-record their discussion, what appears on the tape is unlikely to be exactly as it would have been without the recording, unless the microphone has been in place so long as to be virtually forgotten. What can be taken from such recording is an impression of what pupils can do at their best, and not necessarily an accurate picture of how they normally talk during group work. In listening to such recordings it is important not to bring in expectations drawn from written language. It is perfectly normal in conversation for sentences to be incomplete or to change direction, for much of the meaning to be left implicit and not put into words, for speakers not to make explicit the logical relationship of what they are saying to what has gone before, or for their contributions to overlap. All of this is perfectly normal; if you doubt it I suggest that you tape record yourself and your colleagues arguing in the staff room, and look at that. Once you have taken this into account, and listened to the discussions several times until you understand what is implicit in them, you are likely to find that your pupils have capacities which are not often visible in classwork. The questions which follow relate both to what the pupils are able to do, and to the part which the teacher plays in setting up group discussion.

11. Are your pupils really discussing the meaning of what they are doing? For example, in science are they talking about how their 'experiment' relates to the principle in question, or is their talk mainly at the 'Pass the matches!' level?

12. Can they find problems and formulate them, put forward explanatory hypotheses, use evidence to evaluate alternatives, plan lines of action?

13. Can they cope with differences of opinion, share out the jobs to be done, move steadily through a series of tasks, summarize what they have decided, reflect on the nature of what they are doing?

14. Have the topics which they have chosen or you have prescribed led to useful discussion? Did they succeed in raising valuable issues, or would they have benefited from some help in focusing attention? On the other hand, if you gave them questions to consider, were these helpful or constricting, did they lead to the kind of discussion you believe to be valuable?[1]

---

[1] Issues such as these are analysed in more detail in D. Barnes and F. Todd, *Communication and Learning in Small Groups*, RKP (forthcoming).

Written communication

The emphasis here should fall upon writing as one way in which a pupil increases his understanding both of the world about him and of what is presented to him in lessons. Since most teachers have traditionally treated writing as a way of storing information or of assessing achievement, the accent falls differently here, as a step towards redressing the balance. The central question is: Does the written work being done by your pupils accurately represent the kinds of learning you wish them to be engaged in? In answering this overall question you are likely to consider sub-questions very like the fourteen proposed above for Spoken Communication.

15. In what proportion of the writing being done is the pupil feeding back received knowledge to the teacher solely for the purpose of evaluation? (It has been said that this kind of writing is addressed to The Teacher as Examiner.[1]) Similarly, how much of the writing done appears in a complete and public form, and how much shows the incompleteness, discontinuity, tentativeness, and lack of full explicitness typical of exploratory work-in-progress?

16. How much of the writing is concerned primarily with the copying or paraphrase of information? On the other hand, how much of the writing which begins with such information shows that the learner is reinterpreting, applying, transforming, discussing and evaluating that knowledge?

17. When written work reports first-hand activities or experience, does the writing show that the pupil is interpreting and evaluating?

18. How far do the topics set for written exercises enable and encourage pupils to utilize first-hand experience, and to deal with problems and concerns which they see as urgent?

19. How has the teacher responded to the written work handed to him? Has he only evaluated it? Has he treated it primarily as an opportunity to correct misconceptions in pupils' grasp of subject matter, and solecisms in their use of spelling and punctuation? Is the teacher concerned about appropriate style to the extent of rejecting colloquial formulations? Does he respond to and comment on the ideas put forward by pupils? Is the writing published by being read aloud or displayed on a wall?

My final question relates to all the above, but also to teachers' communication with their pupils at other times than lessons, and in corridor, hall, playground, and street.

20. Do you treat pupils as rational human beings with viewpoints which are different but valid, even when you believe them to be wrong? Do you persuade rather than command? When commands are necessary, do you give them politely and supply reasons? Do you

[1] Britton, *et al., The Development of Writing Ability*, Macmillan, 1975.

treat your pupils' viewpoints with courtesy and understanding, and seek to strengthen their ability to think for themselves?

It would be difficult to supply evidence that answered this last group of questions, yet this is not necessary. We all remember occasions when we have failed to carry out this kind of policy. Teachers work with boisterous young people, some of whom resent school control, and at times talking for learning becomes overwhelmed by a teacher's desire to keep control. It is important however, not to value control for its own sake, to recognize the genuine diversity of children's interests and perspectives, and not to allow 'controlling a class' to become 'stopping young people from thinking for themselves'.

My list of questions implies informal discussions of recorded material and of children's writing. Published systems for the analysis of class-room interaction are clumsy instruments, much inferior to the intuitive understanding of intelligent adults. I am envisaging groups of teachers who share responsibility for a course, and who trust one another enough to collaborate in monitoring the progress of that course. It would probably be best at first not to attempt to listen to enormous stretches of tape or to read large piles of writings, but for one member of the group to select in advance a range of shorter examples characteristic of various kinds of talk or writing.

It can be quite difficult to make sense of a recorded extract from a lesson which one has not taken part in, and teachers will find it necessary to spend considerable time on extracts of no more than a minute or two. Nevertheless, the trouble taken to select an interesting episode, perhaps to transcribe it, to listen to it several times, and to discuss it at length, is likely to be well rewarded by the range of insights and issues that the discussion will reach to. Discussion of these examples should make it possible to define the range of kinds of teaching and learning which the teachers wish to take place and thus help to guide future teaching.

Informal sampling of courses may be all that many teachers can find time to do, but more systematic study would be possible. For example, if special time were made available for this kind of monitoring, it would be possible to analyse recordings of teachers at work to estimate roughly how much of their time was devoted to classroom organization and discipline, how much given to low-level advice about methods, and how much to theoretical discussion of the matter in hand. This last category could then be broken down—again very roughly—into sub-categories such as eliciting description and narration, eliciting reasoning, giving information and explanations. The teachers in the light of such an analysis could consider whether they were carrying out their intentions and decide what changes might be made in their teaching strategies. But it should be emphasized that such a sophisticated approach is far from essential. Every teacher can benefit from even the most informal look at what goes on in his lessons, both at the kinds of communication that his pupils engage in and at the way in which he influences this communication by the part he himself plays.

As a result of this a group of teachers might decide that individualized learning needed to be supported by group seminars, or that their science practicals needed to be accompanied by group discussions to ensure that pupils considered the meaning of the practical work done. They might decide to widen the range of written work to include more occasions when pupils were expected to 'think aloud on paper', or to spend more time in trying to persuade pupils to talk explicitly about what they are learning. They might decide to set up an experimental course with quite different patterns of communication from those currently in use. Thus I am envisaging the beginning of a process of developing new ways of looking at teaching, and not the kind of completeness that 'policy' would suggest. In our time it is beginning to be possible to be much more aware of how we communicate with other people; teachers should be in the forefront of such awareness since communication—in the sense of a two-way attempt by teacher and taught to reach and understand one another —is at the centre of their work.

6. Classroom Technicalities

To make the broader approaches of Chapters 4 and 5 effective, some measure of agreement is also required on what I choose to call 'classroom technicalities'. These are in some ways easier to tackle, for they are limited, instrumental, and technical. I am including a technical vocabulary of language, ways of introducing new vocabulary, approaches to spelling, phonics, punctuation, and the very important study skills. None of these on their own would be of a great deal of value, but they underpin the use of language in a school, and if attention is given to them, they help assure that language growth is effectively assisted by the school.

Language terms

The Report recommends that 'the teacher ought to ensure that in a given period of time the pupils cover certain features of language, and for this purpose he might find a check-list useful. We believe these features should certainly include . . . a knowledge of the modest collection of technical terms useful for a discussion of language' (11.21), and Dr Gatherer confirms that teachers and pupils require what linguists call a metalanguage for talking about language. Whilst it is important not merely to start a game of exchanging jargon, staff need vocabulary for their analysis and discussion. They also need a working vocabulary for helping pupils. This list could serve as a vocabulary for teachers to use with pupils, and thus to be taught systematically:

sound, letter, syllable, word
vowel, short, long, stress, consonant
prefix, root, suffix
upper case, lower case
italic, bold
ascender, descender
comma, semi-colon, colon, full stop, etc.
indentation, paragraph
skim, scan
numerals, Arabic and Roman

headings, sub-headings, running heads
table of contents, index, reference, cross-reference
accent, dialect
phonic
register

Introducing vocabulary

The whole process of education, however we may work to alter it, is a long process of learning more words. The teacher is frequently a supplier of vocabulary. Sometimes one feels the vocabulary is a mere frill, a barrier between the pupil and the understanding. Sometimes a subject like Geology seems to be just a matter of difficult names for easy ideas. If only one could sweep away the verbiage, one feels, one could teach so much more easily. To a small extent that is true, and much learning is made easier if we can work with the pupil's own vocabulary and not drive him too early into words that he cannot himself feel or possess, as I have discussed earlier. However, that aside, there will be a continuous need to work on vocabulary at three levels:

1. Slightly unusual words necessary to a topic and used in their standard sense. The humanities are especially full of such words, e.g. *development, growth, examine, civilized, establish, relations*—and a huge host of others, each used in its normal sense, but not part of the usual vocabulary of a school pupil, in many cases not part of the ordinary vocabulary of domestic and personal talk. These are the words glossaries don't gloss and teachers don't explain.

2. Common words used in special senses, perhaps in their original sense, or core sense, perhaps in a specialized sense, e.g. *veneer, series, contemporary, energy, rotate* (maths), *balance* (physics)—which are also frequently left unexplained. Particular problems, it seems to me, arise out of this second category. Indeed, sometimes the sense of a word can change from subject to subject, e.g. *volume*, which pupils can meet with a different sense in music, science, art, mathematics, and in the library!

3. Technical terms, more or less particular to the subject,. e.g. *ductile, ferrous, taper shank, contour, topology, enfranchize, pipette*. These words are usually explained fairly fully, but sufficient practice in using them may be unavailable.

The sheer size of the English vocabulary is daunting, and we do not seem to have faced up to this in schools. Webster's Third International Dictionary, according to Hunter Diack,[1] contains 450,000 words. If you take 'headwords' only (that is not compounds derived from

[1] Hunter Diack, *Standard Literacy Tests*, Hart-Davis Educational, 1975.

them), the *Concise Oxford* has some 36,000 (with a further 30,000 derived). The first 6,000 are known by an eight-year-old, a further 6,000 are usually learnt between nine and twelve, and a third 6,000 by the 'linguistically adept' between twelve and fifteen. This is a target which will not be met if it is left to chance. As well as the quantity, there is the problem of the unease so many feel with a large part of the vocabulary. We British have a dislike of words which sound unfamiliar and puzzling. Although often referred to as 'long' words, these difficult words are frequently not actually long. It is the unpredictability of pronunciation which puzzles people. They are usually words from Romance origins. Words of Anglo-Saxon origin, though not half so numerous, are used about five times as often.

Not only do others use them—so that we *must* understand them, but *they are the key words for further study, for the exchange of ideas, for debate, indeed for the learning of many school subjects.*

Often the sense of a word is simply hidden because its root is lost by pronunciation changes. 'Significant' is an example. Used in the humanities, literature, and commonly for the early stages in mathematics, it remains vague for many pupils because they have never have pointed out to them the simple fact that it comes from 'signify'—'to be a sign', that is, 'significant' is said of something that *points to* meaning.

The way into these less commonly used words is not merely extensive reading—especially as extensive reading usually means of novels, which have a considerably lower use of such words. Specific help must be given, and this must involve analysing vocabulary.

In a whole-school language policy, I should suggest agreement is reached on modes of introduction, on the technical terms used (as explained in the previous section), and on who in the school will give the basic theoretical teaching about words. There is then the knowledge available to help the introduction of new words, and every teacher is building up the pupil's understanding of how our vocabulary works, and thus making it easier for the pupil to absorb other words later. This again, is a combination of the contextual and specialized language teaching approaches. A policy might include a consideration of five aspects of advice.

Phonic approach and syllabification

If a new word is to be understood and used by a pupil, he needs to see it (I mean actually *see* the structure), hear it clearly, be able to pronounce it himself, and then be able to spell it. Most of us find it difficult to take in new vocabulary, and secondary pupils, struggling with so much at once, often have great difficulty with new words. The first aim is to make the word aurally and visually clear.

Some words offer difficulties beyond their conceptual level simply because of their pronunciation or spelling difficulties. In these cases special phonic care on introduction is worth while, with an opportunity for drilled pronouncing aloud. The 'rh' words are an example. 'Rhombus' is in fact phonically regular, but the 'h' always confuses, as do the other words transliterated from the Greek ρ (*rhetoric, rheumatism, rhinoceros, rhododendron,*—difficult, but all phonically regular). Also in mathematics 'ratio' remains foreign sounding, but need not if related to the other 'sh' pronunciation of 't' in *ration, introduction,* and many other words. I even find that the pronunciation difficulties of 'th' can cause problems, so that 'tenths' and 'hundredths', especially that last one, torture pupils.

A school drawing up a policy might like to consider an agreed procedure for all specially introduced words. It might be based on a list such as this:

1. The word is written on the board or in the individual pupil's book. (The pupil does not yet copy it.)
2. The teacher pronounces it, underlining each separate syllable, e.g. syllabification. The word, therefore, is *not* broken up by lines or gaps.
3. A number of pupils read it aloud, perhaps even the whole class in unison.
4. The pupils are given time to learn it.
5. It is covered up, and the pupils asked to write it from memory.
6. They then check against the original.

In this writing out, reference is made to phonic expectations or irregularities. The pupils are being helped to grasp the sound structure of the word, and the phoneme/grapheme relationship. They are being given a chance!

Word building

In finding his way around the huge vocabulary I mentioned earlier, the pupil is helped considerably if he is aware of two things.

1. The way a word can be changed to give the sense a variety of uses (noun, adjective, adverb—though he needn't know the terms). I call this the family of words. Pupils frequently know only one from each family, and are not used to recognizing, yet alone using, the rest of the family.
2. The way words can come together with other words or special parts to create new compounds.

In other words, pupils need to be taught word building. Sean O'Casey was one of those who put the justification very clearly:

> Ella went over to rummage among the books left behind as unsalable out of her father's fine store. She brought back a Superseded Spelling-book

by Sullivan, who held that by learning affixes and suffixes, Latin and Greek roots, you could net words in hundreds, as against the old method of fishing one word up at a time. . . . (*I Knock at the Door*)

This should not become a fetish—it hardly justifies the learning of Latin, for instance. It also needs to be treated cautiously, for the clues can be misleading. However, the pupil should be taught how words can go together (*sometimes, upstairs*); then they should learn the idea of prefix, root, and suffix. Next, the most useful examples of these should be taught (see Appendix One) by the English teacher in general terms, and the specific ones by subject teachers. Pupils should be taught to tackle an unfamiliar word by examining it for familiar parts. If they know the meaning of that part, they apply it. If not, they deduce the meaning from the known word in which the familiar part appears.

The vocabularies of mathematics and science are rich in regularly constructed Latin or Greek compound words. If these subject teachers can know that prefix, root, and suffix has been taught, they can draw on this knowledge economically to help explain new words, and at the same time widen and reinforce the pupils' grasp of how such words go together. A few examples make the point clear:

divide, division, divisor, equilateral, quadrilateral
polygon, hexagon, diagonal
metre, centimetre, metric
subtract, subtraction
triangle, rectangle
rotate, rotation
convex, concave
energy, energetic
compare, comparison
electric, electrical
mechanic, mechanical
magnet, magnetic
oxide, dioxide, monoxide

The vocabularies of other subjects can be similarly examined by teachers, and vocabulary can be taught in two complementary ways: the basic systems of word analysis should be specifically taught by an agreed department (probably English), but the 'knowledge' must be constantly used and built upon by the range of subject teachers.

Word building is an important process in which all teachers work together to provide the pupils with the necessary knowledge, experience, and confidence. It has the great virtue of de-mystifying the most difficult parts of the English vocabulary, especially those parts used so much in the language of abstract thought and of technical explanation. The knowledge helps spelling, reading, and understanding.

Timing of introduction

One principle which can well be considered by the staff of a school is when special vocabulary is best introduced. At one extreme, the specialized teacher has a 'vocabulary' session. The craft teacher starts the autumn term with a list of tools, metals, and special terms: *ferrous, corrosion, alloy, anneal, ductile,* are listed, explained, learned, tested. Then work starts and the words are used in context. At the opposite extreme, mid-activity, the teacher explains 'And this is called the "drill chuck key"; don't forget that'. Because the demonstration is going on there is no time for explanation, clarification, or discussion of the term.

The first method risks unreality: 'malleable' is explained months before the quality of malleability can be felt. The second risks superficiality: the word is passed over so quickly that it is barely heard, never mind comprehended.

It seems likely that a mixture of approaches is required: the word should be introduced as near as possible to the point of use, but it should be part of a programme of methodical subject vocabulary building, with the word introduced, used, followed up, and then put into a growing subject list by the pupil. Subject teachers should have pupils build up their own subject dictionary. Thus the geographer's personal dictionary will grow during the two years, say, of a CSE option, helping the pupil *possess* the words for himself.

Establishing meaning

In establishing a language policy, a school staff can consider ways of revealing the meaning of words. One principle might be that prior to the teacher's explanation of a word there should usually be a deductive process, to help the pupil work towards the meaning of a word from his own memories of other uses or other parts. I have already spoken in Chapter 5 of the deductive approach to unfamiliar words in reading. I suggest something slightly different could be considered when new vocabulary is introduced orally—somewhat less deductive, more experiential. This both helps the present grasp of the new word and helps the general learning of how one can and should focus the past on newly discovered words.

Relationships of meaning

I pointed out that one of the difficulties for pupils is the special use of words familiar in wider or different contexts. A language policy should face up to this, and establish what can be gained from it. I should suggest that normally it is worth while *making use* of the relationship between the current meaning and others, both to assist the present apprehension of sense and to assist the general under-standing of sense relationships. With many words which are used for subject concepts, the starting point should be the 'everyday' sense: with a word such as *cycle* used in religious studies or biology, for instance, or *series* used in mathematics, the teacher moves from the general to the specific sense, pausing briefly on other related specific senses (the bi*cycle*, the TV *series*).

The reverse process is also important, moving from specific uses to the wider metaphorical sense, for this also not only sharpens the understanding of the moment, but also deepens an understanding of the nature of language. Thus the craft teacher introducing *veneer* would move out to the metaphorical sense.

Conclusion

Certainly vocabulary grows by experience, but not adequately. A whole-school policy of mutual reinforcement, on the other hand, can combine the specific with the contextual approach with great power. Only *use* can help a pupil take possession of words, and the use must be in a variety of real situations. But a carefully co-ordinated series of approaches to the extension and explanation of new vocabulary will help the subject understanding and deepen the pupils' underlying grasp of how words work.

The phonics approach

Ann Dubs

In the English alphabet, letters and groups of letters represent different sounds. Phonics relates these letters to forty-six[1] possible sounds. Many people learn these letter/sound relationships through a non-conscious, seemingly automatic process. Others find this difficult and need to be taught the relationships to a greater or lesser extent.

Many children entering secondary school have acquired a large sight (or whole-word) vocabulary, yet are still unable to read a new word when they come across it. They have not learned how to identify the sounds of the letters or groups of letters in that particular word. This is even more apparent from their poor spelling performance. Nearly every eleven-year-old I have tested from amongst our poorer readers could for example spell the word 'big'; few could manage 'wig'. Both are phonically regular, but 'big' is a word which every

[1] Axel Wijk, *Rules of Pronunciation for the English Language*, OUP, 1966.

child would have known and used since infant school; 'wig' is far less familiar. To spell the word correctly, these pupils have to be taught formally the generalization that 'g' at the end of a word has a hard sound.

The majority of words in English are phonically regular, i.e. they conform to basic rules. The reader who has learned or absorbed these phonic rules does not then have to learn every word separately. Of course, some words do not conform to the basic code and they will have to be learned as sight words. But the phonic system does simplify the spelling and reading process by reducing the number of words that have to be learned. Experience has shown that the pupils whose spelling and reading performance has improved the most are those whose programme has included a very systematic phonic training in encoding and decoding sounds.

At first sight, identifying a word by its component sounds may not appear to be altogether straightforward. 'Cat' is sounded k/a/t/ only because the reader has the foreknowledge of what sounds the letters 'c' 'a' 't' make in that particular word. Yet the letter 'c' can make, amongst others, the sounds /s/, /k/, or /sh/ depending on the letters with which it is associated in the word in question. The letter 'a' can make, amongst others, the sounds heard in 'hat', 'hate', 'halt', and 'hart'. 'T' makes a /sh/ sound when it is followed by an 'i', as in 'patient'.

Nonetheless, the choice of which sounds to select for the letters in the word 'cat' is not an arbitrary one. The reader familiar with phonics knows that 'c' preceding an 'a' has a /k/ sound and that the vowel 'a' in a single syllabled word, with a consonant-vowel-consonant structure has a short sound.

The remedial reading teacher quite often has to explain this process in some detail. The subject teacher would not have time to do this. But, in order to clarify the system of word pronunciation for pupils who are not fluent readers it is helpful if the teacher is familiar with the broad outline of phonics instruction. That teacher is then able to do two things: (1) To identify the irregular words which cannot be sounded out or decoded, such as 'any' and 'does', and which have to be learned as whole words. (2) To indicate certain letter combinations within a word which give a clue to the pronunciation of the word.

Most teachers, as fluent readers, are unaware of the phonic elements of words. They never stop to question the ways in which the letters go together to make the sounds because they take them for granted.

But, an ability to recognize the combinations of letters which produce a particular sound is essential to anyone trying to teach a new word by this method. A pupil baffled, for example, by 'shingle' would be helped if told that the final sound is similar to that at the end of a familiar word like 'table'. He may, at first, need help in isolating the sound made by the letters 'le', particularly if he is learning French!

A large part of the teacher's task is to help the pupil use his secure knowledge of common words as raw material for interpreting less common words. Let us suppose that the word 'phalanx' has come up in Roman History. The teacher points out that the 'ph' is /f/ 'as in "photograph"' and the 'x' represents the /ks/ sound 'heard in "tax"'.

'Scutage' is another technical word which occurs in Longman's Secondary History *The Middle Ages*, a textbook used quite frequently in the lower part of the secondary school. Such a word would obviously first have to be taught in context and with reference to its derivation. It is an extra *aide-mémoire* for the pupil if the teacher can make an analogy with other '-age' ending words in the same course, like 'pilgrimage' and 'Carthage'

It is apparent that there are certain groups of sounds which can cause difficulty to all but the most competent readers. It is the aim of this chapter to indicate (a) some of these sounds, (b) certain basic aspects of phonics which it would be helpful for the subject teacher to be aware of when introducing new words or reading a text with a class. Examples are taken from the secondary school learning vocabulary.

With this purpose in mind, it is useful if the subject teacher lists, over a period of time, technical words or words which are giving difficulty to his pupils. It is not suggested that these words should be taught out of context or that the subject teacher should teach lists of words as such. Once he is aware of the phonic patterns within the words relevant to his subject, he can then make generalizations which will help the slower reader to learn and retain that word.

As some of the simplest identifiable phonic groups are found at the end of a word, a brief list of suffixes that are useful phonically is given first. Those suffixes that are primarily of semantic interest (such as '-by' in place names) are intentionally excluded.

Next, certain consonant groups are mentioned. Finally, there is some basic information about long and short vowels.

The principle of breaking up words into syllables which is essential to the process of phonic analysis is dealt with in the section on spelling.

Word endings

(a) *Examples*:

-*age* (idge sound)   language, mortgage, homage, vintage, hostage, average

-*er*   lawyer, thermometer.
   Note: the same sound can be made by -or, -our and -ar and mainly in the beginning and middle of words by -ir, -ur, and -ear. (For examples, see section 5, long vowels).

-ic democratic, epic, ethic, bionic, isometric, ironic, nomadic, Arctic, Hellenic.
Note: all these words have more than one syllable. A single syllabled word with this ending would be spelled -ick, e.g. trick. (See short vowel section.)

-le shingle, quadruple, portable, double.
Note: to represent the sound heard at the end of 'shingle' it is necessary to use the phonetic symbol /ə/, which designates the murmur or schwa sound, +l.

-y (/ee/ sound) colony, injury, philosophy, savagery, clemency, magistracy, monastery, philology, slavery, astronomy, yeomanry.
Note: the fact that '-y' becomes '-ies' in the plural does not affect the pronunciation.

-ous (/us/ sound) ferrous, ingenious, igneous, monotonous.

-tion (/shun/ sound) elevation, initiation, projection, construction. (These are just a few of the words with this suffix collected from the Longman's text *The Middle Ages* referred to earlier: constitution, population, destruction, publication, execution, Reformation, foundation, excommunication, occupation.)

-sion (/zhun/ sound) confusion, corrosion, division, precision, explosion.

-ssion (/shun/ sound) admission, fission, progression, accession.

-sion (/shun/ sound) pension, mansion.

-ture (/cher/ sound) conjecture, pasture, texture, fixture.

-sure (/zher/ sound) exposure, enclosure, measure.

(b) *Inflected endings* cause many pupils difficulty.
The most helpful generalization, clarifying the pronunciation muddle for children of West Indian origin in particular, is that '-ed' is pronounced /id/ when it is preceded by a 't' or a 'd', e.g. demented, added. In all other cases, '-ed' adds a /t/ or a /d/ sound to the root of the word; it does not add an extra syllable. If more clarification is needed, the rule can be amplified further

ed = /t/ after 'p', 'ck', 'f', 'ch', 'tch', 's' (as in 'sell'), and 'th' (as in thirty).
e.g. hopped, puffed, wished, watched.

ed = /id/ after the consonants 'd' and 't'.
e.g. abated, faded.

ed = /d/ after 'b', 'g', 'v', /j/ ('g', 'ge', 'dge'), 'z', 's' (as in pleasure), 'th' (as in those), 'l', 'm', 'n', 'r', 'w', 'y', 'ng'.
e.g. stabbed, hugged, trudged, clanged.

and following $\left\{\begin{array}{l}\text{vowels}\\\text{vowels} + \text{'r'}\end{array}\right.$
   e.g. fried, stirred.

(*c*) *-s, -es endings*
-s = /*s*/   after the consonants p, t, ck, th, f.
   e.g. crops, currents, rocks, months, giraffes.

-s = /*z*/   after consonants b, d, g, m, n, -ng, -th, v, l.
   e.g. tubes, liquids, drags, diagrams, basins, filings, tithes, dissolves, levels.

   and after vowels +'r'
   e.g. stirs.

-es = /*iz*/   when it has been added to a word whose root ends in a hissing sound:

-ss	e.g. compresses
-x	coaxes
-ch	dispatches
-sh	punishes
-ces	substances

   and to
   -ge   e.g. abridges

Selected consonants

$\left.\begin{array}{l}\text{c}\\\text{g}\end{array}\right\}$ (attached)
gh
   In a final position, 'gh' is sometimes pronounced /f/:
   e.g. rough, tough, enough
      cough
      but not in bough, Slough, through, thorough, borough, where it is silent.

-igh   has a 'long i' sound:
   e.g. fright.

ph   A not very frequent representation of the /f/ sound found in words of Greek origin:
   e.g. amphora, Euphrates, hieroglyphic, phalanx, chlorophyll.

qu   Never separate in English. Together they represent the consonant /k/ sound heard in cheque, oblique;
   or more frequently the (kw) sound in equatorial, aquatic, adequate.

th   has two main sounds:
   (a) as in *th*eme, *th*eorem;
   (b) as in la*th*e.

These sounds are so similar that there is no need to make the distinction explicit.

Its third sound, /t/, is one which many children of West Indian origin use in preference. The most helpful way of clarifying this is to present the relatively few words in which 'th' has this /t/ sound, e.g. *Th*ames, *Th*omas, Neander*th*al, *th*yme, disco*th*eque.

s   makes a /sh/ sound in sugar, sure, surely, insurance. (See also suffixes.)

and a /zh/ sound in usual, casual.

wh   sounds the 'h' alone in *wh*ole, *wh*o

and the combined letters /w/ or /hw/ sounds in *wh*irl, *wh*isky.

x   has a /ks/ or /gs/ sound in phalanx, paradox, exact.

Letter 'c'

The rules relating to the pronunciation of the letter 'c' are practically infallible. They are worth learning thoroughly as their application is very wide and relevant also to the pronunciation of 'c' in French, apart from the minor variation of the soft (ch) sound. (In any case, we have retained the French /ch/ sound in English in several borrowed words, such as 'ma*ch*ine' and 'cro*ch*et'.)

The rules are summarized as follows:

Spelling	Pronunciation	Examples	Exceptions
1. 'c' followed by the vowels 'e', 'i' or 'y'.	's' sound called 'soft c'.	*c*entimetre con*c*entrated *c*ircumference *c*itizen os*c*illate *c*yclone cal*c*ium	(a) Certain words containing letters 'ci'. (See 4 below). (b) Celt, soccer, sceptical (c) Several musical terms borrowed from Italian, e.g. crescendo, concerto.
2. 'c' followed by: (i) the vowels 'a', 'o' or 'u'; (ii) other consonants.	'k' sound called 'hard c'.	college Carthage carburettor collateral statistics crevasse classical clarinet	Soft /ch/ sound as in chur*ch*. /ish/ sound in words of French origin: chef, machine.
3. 'ch'.	'k' sound.	chemist stomach-ache mechanical technical	As above.

Spelling	Pronunciation	Examples	Exceptions
		characteristic chronicle Chrysler saccharine Jericho archaeology chlorophyll	
4. 'ci'.	'sh' sound.	precious ancient racial electrician musician association coefficient	(Most words conform to rule 1, e.g. deciduous.)

Note
(a) The pronunciation of words containing a double 'cc' conforms closely to 1 and 2 above.
  e.g. *soft c* —access, accident, success, accession
  hard c—occasion, occur, tobacco.
(b) (Rules 3 and 4) As with many 'rules', to over-teach them is to risk confusing the pupil by leading him to make false analogies. For example, a pupil who has practised a list of words of Greek origin with a spelling for the 'ch' (/k/ sound) like 'chorus' is liable to apply the same pronunciation to words like 'chimney'. By grouping technical words which contain a similar element, e.g. *ch*lorophyll, *ch*loroform, *ch*lorine, and teaching them in context, the subject teacher can try to avoid this phonic trap.
  In the case of 'ci' words listed in Rule 4, it should be noted that in every instance the 'ci' (/sh/ sound) never appears in the first syllable. It always occurs at the beginning of one of the subsequent syllables. An awareness of this might help to prevent the mispronunciation of a word like 'city'.
(c) When an oddity like the word 'circuit' is met, it is helpful to draw a parallel with the more familiar word 'biscuit', explaining the connection between the hard 'c' and the following 'u', which has no function other than that of preserving the /k/ sound.

Letter 'g'
The letter 'g' has two sounds, soft like /j/ or hard as in *g*un. Basically similar to the 'c rule', the 'g rule' works on the principle that the 'g' has to be followed by 'e', 'i', or 'y' to have a soft /j/ sound.

Not all words containing the letter 'g' conform to this rule. However, since the list of exceptions includes quite a lot of very familiar, over-learned words—e.g. get, girl, gift, begin, geese, give, etc.—it is still worth teaching the basic rule nonetheless.

Note
(a) If the /j/ sound is to be written in a word, the 'g' has to be followed by 'e', 'i', or 'y', e.g. *G*eorge, *g*orgeous, pi*g*eon, reli*g*ious, inte*g*er.
(b) The letter 'u' indicates a hard 'g' in a fairly large group of words containing the letters 'gu', e.g. *g*uarantee, *G*uardian, dis*g*uise, *g*uidance, *G*uernsey, *g*uest, *g*uild, *g*uillotine, *g*uilty, *g*uitar,

demagogue. Once this is appreciated, the pronunciation of these words becomes much more logical.

(c) When the letters 'gg' appear in a word their sound is always hard before 'e', 'i', 'y', e.g. Rugger, Maggie, stagger, carpet-bagger; exceptions: suggest, exaggerate, Reggie.

(d) 'g' as a final letter is always hard, e.g. wig. (Reg is an exception because it is a shortened form.)

(e) For -*age*, see section on suffixes.

Consonant Combination

	Initial			Final		
br	bl	sc		-lb	-ft	-nk
cr	cl	sk		-ld	-lt	-lk
dr	fl	sm		-nd	-nt	-sk
fr	gl	sn		-lm	-pt	
gr	pl	sp		-lp	-st	
pr	sl	st		-mp	-ct	
tr		sw		-sp		
		spr				
		scr				
		str				
sh				-sh		
shr				-ch		
th				-ng		
thr						

Some pupils find it difficult to blend consonants together. Although the word *scratch*, say, might be familiar as a sight word the pupil could still have difficulty in reading *scree*. The stumbling block would most probably be the first three letters, and practice is required.

Many slower readers find it hard to articulate the different sounds in the initial blends group, such as 'cr'/'gr' and 'cl'/'gl'. This is confusing when they are attempting to read a word which is unfamiliar.

Silent consonants

Teachers will wish to devise their own lists of words which contain silent consonants. It is worth mentioning here, however, the fact that such letters are rarely silent in isolation.

For example, 'k' is silent when combined with 'n' in 'knowledge', 'Knaresborough', etc.; 'b' is silent before 't' in 'subtle' and 'debt' and after 'm' in 'thumb', 'catacomb', 'succumb'; 'p' is silent when it precedes 'n', 's', and 't' at the beginning of a word, e.g. pneumonia, psychology, Ptolemy. 'H' is an exception: honour, vehicle.

Stress

Before considering short and long vowel sounds, the effect of stress on the pronunciation of words must be mentioned.

The generalization many primary school children are taught that the final 'silent -e' has a lengthening effect on a preceding vowel (e.g. pine) is a very useful one to know. It does not, however, apply to multi-syllable words like reproduct*ive*, determ*ine*, clim*ate*, because the sounded vowel in these final syllables is unstressed and pronounced with the schwa (ə) sound. The long vowel effect is heard in multi-syllable words where the relevant syllable is stressed, or has a secondary stress, e.g. stam*pede*, consti*tute*, Constan*tine*.

In very broad terms, the position of the stressed syllable can be determined in these ways:

1. The first syllable of a two-syllabled word is frequently stressed, e.g. *mu*sic, *pic*nic.
2. The root of a word is generally accented, e.g. im*pos*sible, un*do*ing. Though if a two-syllable word begins with a prefix it is then the second syllable which is stressed, e.g. re*call*.

In longer words vowels tend to have a short sound in a closed syllable (i.e. one that ends with a consonant) and a long sound in an open syllable (i.e. ending with a vowel). The stressed syllable is underlined in the examples:

vol *ca* no
u ni form } (open syllables)

clas sic al
re *jec* ted } (closed syllables)

Of course, children cannot be taught to decode automatically by this method because there are so many exceptions. In many words, the vowel is in a stressed syllable with only one consonant following the vowel yet the vowel is short, e.g. *ra*dish, *li*nen, *me*rit, *vo*luble.

While it is important for a teacher to appreciate the distinction between open and closed syllables so that he can syllabify a word in the most helpful way when presenting it to a class, it is not necessary for pupils to be taught the distinction in these terms.

A simpler method of explanation is suggested at the beginning of the section on long vowels. (1b)

Some teachers might find it useful, in addition, to familiarize their pupils with the diacritical markings in a dictionary, emphasizing the role of the stressed syllable.

However obvious, it must be restated that the most important factor in helping children to decode apparently unfamiliar long words is the teacher's consciousness of the need to enlarge their listening and speaking vocabulary. The pupil who looks blankly at a word like *concentration* will decode it successfully by combining a little phonic knowledge with an oral/aural awareness of that word.

Short vowels

The short vowel sound is the sound heard in had, red, bit, pot, cup.
1. (a) In consonant-vowel-consonant (cvc) monosyllables the vowel is short. Such syllables are known as 'closed' syllables.
Exceptions occur when the vowel is followed by '-r' or '-w'. (See section on long vowels.)
(b) In words of more than one syllable all five short vowel sounds may have a schwa sound (ə) in unstressed syllables. This is the sound heard in: alone, between, cannon.
2. A vowel is short when it is followed by
(a) a doubled consonant, e.g. mammal, narcissus, pollen, mussel, buzzing, potassium.
Note: in a monosyllabic word or root after a short vowel:
　　(i) The /k/ sound is spelled -ck, e.g. trick.
　　(ii) The /ch/ sound is spelled -tch, e.g. pitching.
　　(iii) The /j/ sound is spelled -dge, e.g. abridged.
　　　　(Exceptions: much, rich, such, which.)
(b) two consonants, e.g. crank, cling, temper, pond.
(For exceptions, see long vowel section.)
3. Stressed cvc forms + suffixes become cvcc + suffixes.
In the following type of structure a doubled consonant marks the short vowel sound, e.g.

	Root + suffix	
-ing	drag + -ing	= dragging
-ed	stun + -ed	= stunned
-y	wit + -y	= witty
-en	trod + -en	= trodden
-er	dim + -er	= dimmer
-est	thin + -est	= thinnest

The same principle applies broadly to these suffixes:
　　-le The short vowel sound is heard in battle, settle, preceding the doubled consonant, while the long vowel sound is heard in sta/ble, etc.
　　-ow e.g. bellow, follow
　　-et 　　puppet, bonnet
　　-el 　　kennel, funnel.
This generalization, though helpful, is far from infallible as exceptions like 'city', 'coming', 'model', and 'panel' show.
4. When the letters wa, swa, qua, squa are followed by '-a', that 'a' generally represents the short o sound, e.g. wasp, wallet, wander, swab, swamp, quantity, quarry, squatter, squabble.
Exceptions: most cases where a + r is found, e.g. warning, quart, swarthy.

5. The letter 'y' is sometimes a vowel and sometimes a consonant. It is only a consonant if it is positioned at the beginning of a syllable, e.g. yeoman, youth, yacht. When 'y' is found medially it generally has a short /i/ sound, e.g. synthesis, cylinder, mythology.

In the Greek prefix forms hyper, hypo, the 'y' has generally a long /i/ sound, e.g. hypothesis, hypotenuse, hyphen, hydrogen, hyposulphite, though it is short in hypocrite and hypnotize.

Long vowels

1. (a) Where there is one vowel in a syllable and it comes at the end of that syllable the vowel is frequently long. The syllable is then called an open syllable, e.g. me, so/lo, mu/sic.

(b) There are so many exceptions to this rule in multi-syllabled words that it is more helpful to children decoding long words to tell them that a cvcv syllable may be either open as in i/vor/y, i/tinerant, or closed as in ac/tiv/ity, hos/til/it/y.

In simpler terms, the instruction, 'Try /ĭ/ (short i) then /ī/ (long i)' is adequate in most cases. By breaking down words syllabically the pupil will probably reach a more accurate version of the word than he would by guessing.

2. *Silent -e*

(a) A 'silent -e' at the end of a word makes the vowel in front say its own 'name'.

In cvc+e structures the first vowel is long, e.g. evapor*ate*, extr*eme*, ag*ile*, micr*obe*, cath*ode*.

The final '-e' is dropped when all suffixes except '-ly' are added,

    e.g.   hope—hoping
            shine—shiny
            ride—rider
but   time—timely

This generalization can be applied to a few cvcc+e structures, like *taste*, but usually the short vowel sound is heard as in *hinge*.

The most common exceptions to the 'silent -e' rule are well known as sight words to most secondary school pupils. They are:

    have, give, love, shone, above, dove, glove, come, some, done, none,
    gone, shone, lose, whose, move, prove.

(b) When a vowel is followed in a syllable by '-re' the sounds are modified in this way:

    -*are*   compare, share, rare, spare, scare, etc.

    -*ore*   sh*ore*, sp*ore*.

3. *Some vowel digraphs*

(a) -*ee*, -*ea*, -*ai*, -*oa*.

In these combinations, the sound is the name of the first letter, i.e. the long vowel sound,

e.g. ch*ee*tah, discr*ee*t, st*ee*l

    b*ea*ker, guin*ea*, sh*ea*ves, incr*ease*

retain, appraise

groat, moat, approach.

(There is also a large group of words in which *ea* has a 'short e' sound, e.g. weather.)

(*b*)

'ai', 'ay' have the same sound in plaintiff and dismay, but the '-ay' spelling only appears at the end of the word or root.

'oi', 'oy' have the same sound in celluloid, rhomboid, and alloy, but the '-oy' spelling is generally found at the end of the word or root.

(Exceptions: oyster, royal, voyage.)

'au', 'aw' have the same sound in dinosaur, inaugurate, automatic and the *augh* group, trawler and sprawl.

'or', 'our' can also make the same sound:

e.g. storage, courtier.

'ow' has two sounds:

the '-ow' sound in owl and

the '-ow' sound in yellow, growth.

'ou', 'ow' have the same sound:

e.g. doubt and cowl.

'ui', 'oo' have three very similar sounds:

(i) as in fruit juice, bruise, suitable, cruise, sluice, and stoop, harpoon;

(ii) as in nuisance, pursuit,

(iii) as in partook, shook.

'ew', 'ue' have two closely related sounds:

(i) as in clerihew and argue;

(ii) as in shrew and glue.

'ie', 'er' have a 'long e' sound:

e.g. fief and plebeian.

4. '-y' in a final position in a one-syllabled word usually has a 'long i' sound, e.g. spry, by.

(Final '-y' in a word of two or more syllables generally has a 'long e' sound. See suffixes.)

5. *Exceptions to rule of short vowel in medial position.*

(*a*) When letter 'r' follows a single vowel in a syllable the combined vowel +r can produce one of three sounds:

(i) /ar/ sound   as in Carthusian, argument.

(ii) /or/ sound   as in swarthy, warden.

(iii) /er/ sound   as in burglar, Cistercian, emerge, girder, thirtieth, donor, vector, carburettor, bursary, sturgeon, survival.

Note: in addition to the five spellings in (iii) above the /er/ sound can also be found in these forms:

wor   as in Worthing, world, worship.

ear   as in earth, early, learn, heard, pearl, search, rehearsal.

our   as in vapour, labour.

(b)
$c+a+ll, ld, lt, lk$ e.g. enthrall, bald, malt, walk.
$c+o+ll, ld, lt$ e.g. poll, scold, bolt.
$c+a+ft, th, sh, sp, ss, st, etc.$ e.g. craft, bath, surpass, task, clasp, cast.
$i+nd$ e.g. find, blind.
$i+ld$ e.g. wild, child.
$i+gh$ e.g. sigh, flight.
$i+gn$ e.g. sign, design.
$a+w$ e.g. law, crawl.

Although this chapter is concerned with the teaching of phonics it must be read within the framework of the general thesis of this book. Poor readers have to use their phonic skills in the context of cues provided syntactically or semantically by the text. To concentrate too heavily on word analysis is detrimental to comprehension. The purpose of phonics teaching as defined here is to simplify the pronunciation process for poorer readers and to help those who have difficulty in decoding and retaining unfamiliar and technical words.

Spelling

There is no doubt that spelling remains a bugbear, even if you try to avoid making an issue of it by playing it down. On the one hand it can be argued that it becomes more of a burden to the pupil when not enough attention is given to spelling, and on the other it is possible to argue that our society places too much emphasis on spelling. Many critics have attempted to demonstrate that concern for correct spelling is misguided, belonging either to class snobbery or to capitalism's wish to stratify people.

These arguments, I have found, convince neither parents nor pupils: they *know* that uncertainty over spelling embarrasses them at a number of moments in life. The examples of the words that can be spelt wrongly without impeding communication given by most theorists are irrelevant. It is true of a few examples a few times, but not logically true if extended to many words for a long time. To argue from some apparently non-significant variations to the low importance of spelling in general is weak. Those who argue thus do not seem to have in mind the real difficulties of the poor-spelling pupil, who is wandering amongst a forest of lost letters, and to whom writing is an agony.

I have little sympathy anyway with those who dub correct spelling merely the equivalent of literary table manners, to be fussed over only by the socially squeamish. These people overlook the contribution that a grasp of spelling makes to an understanding of *sense*.

The two are closely linked. The importance of helping the pupil's vocabulary growth depends heavily on the pupil's understanding of the way words are what they are. For many pupils, vague teaching for ten years has given the impression that words are chance collections of letters, with a vague meaning loosely attached. Attention to spelling, on the other hand, is attention above all to form and therefore to meaning. For instance, the pupil confused about the number of 'r's in 'interrupt' has lost, or never found, the relationships *inter—between*; and *rupture—to break*; hence *to break between or into*. The pupil missing the first 'a' in 'extraordinary' has lost the sense, of *extra—beyond*, hence *beyond the ordinary*.

The Report stressed not only the social disability of poor spelling, but the fact that 'confidence in spelling frees the child to write to fulfil his purpose' (11.43). Spelling is really part of the concern for vocabulary, which I have argued must be part of a whole-school language policy. The school must pay regard to the needs of a pupil's spelling because it helps him to write freely, it helps with an understanding of vocabulary, it gives a useful occupational and social skill.

The difficulties in spelling come from the diverse roots of our vocabulary, and the layers of confusion that they introduce:

1. Similar sounds from different roots have been transliterated differently, e.g. especially the Greek sources, where the different alphabet has led to the 'ph', 'rh', 'ps' combinations. Homophones are inevitable in a spelling that retains the etymological origins of words (e.g. *tied* and *tide*).

2. Pronunciation shifts alter the sound of root words, e.g. *phótograph, photógraphy*.

The result is a vocabulary in which there is a constant tension between phonic and etymological pressures on the spelling of a word. What should the school do? There are two extremes in teaching techniques, both impossible not to label pejoratively: (*a*) Issuing word lists, and requiring them to be learnt; (*b*) Leaving the learning to be 'picked up' as words are met over the years. In overall curriculum terms, there is also the choice of putting the responsibility on one Department, or leaving it to all. Virtually nobody who has done research into or written on spelling or vocabulary considers that either is adequate. Certainly the learning of de-contexualized lists has been shown to be ineffective, as has mere leaving to the chance of picking up. The latter seems almost impossible for most pupils given the comparatively small amount of reading and writing that they do.

Indeed, learning for recognition in reading is far from providing most people with the complete mastery required for writing, in which a complete recall is necessary. This is why, of course, contrary to much popular opinion amongst teachers of English, extensive reading is not itself sufficient to help pupils spell adequately, though it is a vital background and helps some children extremely well.

We do not know which pupils are likely to learn best by which methods, but it does seem that the less able and those from less educationally supportive homes are the most in need of specific teaching. Phonic analysis is necessary even for what seems to be sight reading, and those from less favoured homes have had less home training in the underlying skills, an idea which I suspect underlies the Report's survey results: the poorer readers increasingly come from the families of the lower socio-economic classes. It is for these kinds of reasons that the Committee said of spelling that: 'less favoured children need to be taught, and taught rationally and systematically' (11.48). A witness to the Committee from an EPA area stressed: 'An incidental learning approach (to spelling) is hazardous for all children, but particularly so for those from disadvantaged homes' (11.48). The Report therefore called for 'an agreed policy' between subject teachers (11.13), but did not specify in detail what this could comprise. However, a school planning a language policy would clearly wish to arrange a co-ordinated programme of specialized attention in the English class, and attention in the context of all subjects, especially as far as the special vocabulary of each subject is concerned.

What should that attention involve? Not mere list learning—indeed the sheer size of the task would obviously make that unsuccessful. A more careful approach is required, for the heart of helping a pupil to spell well is helping him to concentrate on the internal structure of words.

Children talk about 'difficult words'. It is helpful if they can be brought to think instead of 'difficult syllables'. For most words that offer difficulties there is a vague sense that the whole word is a muddle: the pupil just cannot embark on it. The first task, therefore, is to break it into its constituent syllables. Obviously to do this the nature of syllabification must be taught specifically and used generally.

Having syllabified a word, itself a phonic activity which concentrates attention on sound and structure, each syllable needs to be examined according to the phonic expectations. (Thus the simple phonic patterns of our language should be ready at hand for all teachers.) Then the phonically irregular or difficult syllables can be focused on, and the difficulty isolated, and, incidentally, reduced in size. Thus the teacher facing a pupil or a group having difficulty with the following words would syllabify them:

separate	separate
corridor	corridor
necessary	necessary
analysis	analysis.

Before asking the pupil to memorize the correct spelling, the teacher would indicate the syllable that needs special care. *This* is the one that requires learning, for this is the one that is phonically irregular or for which there are reasonable phonic alternatives. These syllables are then specified, thus:

separate	with an 'a' in the middle
corridor	with only one 'o' at the end
necessary	with double 's'
analysis	with a 'y' in the middle.

Many weak spellers in secondary schools simply have not got the knowledge of the basic regular phonic patterns. There are pupils in secondary schools who are vague about the sound possibilities of the single vowel sounds (e.g. they confuse 'i' with 'a' in short sounds) or blends (e.g. 'dr', 'spl', etc.). The teacher coming across spelling difficulties must (a) recognize the cause of the difficulty, (b) refer the pupil to the phonic generalization, and (c) if possible add other examples. *Merely to correct gives the pupil no power over the spelling: he cannot re-use the correction on other occasions.* Therefore a school-wide knowledge of simple phonics is vital. This involves being able to draw on a commonly agreed pool of phonic teaching. Again, taught specifically, and used throughout.

Phonic generalizations, however, rarely stand up to the full range of examples because of changes of pronunciation or because of alternative less common phonic patterns. The report quoted the letters 'calmbost'. This group of letters could be pronounced as 'chemist' by applying a set of sound values, each of which is used in English (candle, many, silent 'l' as in calm, harm, silent 'b' as in lambing, women, and so on). The simple phonic rules do not cover all of the vocabulary, for as many as 1 in 3 words may be irregular. America's experience of burdening children with a vast number of such generalizations has shown that it does not help, and after a point can be counter-productive. Furthermore, some of the rules are inevitably complicated in expression and use terms which are unfamiliar to the pupils. The rules are more difficult than the points being covered. It is therefore only the firm central phonic principles that should be taught and used, and a school drawing up a language policy should commission its own list, and agree it as a full staff.

Clues

Apart from this phonic analysis, clues can be offered to help the concentration on the key difficult syllables. The main ones are:

1. *Derivations.* Clues come from a knowledge of structure, and this means considering what a word comes from, and what it does not come from. For instance, 'corridor' seems to have something to do with doors, and many writers want to put two 'o's at the end. In fact it comes from an Italian word (via French) meaning a runner. The teacher would therefore point out that the word has nothing to do with doors and comes from quite a different word altogether.

2. *Patterns inside words.* There are a number of letter groups that are used often as alternatives to the simple regular phonic pattern. These should be taught, and when a subject teacher comes across a word that uses such a combination he should refer to the similar ones, such as the examples below:

'ph' is pronounced like 'f' sound (except in Clapham): photograph, etc.

'aught' group: caught, taught, naughty, daughter.

There are approximately 1900 words in English which end in '-tion'. It is useful to point out that this suffix is pronounced 'shun'.

'-uit' group: fruit, suit.

3. *Unstressed sounds.* One of the hardest parts of spelling is choosing the right form for the unstressed syllables in each word. Most errors occur in such places (e.g. sep<u>a</u>rate, poss<u>i</u>ble, opp<u>or</u>tunity). There are only three approaches:

(*a*) Pure visualization. Some people learn them best by learning the look of the difficult area.

(*b*) Artificially stressed pronunciation.

(*c*) Knowledge of etymology to trace the source of the indeterminate sound.

4. *Homophones.* With homophones, pronounced the same but having different meanings and different etymologies, research[1] shows that for secondary-age children they are best taught separately. It would seem wise to tie each word to its etymological family of words.

5. *Spelling rules.* The Report commended a comprehensive school that used a list of spelling rules. The trouble is, as I said of phonic generalizations, on the one hand there are often exceptions, and on the other hand some rules are extremely difficult to comprehend. However, those rules that are simple and fairly reliable should be known, agreed, taught, and used.

The maths teacher, for instance, should know the 'silent e' rule, and have that knowledge available if any of his pupils are having a moment's difficulty over any of the following words, all from a first-year maths syllabus:

rotate

divide

spike (a spike abacus)

prime.

To do this he must also be sure that the rule has been taught elsewhere specifically. Only if that is so can he call on the rule to help him, without risking having to embark on a full-length explanation of the rule. Of course, by making this reference in context, especially with a central word like 'prime', over which he is likely to linger, the maths teacher is reinforcing the pupils' knowledge of the rule and ability to use it readily as required.

[1] K. C. Harder, *The Relative Efficiency of the Separate and Together Methods of Teaching Homonyms,* Journal of Experimental Education, VI, 1937.

Methods

Because some people have less good visual perception and memory, much bad spelling results from not having ever really *looked at* the letter patterns. Thus mechanical copying is as bad as rote learning. The pupil given a written word to help him with his spelling should be encouraged by all subject teachers to follow the routine: look—cover—write—check. There is ample evidence that for most pupils if this drill is not followed, and if mere copying is allowed, the pupils will usually *use* the word, and as promptly let it go from their minds in the same way that most people promptly forget a telephone number recently looked up the moment they have dialled it.

Reading aloud has been shown to help spelling (as I believe it helps punctuation and grasp of argument). This is because a fuller articulation of the word is required for a vocalized reading, and this fuller articulation involves auditory analysis (e.g. in the words 'multiply', 'analysis', 'elasticity', 'corrosion').

The key question for the teacher is not 'Is this pupil a bad speller?', but 'What kind of mistakes?' does he make? A continuous professional curiosity is vital—the too-common attitudes of hopeful optimism or hopeless despair are useless. Teachers of English can not teach a general skill, 'good spelling'. However, they can give the familiarity with words, their structure, the phonic basis, and methods of analysing words which provide the necessary foundations. Each subject teacher then uses this in handling the vocabulary of his subject and the writing in it.

A school language policy might, then, involve attention to the following ideas:

1. Phonic introduction of all new, or specially important words in each subject.

2. Words always to be syllabified, to help the understanding of structure, and to make it possible to focus on the difficult syllables. Draw attention to the syllable to be specially remembered.

3. A school-agreed list of spelling rules to be drawn up and promulgated. When commenting on pupils' spelling these to be referred to.

4. A list of some of the main word patterns to be similarly drawn up. This is likely to be specially important in schools with less able intakes where many of the simple patterns (e.g. 'aught', '-uit', or '-tion') have not been learnt at younger stages.

5. A list of useful prefixes, roots, and suffixes to be used throughout the school for reference when introducing new words or correcting old.

6. Always giving a spelling in writing, never letter by letter aurally. Usually then have the pupil memorize it before writing it. Discourage copying.

7. Each subject teacher should produce a word list for a sequence of work, and these words should be tested and taught.

Punctuation

Punctuation is an integral part of writing, arguably the most important technical aspect, and one that has been a least successful part of formal education. The paradox is that if we are to help those that can take least theorizing, *our* theory of punctuations has to be so much better, so that what instruction we do judge fit to give is not only accurate, but also useful because it refers to fundamentals and can thus be applied on other occasions.

The teaching of punctuation has suffered from a number of fallacies:

1. *That as the teaching of grammar is out, punctuation should not be taught.* This rests on a misunderstanding of the definition of grammar and the arguments against teaching it. As Dr Gatherer has made clear in this book, punctuation is not part of grammar as taught, so that the arguments against the teaching of it are in most ways irrelevant to the teaching of punctuation. The available research shows that punctuation is used best by pupils who have been taught it.[1]

2. *That punctuation is all a matter of personal taste, and so cannot be taught.* It is true that there is considerable latitude for personal taste in, for instance, the use of the comma. In most sentences the absence or presence of one comma alone is unlikely to make that sentence either 'wrong' or 'right'. However, the fact that we could leave that verdict to personal taste judged against stylistic consistency with the rest of the passage should not obscure the vital fact that there is an underlying system for the use of the comma. There are thus frequent occasions when its presence, or absence, *is* indubitably wrong judged by the sense needs of the writer's intention, and all teachers will be able to think of examples from pupils' texts alone. Thus even on those occasions when we agree to differ, we differ only in our judgement of the effect of the application of certain aspects of the system on any particular occasion, not on the system itself. Whatever choice we make is not a matter of pure taste, such as whether you prefer Double Gloucester or Wensleydale cheese. The fact of there being a choice depends on our understanding of an underlying system of punctuation: thus the choice is better exercised if the system is understood.

It is also misleading to invoke reaction against the teacher's personal taste in the argument, since we do teach many matters in school which are aspects of personal taste. Clearly the approach and aims are different from the more objective parts of education, but we do not consider them beyond the scope of our work.

3. *That concentrating on mechanical aspects of language inhibits creativity, and therefore punctuation should not be taught.* Apart from the way the use

[1] W. G. Heath, *Library-Centred English*, University of Birmingham Review, 1962.

of this kind of metaphor preempts the argument, this is more a point about the method of teaching than against the need to teach punctuation.[1] Even the teacher least willing to explain a point of punctuation is eventually driven to object to eccentric punctuation, for eccentric punctuation inhibits not merely communication but the actual articulation of thought. The Report commented that: 'It was not uncommon to find the despairing comment "Your punctuation must improve" on the writing of pupils who seemed to have received little or no specific instruction in it' (11.21). *This* broadside and blunt criticism certainly inhibits creativity, and, what is more, it does not in any way help the pupil with his punctuation.

4. *That for the less able it should be 'first things first', and any consideration of punctuation is an excessive burden for such pupils.* This point of view has an honourable lineage, and, indeed, did a great deal of good, helping to get pupils away from pointless exercises (e.g. *Visual English*, all little cartoons and error correction) onto writing. David Holbrook approves of this realignment from exercise to doing (although it is worth reminding teachers who romanticize his approach that he includes 'drills' in punctuation in his philosophy.[2] This approach of putting writing before punctuating was well put by the Newsom report when it tartly commented:

> In practice many of the weaker pupils never seem to reach the point at which real English begins.[3]

However, this should not deny pupils the opportunity of sharpening their understanding and practice of punctuating, but merely ensure that the tuition is kept in its place—that is, learnt in sufficient depth and detail for the writing in hand. We can accept that less able pupils cannot be taken through a decontextualized systematic analysis of all aspects of punctuation (indeed, should any normal secondary pupil?). However, the more we argue for reduction the more important it is that what is brought before the pupil is both accurate and above all usable on other occasions, when the advice may come to mind. Now this patently has not been done so far. Inaccurate half truths, of little use for extrapolation, have been the sop for the less able: 'A full stop marks a big pause, and a comma a short one'; 'never start a sentence with "but"'; 'all paragraphs start a little way in.' These rough 'rules' are not 'ready' because they are neither accurate nor useful. They commit the first sin of curriculum building, losing essential concepts while accommodating ourselves to the less able. Indeed, I should argue that it is precisely by sifting out the essential concepts, clarifying them, and then concentrating our practice on those that one works with the less able.

[1] cf. Neville Bennett, *Teaching Style and Pupil Progress*, Open Books, 1976.
[2] cf. *English for Maturity*, CUP, 1961, and *I've Got to Use Words*, CUP, 1964.
[3] *Half Our Future*, para. 462, HMSO, 1963.

5. *That punctuation is unnecessary in any case.* Those who attack the need for punctuation ('Bernard Shaw didn't use the apostrophe') use phrases such as 'middle-class', 'only snob value for big business'. Their attack is usually illustrated by a lengthy diatribe against the pointlessness of the apostrophe – somewhat similar in my mind to attacking the value of the vital organs of the body by criticizing the appendix.

A theory of punctuation

The Report definitely recommends (11.21) 'explicit instruction' in punctuation. The Report's twin approaches of general use and specific intervention fit especially well. Within the writing policy, pupils should be given as many and as varied assignments as possible, but:

1. Some named sector of the school should be responsible for relevant explanation in context. This is likely to be the English teachers, but could be any Humanities Department.

2. Every teacher should work to the same principles, agreeing on punctuation priorities, and knowing the methods of explanation of others. This general support should concentrate on the fundamental principles.

There seem to me to be two problems—firstly of classification analysis and secondly of terminology. In other words, a theory of punctuation is required which is adequate to cover the complexities of real language, but simple enough for use in all subject classes. I know of nothing published to help us in the classroom, and I therefore offer my own approach.

Punctuation should be analysed by function, not by sign. The common analysis by sign (e.g. 'a capital letter is used to start sentences and for the names of persons or places') confuses quite separate functions. The full stop ending a sentence is quite different from the one showing an abbreviation. Instead of the uses of the semi-colon, we need to show seven ways of marking off a sense group, or the three ways of inserting interruptions (commas, dashes, brackets), or the three ways of indicating a word or phrase has been borrowed from a special use. In this analysis the emphasis would be on the way we have to group words to make sense. Ungrouped words are mere puzzles, and wrongly grouped words are nonsense. These ways of grouping are by use of:

the comma
the semicolon
the bracket
the full-stop with space and upper-case letter
the paragraph indentation (or the extra horizontal space)
the space or signs for section divisions
the chapter-ending space

These signs can be seen as a hierarchy of sense groupings from the phrase to the chapter.

I am suggesting that space should be seen as a sign, for it is used to signal meaning to the reader. The paragraph was originally marked indeed by a sign in the margin. Only in modern times was the indentation used instead (now frequently replaced by a wide horizontal band of space).

Another aspect of this functional analysis is that the full stop is never on its own. As a major, *the* major, sense-group marker it is always followed either by an upper-case letter or by space to the end of the line (to show the end of a paragraph). Thus the combination of full stop and upper-case letter should be taught from the first, and explained as a *pair* dividing off sense groups.

If punctuation is to be taught, both in a specialized and contextualized way, and if an analysis by function is to be used, the points made must be accurate and consistent. The theory of stops most commonly taught, and indeed the one which seems to rise to the teacher's mind when he is obliged to point out a failing and thus to explain, is surely inaccurate. It is often said to help young pupils in schools that the main punctuation marks (, ; : .) indicate pauses of varying lengths. This is so commonly said that you may well have come across it. It is a limited definition of their function, since the statement applies only to prose read aloud. Even then it is only partially true; experience shows that pauses are almost entirely left to the reader's own choice. The strongest pause signs are paragraph indentations but the others are flexible. If you listen to yourself or any other reader you will soon notice that the *length* of a pause has almost no relation to the punctuation marks. Pauses are part of a speaker's effect, but in prose read aloud do more to control the impression and atmosphere, than the sense, which is indicated far more by the intonation or pitch of the voice. Consider this remark and try reading it aloud as dramatically as possible.

> I want you to realize this is your last warning. If you come late again, you're fired!

Where have you made your longest pause? It could be at the full stop after 'warning', but there are many other places where a speaker might make a long pause. One could occur at the comma after 'again', but is just as likely even where there is no sense break, e.g. after 'realize' and after 'you're'. It could be said like this:

> I want you to realize [*long pause*]
> [*very fast*] this is your last warning. If you come late again, you're
> [*long threatening pause*]
> [*slowly*] fired!

That would be a natural and effective way of reading it. The pauses would be for dramatic effect, and not for basic sense. If you listen carefully, you will notice that the sense groups marked by the punctuation marks (the full stop and comma) *are* used by the reader. *Punctuation helps control his phrasing and intonation, while the reader's speed and length of pauses are left up to him.*

It is also important to put across the idea that the 'signs' do not include only the familiar stops, but space, and the change of letter-shape to upper case, and italics or bold type.

Terminology

Finally, there is the question of terminology. When analytic grammar was taught, 'phrase', 'clause', and 'sentence' could be used, as could the remainder of the Latin terms. A school must establish its agreed vocabulary, what the report calls 'the modest collection of technical terms useful for a discussion of language'. 'Sentence' would no doubt be retained. I have already mentioned that I find the idea of 'sense groups' useful. Here are two other important things to teach when explaining punctuation conventions.

1. *Defining or non-defining clauses*, the correct use of which is especially necessary in transactional writing. Teachers will have their own methods for explaining these to pupils. A useful question is: 'Does the extra part tell you which or tell you more?'

A word or phrase in apposition would also be described as one 'that tells you more', or 'a second naming' and taught similarly in both reading and writing.

2. *Vocative of address*, very frequent in story writing, autobiography, and plays. It is best to say: 'Always put a comma before the name of the person you're speaking to, never before the name of the person you're speaking about.' Thus: 'Come here, John'; and 'I don't agree with John'. In all these examples the emphasis should be on punctuation as it indicates *intonation patterns*, not pauses.

Conclusion

A school, then, needs a theory of punctuation. It needs an agreed analysis and set of terms. A named department should have the explicit, specific teaching function, probably with check list and record system to ensure that opportunities are found for all aspects to be explained. The system should be agreed on by the whole staff, expounded, and circulated, with a list of priorities. Each teacher is then in a position to reinforce the teaching within the context of the writing requirements of his or her learning activity.

Study skills

We award the highest academic accolade to a student who can see a question, focus it into an enquiry, trace sources, find relevant information in those sources, collate the information, reorganize that information in a way that meets the question posed, and write up the reorganized material as a report. To those who achieve that pinnacle of scholarship we award a Ph.D. This same process is the one we have adopted as the main teaching method for the less academic and less well motivated school pupil, almost completely in the humanities, and extensively even in the crafts and sciences. Yet we often give no specific help. The 'project method' is felt to motivate inherently, and this motivation is supposed to solve all learning problems. No wonder projects are often so feeble.

The spread of CSE brought the spread of the project, sometimes even called the research project. From the earliest days perceptive teachers were worried about the quality of the pupils' work. What looked to a superficial glance like involvement, perhaps because of the cover and presentation, proved on a closer look to be undigested and uninvolved copying. In history it was reported that 'it became apparent that many pupils had copied sections from their source book, sometimes showing evidence of lack of understanding'[1] and examiners in geography similarly reported: 'A large number copied enormous amounts or made thinly disguised paraphrases, and this led to over-lengthy accounts'.[2] I came across one example of a page of pupil's project writing that included the words 'see Fig. 24' — but the reference was to the original, slavishly copied, text. Too often the pupil was not merely not understanding the subject matter, he was not getting experience in writing, for he was not sufficiently master of what he wanted to say to be handling the ideas in his own language. Indeed the project or research folder too often provided the antithesis of the writing approaches recommended in Nancy Martin's contribution to Chapter 5. Ironically some teachers, having recognized these weaknesses, have tried to avoid the problem by recommending that project topics should be less factual. In other words they have fallen for the temptation yet again of recognizing a linguistic difficulty, but, instead of facing it and attempting to solve it, are seeking to avoid it.

The process has slid down the years, because of a peculiar logical flaw of assumptive teaching. The teacher first conscientiously notes that the next stage of education is going to require some particular ability; he then decides to prepare for that next stage; then, instead of teaching *for* it, he merely presents the new situation, assuming that

[1] Schools Council, *The Certificate of Secondary Education: The Place of the Personal Topic – History*, Examination Bulletin No. 18, H.M.S.O., 1968.

[2] Schools Council, *The Certificate of Secondary Education: Trial Examinations – Geography*, Examination Bulletin No. 14, H.M.S.O., 1966.

the pupil will develop the required abilities to tackle it. Thus projects go further and further down the age group, with as little tuition in reference skills as ever. Many textbooks habitually throw hopelessly huge investigations at pupils, and many teachers accept from pupils hopelessly vague project titles. Typical assignments from a religious studies book, clearly designed for average pupils of fourteen or fifteen, include such bald statements as: 'Make a detailed study of Yoga'. Where would a pupil start? What kinds of things would be appropriate to include? Where might he find information?

As any school Librarian will tell you, usually pupils embarking on a project go to the library for a book. The famous 'use of initiative', 'discovery method', and so on, often consist of 'ask Miss', and 'Miss' is asked for a book of the precise title of the project. 'Please, Miss, have you a book *on* "Pollution Today"?' What knowledge of the library was taught in Junior schools is often lost in the early years of the secondary school.

The study skills are yet another part of the policy that requires both a specialized and a disseminated approach. At certain moments in the school, agreed in the policy, some department must be responsible for specific teaching of the basics. All subjects must then draw on this foundation knowledge, add the key data for their own subject, and give the study skills scope by guided experience in their subject field.

The study skills could be listed as:

1. Some idea of the range of kinds of sources available, which includes non-book sources.
2. Some idea of the kinds of books from which knowledge can be gained, that is, not only from full-length books on the subject being explored but reference books, encyclopaedias, compendiums, books ostensibly on other subjects, etc.
3. A growing understanding of how a library is organized, with special emphasis on the use of the catalogue.
4. Familiarity with the conventions of books, including table of contents, index, etc. It is especially important to understand the value of an index for freeing the reader from the author's organization of ideas, and allowing a different means of entry into the book.
5. Training in flexible approaches to books, especially skimming and scanning and searching for the points required.
6. Training in noting information and ideas.
7. Guidance in reorganizing and synthesizing—possibly the most difficult task.

Underlying all these sub-skills is the central need to focus the enquiry in such a way as to make it a definite enquiry and not merely a wander through any facts which come up.

It would help if more subject teachers gave specific lead-ins to any enquiries that they are mounting. Below, is an example of a worksheet prepared for third-year classes to help them handle an enquiry into a religion.

Report on a religion

These notes give you advice on how you might gather in information for, and write up a report. It is only *advice*, and you should use your own ideas as much as possible.

1. *Choose the religion you will study* (n.b. if you are yourself a member of a religion, it is advisable to choose *another* one.) Hinduism, Sikhism, Judaism, Buddhism, Islam.

2. *Sources.* You may use any of these ways of finding out. It is best to use a number of ways. When you are finding out, make notes as you go along. When you are using books, do *not* copy—put the ideas into your own words.

(*a*) Class book box.

(*b*) School Library.

(*c*) Public Library.

In libraries the classification numbers of books on religions are:

Christianity	230
Comparative religion	290
Judaism	296
Islam	297
Hinduism	294
Sikhism	294
Buddhism	294

You can also look up the religions under their alphabetical position in Encyclopaedias (classified as Reference). Books on peoples' lives are 920.

(*d*) Newspapers—cut out any news item from local or national papers.

(*e*) TV—note any news item, and watch any documentaries.

(*f*) People—ask friends, relatives, neighbours.

(*g*) Places—try to visit at least one religious building. Look carefully and make notes.

3. *Questions.* These are the sort of questions you might ask: When was the religion started? By whom? Where? What are the main beliefs? How do these beliefs affect ordinary life? How do they affect families? Could you make this clearer? What are the main rituals? What do they stand for? Describe one in detail. What are the main festivals? Describe one in detail.

Does worship take place at home or in a special building, or both? Is the home or the building more important for worship? What is one special building like from the outside? If you went inside, what would you find?

Is there a special person in charge of worship in the building? What is he/she called? What does he/she do?

All religions have stories (myths) about their god/gods or their founders. Re-tell one of the stories which plays an important part in the religion you are studying (e.g. if you were studying Christianity you might want to tell the story of the death and resurrection of Jesus).

4. *Writing it up.*

(*a*) Make notes as you find out facts, opinions, etc.

(*b*) List the main headings under which you will write. (For example: history; main countries; main beliefs; rituals; one in detail, etc.)

(c) Re-arrange the points in your notes so that you write about all you know on each part of the subject.

(d) When you have spoken to someone to find things out, quote his or her actual words.

(e) Only quote key phrases or sentences from books, and put them in quotation marks. (Put everything else in your own words.)

Use diagrams, drawings, pictures, photographs, to illustrate what you have written.

Previewing a book

Amongst the problems of individual study, possibly the hardest is how to get from a book what is wanted without a complete read from page one onwards. Colin Harrison has described in Chapter 5 how even quite advanced students start at the beginning of a book, and are unable to grasp its whole, indeed to establish whether it is likely to include the information being sought. It is helpful if they are specifically taught the clues which can be used in previewing. If a whole class is to spend some time using a main source book, the teacher would be well advised to preview it with the class, thus both helping the pupils use that book throughout the year, and giving a specific setting for the knowledge of book structure that will be taught specifically elsewhere. Such previewing involves considering:

1. *Title and sub-title.* Particularly the latter gives clues to the angle, focus, or purpose of a book.

2. *Blurb or preliminary summary.* The book jacket blurb, frequently reprinted on the half-title page is, as it were, an extension of the sub-title.

3. *Editorial details.* Almost always found on the reverse of the title page, these establish the date of publication, and any subsequent revision. (The difference between a new 'impression' and a new 'edition' should be taught.) Even the youngest pupils should learn to consider this for all information books.

4. *Table of Contents.*

5. *Introduction.*

6. *Summaries* or terminal aids. Many books have conclusions, summaries, or sets of questions that review the entire book.

As more and more subjects use individual enquiry as a teaching method, it is vital that the school as a whole helps pupils develop what have come to be called the study skills, as the ability to do this is of lasting value to the adult. The need is put clearly and forcefully by this American writer:

> What is the teaching of a particular subject such as social studies, science, literature, or home economics other than that of teaching the pupil or student to recognise and face issues, questions or problems inherent to

that body of content; to locate appropriate information content; and to devise from that content ideas, generalisations, and principles that will help him answer his questions, resolve the issues, or form valid bases for opinions or judgements? It would seem, then, we are saying only in another way that *the teaching of a particular subject is the teaching of the study of that subject*; and that makes inescapable the fact that every teacher is a teacher of reading and study.[1]

[1] A. Sterl Artley, 'Effective Study—Its Nature and Nurture', J. Allen (ed.), *Forging Ahead in Reading*, International Reading Association, Newark, Del., 1968.

7. School Organization

To be concerned with language development in the secondary school is to be concerned with the total school context, and that means matters of school organization. The classroom approaches outlined so far need backing up by a number of school-wide organization points. In this chapter we shall consider the main ingredients for a language policy which are aspects of school organization.

Liaison between schools

One of the romantic vaguenesses peddled frequently in educational exhortation is that we should create a closer liaison between educational institutions. In fact it is one of the hardest things to achieve, and can consume considerable time to little effect. Hadow in 1926, Plowden forty years later, and Bullock again have all echoed the need. Certainly the comprehensive school has usually built some form of liaison into its pastoral system,[1] but too often this is the responsibility of one Head of Year, and whilst something is achieved, it cannot be said that there is a functional continuity. Usually, even at its best, this is only a pastoral continuity; it rarely has a curriculum dimension, and frequently has only the sketchiest reference to language.

Where such pastoral continuity so often goes wrong is in its emphasis on a *personal* relationship between a limited number of teachers. This in practice must mean a relationship between a very small number of people, and thus very little real information is passed on to those who need it most. The records most transfer systems use are inadequate. Suspicion of measured reading data is made worse by the erratic and incompatible systems used even within one group of schools. From a language point of view this lack of continuity means work is not built on, needs are not prepared for, there are confusing conflicting explanations for pupils, and the language environment can change awkwardly and unprofitably from school to school. A secondary

---

[1] cf. Michael Marland. *Pastoral Care*. Organization in Schools Series, Heinemann Educational, London, 1974.

school needs to know what has gone before, generally and specifically. Whatever other reasons there are for policies of continuity, a secondary-school language policy depends upon one. A good liaison policy is likely to consider:

1. Shared knowledge of the learning activities, materials, and styles.
2. Specific information about individual pupils, their reading and language skills (this will include the interchange of samples of writing).
3. An agreed policy on at least some aspects of a language policy across the schools.

In my view there is no harm in there being stages in education, marked by distinct breaks. Rather than energy being devoted to removing or smoothing over these breaks, the pastoral side of schools should be devoted to helping the growing pupil *use* them. There is therefore a need for preparation. However, the pupil needs to know that he is known, and that his achievements have not been left behind him. The receiving staff need to know what experiences he has had, what his strengths are, and what special help he requires. This information is ill-supplied by many current LEA transfer systems. What is clearly required is four-fold:

1. A language 'profile', that is a description of each pupil's language use.
2. Standardized reading data, measured by the same tests throughout the group of schools, preferably in the form of informal reading inventories or diagnostic test data.
3. Titles of books read.
4. Samples of various kinds of writing by the pupil.

Knowledge either of schools or of individual pupils is not itself enough, unless it leads to action. Part of the action will be the adjustment of the teaching process and learning materials to take cognizance of the complementary approaches of the other school. I do not mean by this a wholesale change, indeed it would be ridiculous to try to reproduce, say, primary methods for thirteen- or fourteen-year-olds. It is, though, possible to retain the differences between school types, and between subject experiences within the secondary school, and still to respect and react to this knowledge. This action is likely to be on a number of planes:

1. Agreement about terms to be used amongst the community of schools (see the first section of Chapter 6).
2. Agreement about definitions and methods of explanation (e.g., 'What is a full stop?').
3. Agreed approaches to the hierarchy of reading skill.

4. Broad agreements about the language demands reasonable at various stages, and of the preparation that will lead to those stages.

Thus liaison would result eventually in a community of schools creating a broad plan, within which specific details were agreed, and individual pupils' work known. This would allow each of us to work at our specialism in its own way and within our own approach knowing what we build on, what we are expected to contribute, and the common terms with which to do it.

Testing policy

There is a tendency in education for crusaders to be left fighting enemies that have all but disappeared. If they stopped to consider, these crusaders would often find the battle has been won, and that the enemy they are now hitting at is not the original target, and does not deserve their animus. So it is with testing. Those who fought against selection and for the common school had in the course of the war to wage a battle against selection, which involved attacking and destroying the selection tests that were part of it. There are those who are still fighting tests as the agents of selection, even when selection has been defeated.

More sophisticated objectors to test programmes are those who have read about the labelling effect of testing. 'Teacher expectation and pupil performance' was the telling phrase that Douglas Pidgeon used as the title for his important study of the possible bad effects of the misuse of crude test data.[1] This was, and still is, an important argument: crude labelling of pupils' abilities can lead teachers to teach to and keep children to the levels declared by the labels, and at the same time can lead pupils to live up to, or down to, those label levels.

Thus educational testing and measuring of all kinds became discredited in the eyes of many teachers who were anxious to encourage high achievement, and who regarded themselves as progressive. In fact, however, such blatant condemnation of testing does not stand serious consideration, nor should the possibilities of mis-use of data logically lead to the abandoning of the data if it can have good uses.

The Report retains a healthy scepticism about much testing, and notes all the possible fears, even that testing time uses up precious teaching time. However, it establishes a new positive approach to testing, given the main point that 'Testing should never be carried out without a real purpose' (17.15).

[1] Douglas Pidgeon, *Teacher Expectation and Pupil Performance*, NFER, 1970.

The Report recommends testing programmes at three levels: national monitoring, local authority screening, and in a school. It is this last that is part, as I see it, of a whole-school language policy. However, it should be seen in relation to the first two. National monitoring by a rolling programme of light sampling based on frequently revised item banks is carefully worked out by the Committee as the only sane, reliable, and valid way by which the nation can know how its education is doing. The community is, quite rightly, bound to want an answer to that question. A method such as the committee recommended avoids the panic that blows up from time to time without it, and the inefficient measurement that results. The case for local authority screening is equally strong. By the use of appropriate measures at younger ages, pupils whose language deficiencies need attention can be identified.

My concern, however, is with in-school programmes. The argument I shall develop grows out of Bullock, but in some ways goes further than the Report, which does not go into great detail in the secondary phase.

The central purpose of testing is to obtain knowledge that can be used to help the pupil. There is plenty of evidence that we have insufficient knowledge to provide the help that is required. One interesting example was produced by the Committee's survey, and it is an example which could be replicated with teachers of older children also. For me it epitomizes one level of knowledge requirement. The teachers of six- and nine-year-olds in the sample were asked to estimate how many of their own pupils would be placed in the top 25 per cent ('Good') of the *nation*'s readers, the middle 50 per cent ('Average'), and the bottom 25 per cent ('Poor'). Obviously when totalled and expressed as a proportion of the total number of children in the whole sample, the number of children in each category should roughly equal the defined percentage of each category: 25 per cent, 50 per cent, 25 per cent. In fact the teachers of nine-year-olds over-estimated the ability of their children by putting no fewer than 41.1 per cent of the pupils in the top 25 per cent, only 37.4 per cent in the middle 50 per cent, and only 21.5 per cent in the bottom 25 per cent (Survey Table 43). My impression is that similar over-estimating of reading ability is common in many schools, though it is difficult to get firm evidence.[1] Secondary schools have a

---

[1] cf. the report on *William Tyndale Junior and Infants School Public Inquiry*, ILEA, 1976, p. 59. ILEA Junior schools are obliged to place pupils in similar percentiles (25, 50, 25) for transfer to secondary school. This table is interesting quite apart from the purpose for which it is quoted in the Inquiry Report (i.e. to assess the levels of the pupils' achievement). The figures show the teachers' assessments, in the centre column, as widely discrepant from the NFER-monitored test results for the school, shown in the right-hand column:

way of reacting with shock after the first screening tests that they administer—a shock that can be explained only by the inaccuracy of the previous estimates. I therefore believe a reading testing programme is necessary as part of a whole-school language policy:

1. To monitor the overall curve of the ability:
(a) to establish changes from year to year, or following alterations in the transfer procedure.
(b) to help consideration of the curriculum.
(c) to help consideration of pupil-grouping methods.
(d) to help the disposition of all forms of resources, teachers, rooms, money.
(e) to contribute towards the evaluation of the curriculum and teaching by comparison with follow-up figures.
2. To identify those pupils who seem to be deficient and who need help.

Clearly individual pupils would then require detailed diagnostic tests to establish what kind of help they require and to suggest a teaching programme in reading for them. (There seems no doubt that pupils who need such help can go unnoticed by class teachers without this kind of check.)

Ideally such a programme would be established over a community of schools. If, as the Report suggests, all contributory schools and the secondary school worked to a standard test programme, there would be no need for early testing, only for later follow-up testing. Such a co-ordinated programme, however, seems too difficult for most schools to achieve, and in that case I believe the secondary school must initiate its own programme.

	Authority-wide proportions	School assessment of 4th-year pupils	ILEA's banding of them
Above average	25%	25.8%	16.4%
Average	50%	38.2%	54.1%
Below average	25%	36%	29.5%

The differences between the centre and the right-hand columns are the measure of how far out teachers can be in their unmoderated assessments! Although there are reasons for believing that this group of teachers had greater discrepancy than most.

Planning a screening test of reading attainment at lower secondary level

Colin Harrison

The aim of conducting a screening exercise is straightforward—to identify at an early stage those children who urgently need extra help in reading. This aim is much more general than that of determining exactly what kind of help is needed, which can only come from a closer diagnosis. Individual diagnoses and specific remedial attention would then need to be made available, but initially for an intake of two or three hundred an individual diagnostic reading test would probably be impracticable and inefficient. What is necessary is to pinpoint those in need of extra help during the first half-term, and not allow a situation to develop in which nearly a year passes before a child who had been considered simply quiet or lazy is found to be a very backward reader.

The field of mental testing is a highly complex one, and many teachers who use reading tests are blissfully unaware of the fact that their method of administrating and scoring a test may totally invalidate its results. Most of us have very little understanding of the private world of the statistician and psychometrician, but this does not stop us from borrowing their tools. The kind of error we make may be comparable in magnitude to that of the child who tries to measure the length of a line with a slide rule, but we obtain a result—a number, a score, or better still a quotient, and we make the assumption that it has given us accurate information about the child's current level of ability. We may indeed put much more faith in the score, which seems objective, than in the evidence of our observation and experience, which are necessarily subjective. Every test ever published has weaknesses, and each test result gives only an approximation of the child's true ability. Most test norms have a standard error associated with them, which indicates the range within which the true score is likely to lie. This may be for example plus or minus two months in terms of reading age on a well-constructed test, which represents a range of four months. Even this range would only take account of 68 per cent of the results; the standard error concept implies that 32 per cent of true scores will in fact be outside that range. The need therefore is for great caution in interpreting test scores; the teacher must try to ensure that the results are used to help children, rather than simply to brand them.

The very concept of reading age, so cherished by teachers in their urgent need to classify those in need of help, is not accepted as meaningful by many of the most eminent experts on testing. The manner in which we divide reading age by chronological age in order to produce a reading quotient is similarly regarded as unsound. It presupposes that development in reading ability is linear, i.e. that it tends to follow a straight line, as in graph (a) of Table I. Such a notion is highly speculative, and would indeed be challenged by many teachers. Progress in reading probably conforms more to graph (b), or even graph (c), either of which would be wholly incompatible with the usual manner in which we obtain a reading quotient.

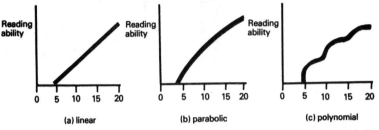

Table I: *Graphs illustrating three possible types of relationship between chronological age and the development of reading ability*

The fact that the field of testing is a complex one need not deter one from using tests, but rather should alert us to the dangers of misuse and misinterpretation. In the present case, where the aim is to carry out a general screening exercise, heeding such a warning may in fact save a great deal of time. In particular, if all we wish to do is to establish which children are not reading well, there is no real point in changing every child's raw score (e.g. 37 out of 53 items) into a percentage, then obtaining a reading age, and moving from there to a reading quotient. If the test has pinpointed the five poorest readers in a class it has done its job. If four of those children are younger than average, using raw scores has not made allowance for the fact, but this could be considered an advantage since they are still the children who are reading most poorly and this would have been masked to some extent by working from a reading quotient.

Choosing a test

An excellent and comprehensive review of the reading tests available in this country has appeared, which tabulates details of ninety-one British and overseas tests as well as offering an introduction to some of the theoretical and practical aspects of testing. This is a monograph in the UKRA Teaching of Reading series, called *Reading : Tests and Assessment Techniques*. Its author is Peter Pumfrey, and the book is in the Hodder and Stoughton Unibook series.

The tests in Table II overleaf could all be used at first-year secondary level. Those cited are necessarily a small selection of the ones available, but represent some of the most widely known and most easily accessible, together with one or two which are less well known

Table II. Reading tests suitable for lower secondary level

Name of Test	Chronological Age range	Test mode	Publisher	Skills examined	Time	Remarks
Burt Word Reading Test	5:0–12:0	Individual, oral	Hodder and Stoughton Educational	Reading single words aloud	≤ 10 mins	Unreliable at upper and lower limits.
Neale Analysis of Reading Ability	6:0–12:0	Individual, oral	Macmillan	Oral reading and comprehension of brief passages.	≤ 20 mins	Gives remedial teachers valuable diagnostic information.
Schonell R1 Graded Word Reading	5:0–15:0	Individual, oral	Oliver and Boyd	Reading single words aloud	≤ 10 mins	Unreliable at upper and lower limits; used so frequently that it must be known by heart by some children.
Vernon Graded Word Reading Test	6:0–18:0	Individual, oral	Hodder and Stoughton Educational	Reading single words aloud	≤ 10 mins	Unreliable at upper and lower limits; less widely used at primary level than Burt or Schonell R1, which could be an advantage.

Table II—*continued*

Name	Age range	Type	Publisher	Skills	Time	Comments
Brimer Wide-Span	7:0–15:0	Group, silent	Nelson	Comprehension (sentence completion)	⩽ 30 mins	Gives diagnostic information; test booklet can be re-used.
Edinburgh Stage 3	10:0–12:6	Group, silent	Hodder and Stoughton Educational	Various subskills of comprehension	⩽ 30 mins practice part 1:30 mins part 2:25 mins	Attempts to assess individuals' attainment on separate sub-skills such as comprehension of main points, vocabulary, etc.; time-consuming to mark.
Gap	7:0–12:0	Group, silent	Heinemann Educational	Reading comprehension (type of 'cloze' test)	15 mins	Easy to administer and mark; less well researched than most standardized tests. Easy to copy answers—teacher must separate pupils.
Gapadol	7:3–16:11	Group, silent	Heinemann Educational	Reading comprehension (type of 'cloze' test)	30 mins	Similar to Gap test; somewhat stronger in its statistical framework. Again, teacher must be sure to prevent possible cheating.
NFER DE	10:0–12:6	Group, silent	NFER (Ginn)	Various subskills of reading comprehension and vocabulary	⩽ 45 mins	A relatively recent test (1967) which is similar in concept to Edinburgh 3, but simpler to administer.
Schonell R4 Silent Reading B	7:0–14:0	Group, silent	Oliver and Boyd	Reading comprehension (multiple choice)	15 mins	Straightforward to administer; quite widely used although supported by no details of reliability or standardization.

221

but which merit consideration. A point which must be made in these days of financial stringency is that some tests are very expensive. The Edinburgh Reading Test Stage 3 (at the time of writing) costs about £24 per 100 pupils, and this is by no means the most expensive on the market. It is not surprising that many teachers pirate tests or at least try to re-use the answer books. Generally speaking the cost of the test is proportional to the amount of diagnostic information gained from it; one should also note that longer tests (i.e. those with more items) tend to be more reliable. Nevertheless the word reading tests should not be dismissed out of hand, simply because they do not require the child to read a whole passage and answer questions on it. The point is that the ability to articulate single words on a list has consistently been found to correlate highly with overall reading attainment, and although they give very little diagnostic information such tests can be useful because they are straightforward and quick to administer. Finally, some warnings are offered.

Warnings

1. Any group test causes stress and anxiety for children, but this can be reduced to some extent:
(a) Do not give a screening test in the first fortnight of the school year.
(b) Ensure that so far as is practicable any group test is administered in an atmosphere and environment similar to that of a normal class lesson.
2. Make absolutely sure that each class or group does the test with the same instructions, and is allowed the same time.
3. Make absolutely sure that the results of each class or group are scored consistently. If the questions are multiple-choice, no divergence from the published answers can be accepted. If there is any element of subjectivity of interpretation in the scoring, then the only safe course is to have every child's answers marked by the same person. If a test has subtests, different teachers could share the load by marking every child's answers on one subtest.
4. Do not give children exactly the same test again for at least nine months. If you do, results may be enhanced by a practice effect rather than by any gain in the skills the test measures.

Pupil grouping

How pupils should be grouped is one of the technical issues of education which has become a subject for popular debate, not only throughout the teaching profession but in the general press and amongst parents as well. Its socio-political aspects have given it this status, and everyone feels qualified to opinionate. There are more assertions than considered arguments and many factors are omitted. 'Mixed-ability' versus 'streaming' is the popular form of the argument, an over-simplified polarity which blunts the points being considered. In many ways it was unfortunate that throughout the 'sixties the battle raged so long and so fiercely that the details of teaching method were too frequently lost sight of amongst the broader sociological arguments.

However, there are some issues which focus very sharply on language use, and although this would not be the right place for a full discussion I must try to outline the language learning aspects.

The outline arguments are now well known, but have a new look if you consider them for the moment purely in terms of language use: however careful you are, classifying pupils makes different pupils in the same group seem more similar than they are, and similar pupils in different groups more different than they are. The arguments for not attempting to separate (a formulation that I find better than 'for mixed-ability') are normally stated in the negative in that this, that, or the other disadvantage can be avoided. The positive point is a hope that the differences between people can be used advantageously.

The argument will have to consider the particular circumstances of the school: its intake curve, the local area, the size of a year-group to be divided, the size of teaching group, the classroom size, the resources, and the teachers. Some people speak as if the only problem is a resources problem, and that a mass of work cards, junction boxes, individual assignment cards, and the like will solve the problem.

There are various patterns of grouping for the remedial pupil, and from the language point of view each has its problems as well as advantages:
homogeneous remedial class;
full mixed-ability group;
partial mixed-ability group;
individual extraction from the mixed-ability groups;
visiting remedial specialist.
Obviously the first three can be used for part rather than all of the work in a variety of combinations.

The Report listed the criteria by which a school should make its decision, and gave a general view that mixed-ability work was likely to be most favourable for English as such (15.1). It did not debate the wider issue of pupil grouping throughout the school. The Committee did specifically worry about the less able reader—how would he or she get help?

There are two respectable philosophies—both have truth in them, and it is futile to think that debate will lead to any overwhelming agreement that all right is on one side.

One is that pupils need individual care and different treatment. Indeed, to give them all the same diet is unfair, and opportunities for specific teaching are possible only if the differences are recognized. As truly individual care is impossible with school staffing ratios (indeed pupil–teacher ratios are one of the main reasons why we have groups at all), some measure of grouping by needs is vital. This argument would have to include an analysis of teaching activity to show that extensive and significant individual teaching is not possible in school classes.

The other philosophy is that interaction is a key element in education, especially in language, and that there is a gain from keeping pupils together. This allows common humanity to be shared, and it allows language to grow more naturally.

The language situation of the backward reader in a fully or partially mixed-ability group needs thought. The conscientious teacher is usually anxious to supply a range of activities. To meet the needs of the less able, many of these will be non-verbal tasks: colouring, drawing, filling-in, modelling. If there is not very great care, the clever backward reader will weave his way through the tasks, going for the less verbally difficult time and again. Indeed it is possible for a pupil of twelve or thirteen with mixed-ability English, humanities, and science to go for weeks without reading—or even doing much writing. One count was 300 words read in six weeks!

Writing of mixed-ability work in Science, J. Dark, Science Adviser for Devon, recommends sub-groups of about four, which, he declares, is 'the best size of group to accommodate non-readers'.[1] What does 'accommodate' actually mean? Too often it means allowing the pupil shelter from having to read, and thus denying him practice in this setting. Such a pupil is probably extracted from his class to be given a couple of periods of reading teaching each week. Valuable as this is, it suffers from breaking the continuity of the lessons precisely for those who need the continuity most, and even the reading teaching is discontinuous. The worst fault is the isolation of the teaching of reading from the other learning processes. Dr Ruth Love Holloway, Director of the US Right to Read effort, shrewdly calls this the 'band-aid' approach.

A version of this for older pupils which we developed at Woodberry Down School is called Support Option. Pupils who require such help choose one fewer option in their subject choices. In the time thus freed they are given support teaching—reading and language work (as well as other skills, of course) through the medium of the materials they are using for their other subject choices. This is an attempt to produce proper integrated teaching for the less able. It is, however, very difficult to get the integration fully effective.

The visiting specialist is a tempting device, and seems to me

[1] In E. C. Wragg (ed.), *Teaching Mixed-Ability Groups*, David and Charles, 1976, p. 147.

effective where there are sufficient staff. I have for instance seen it effectively used for immigrants when the LEA has specifically put in additional staff for the purpose. The remedial or reading specialist visits classes to a timetable. When there he assists those who need help, giving reading and language tuition through the medium of the materials actually in use. This point-of-need tuition can be very effective. It suffers from three problems:

1. There is not always, or indeed often, time for the specialist to give real tuition, only to offer short cuts.
2. The identification, which those who recommend mixed-ability wish to avoid, is re-emphasized.
3. Usually the remedial specialist can visit for only a few of the time-tabled lessons. Thus the help is sporadic, and sometimes hardly worth it.

From the point of view of language development, pupil-grouping should be looked at carefully, and the decisions based on the following criteria: Are groups sufficiently often sufficiently mixed for language interaction? Do those who need specific help get sufficient time with sufficient concentration? Where separate specific help is given, is it related functionally to the other subject teaching?

Print availability

I have stressed throughout that the Committee came to see growth in language as requiring attention to the whole of school experience, and thus involving school management. This is true in the provision of books, what I would like to call a print-use policy. The individual teacher cannot himself provide the range or integration of reading material that is required.

Traditionally each Department selects, orders, and organizes books, and more recently other print sources, in its own way, and a library provides different kinds of books with a totally different method of availability. The pupil himself, with no guidance whatsoever, is expected to integrate the various sources in his mind, know where to go for what, and to make up the deficiencies himself—presumably from public library or home. No wonder education is class differentiated, for this burden is intolerable.

Apart from the question of availability to pupils, there is the question of the economics of supplying books. Are we making best use of limited resources? Money for books is extremely tight, perhaps tighter than staff-room complaints realize. An analysis I did recently showed that the average figure available for spending per pupil per period on learning material for the country's secondary schools was 1p!¹ Consider what materials you can buy for that money. How little is likely to be left for books? Take the 1p per period for a secondary school. A small part will have to be kept by the school for central spending on circular letters home, exam papers, duplication for school administration, etc., say £1.40 a pupil for the whole year or 0.1p per period. Some will have to be put aside for library books. The Association of Education Committees recommend £2.25 as the minimum for secondary pupils up to 16. That would require a further 0.16p per period. This leaves 0.74p to spend on all learning materials for each lesson.

Look next at the costs of a few lessons. I have costed a sixth-form dissection in biology at 69p a period for each pupil, a fourth-form heat structure investigation at 9p, and a particular fifth-year physics lesson at 1.65p a period. A school must run very many chalk-and-talk lessons to save money for these. A page of an exercise book costs about 0.1p—so it is expensive if pupils are too prolific in their writing.

It is not usual to cost printed books on a period basis, but when one does it is clear why schools feel short. Take, for instance, an anthology of plays at £1. It might be read for six weeks for five periods in each week, and three classes might do that in the year. If the books last for six years (the Association of Education Committees' recommendation), the book would cost 0.2p for each period. Even the book cost of an apparently cheap subject like mathematics works out at hardly less. However large the global sum seems, the actual amount of money to spend for every lesson is not adequate for the materials required, certainly not to carry a stock of additional follow-up material.

There is considerable evidence, what is more, that spending on books is decreasing not as a result of conscious educational decisions, but under the pressure of other insatiable needs.² Bullock put the worry clearly: 'Since expenditure on books is part of the larger block of non-teaching costs, it is in competition with other demands, and books are in a particularly vulnerable position in times of financial stringency.'

---

¹ Michael Marland and Ian Leslie, 'A Penny for All Their Thoughts', *Times Educational Supplement*, 26.3.76.

² Details of book spending were given by the Secretary of State to Parliament on October 27, 1975. These show that in the four-year span 1970–71 to 1974–75 the ratio of cash spending on books to that on other materials had steadily worsened from 1:2.1 to 1:2.3 in primary schools and from 1:2.3 to 1:2.6 in secondary schools. In real terms the worsening is even greater. The same parliamentary answer showed that the retail price index increase during the same period had been greater for books than for equipment by a factor of 1.16:1. This means that in 1970–71 schools were spending about a third of the real value of their allowances on books, and by 1974–75 they were spending only a quarter.

A school studying its costs finds that home economics, needlework, science, and crafts require so much consumable material if the subject is to continue that book purchase in those and other subjects declines. Those subjects that can just about get by on old sets are forced to do so.

There are also problems of level. Many school libraries are sixth-form orientated. Many lack simple enough books on topics of interest and importance to pupils of the younger ages.

We need therefore to look carefully at schools' policies for books. Too often there is a muddled selection and ordering procedure, frequent changing of courses without adequate planning, and sheer inefficiency in stock control. A ratepayer seeing the losses, or the piles of disused books in jumbled corners in some schools, might well protest that he does not want his money spent on that kind of waste. Certainly an integrated print-availability policy could be part of a school's language policy.

In summary, then, a print-availability policy would look at: *relevance* to the pupils' lives and to the school courses; *suitability* in levels and style; *availability*; and *integration*, that is, stocks with stocks and stocks with publicity.

The first three are probably self-explanatory, but the last might need some expansion. The interrelated provision should consider each of the following supply sources:

1. *In the classroom*: full-class 'textbooks'; classroom small sets or topic books; classroom individual copies.
2. *The library*: the basic collection; special collections; reference books; non-book material.
3. *The school bookshop.*[1]
4. *Outside the school*: children's library; main library; local bookshops.

The stocking of 1, 2, and 3 should be considered overall. For instance, titles for further reading or research recommended in books available in 1 should normally be provided in 2 and sometimes 3.

Arising out of this is the question of publicity. Book publicity should have at least some measure of co-ordination. Library and bookshop should mount publicity campaigns that are related to topics or themes currently involving at least some of the classes. Displays should be built up in lockable glass cabinets and outstations of both. School and sectional assemblies should feature or refer to available books. (At Woodberry Down, for instance, the list of Assembly Themes always gives the appropriate Dewey classification numbers.) Special units within a school, like the careers room or music department, should have displays and collections integrated with the central library. The school should thus present books in a varied, vigorous, and inter-connected way to all the pupils.

[1] cf. Marilyn Davies, *The School Bookshop*, in Owens and Marland, *The Practice of English Teaching*, Blackie, 1970, also *School Bookshop News*, Penguin.

PART III

Putting It Into Practice

8. Initiating and Implementing a Policy

Throughout this book, whether in analysis or recommendation, my contributors and I have assumed the need for action; we have pictured a school which wished to act, to build on its earlier discussions and thoughts, and to start up a coherent policy. In this chapter we chart some possible ways of starting that action, and of keeping it going. For a school with a tradition of discussion, joint planning, and school-based in-service training the start should not be too difficult. For a school that has not such a tradition, the challenge of Bullock may be a way of stimulating staff discussion.

Initiating
Nancy Martin

Many teachers, of all subjects and at all levels, are extremely puzzled by what a language policy across the curriculum means. Gerald Haigh's parody of the situation in an article in the *Times Educational Supplement*[1] is not so far from the truth and lays bare the confusion with inspired mockery. Let us quote from him:

> *Monday*. Arriving at school in a decisive mood, I wrote on my 'Things to do' pad:
> 1. See the caretaker again about that funny sticky stuff behind the radiator in room three.
> 2. Remove the outdated notices from the board in the corridor.
> 3. Institute a language policy across the curriculum.

This could be a comment on the comic diversity of a head's responsibilities, but it is equally a reflection of the fact that many teachers and administrators and heads do not understand that getting a language policy going is a new, difficult and long-term job, not at all the same kind of operation as, say, a policy for school uniform.

---

[1] *Times Educational Supplement*, 26.3.76.

Monday (continued)

During the morning Marston, my deputy, came to see me. He wanted to know on behalf of the staff, what I meant by a language policy. If it means a crackdown on swearing, they are on my side.

Tuesday

Read Marston's paper on the language policy. It consists of various standardized symbols to be used by teachers in their marking.

One laughs, yet both points are concerns of teachers: what makes them ridiculous is that they are so remote from the major issue of the active role that language could play in learning.

Again Gerald Haigh's fictional headmaster writes:

. . . Johnson is not taking to the language policy. Several of his pupils are citing Bullock as evidence that they are allowed to talk in class, and he is now asking me to define the boundary between insolent and non-insolent talking. How difficult it all is! In my day all talking was impertinence; all argument rebellion. And we were all literate to boot!

Thursday

. . . Parents are beginning to complain about the language policy. Mrs Flotbury wanted to know why her daughter is being taught by a tape-recorder and a pair of earphones. 'I pay my rates and I demand a proper teacher,' she said. . . .

The confusions ridiculed here are painfully near the bone, and are characteristic of all of us when we half get hold of an idea.

Let us look at some reports from real heads of department and real teachers.[1] First from a head of Home Economics in an 11–16 comprehensive school.

Approach to head—interested, though not willing to help formulate any policy or discussion meeting at this stage.

Approach to head of English—agreed, with some doubts, to a discussion meeting with other interested members of staff.

Approach to deputy head—interested and sympathetic—organised an informal staff meeting for interested members of staff.

Informal staff meeting held. Report on the area DES/ATO Course on Language across the Curriculum from me. Discussion that followed mainly concerned with grammatical errors in pupils' work. Only two heads of department attended, therefore head of English suggested a formal heads of departments meeting to discuss their ideas on a language policy. Probably later this term.

Here is a classic situation. A head of department who has a good idea of what the Report is about in its most general educational implications, a doubtful head; a doubtful (at first) head of English; most other departmental heads not seeing it as anything to do with them; and a sympathetic deputy head who occupies the right place in

[1] N. Martin, B. Newton and P. D'Arcy, *Times Educational Supplement*, 26.3.76.

the power structure of the school to organize a meeting for those interested. What is significant is that, once someone organized a meeting, the head of English overcame his doubts and moved into action with a heads-of-department meeting—which would seem a crucial step. And how necessary it is that someone, like the head of home economics in this case, should be explicit about the real issues involved, otherwise the discussion might well stay at the level of 'standardized symbols to be used by teachers in their marking'.

Next a report from the head of English in an 11 to 18 comprehensive school.

> After the DES/ATO course, reported to head who fully supports the notion of moving towards the long-term goal of a language policy. The head feels this will best be achieved by a series of contacts between the head of English and other departments. I myself have come to feel that the process should be institutionalised in the form of a committee. Agreed to discuss the matter in a meeting of the academic committee and determine a strategy with other heads of department.

> Asked the director of studies for an extraordinary meeting of the academic committee. This denied, but the matter was placed on the agenda for the next regular meeting.

> At this meeting I outlined the concerns of the Course and emphasised that there was little recognition of such issues in the school, and that all departments were involved, not simply English. It was stated by some that departmental heads were aware of such matters as the need for a variety of written exercises and were capable of making appropriate judgements themselves. I replied that there were many matters of common interest, and my initiative must not be misinterpreted as an infringement of each department's autonomy, but should be seen as a search for solutions to school problems by group efforts. It was felt that methodologies were at issue as much as language. It was agreed to discuss the matter further at the next meeting.

> Since the meeting I have approached the head of Mathematics, who is interested in linguistics, and asked him if he would be prepared to chair a committee concerned with methodology and language. He agreed to consider the matter.

> It is clear that 'language' is a highly sensitive area where teachers are concerned. Perhaps it would be better to emphasise 'methodology' instead?

In this case the head fully supports the idea though there is some disagreement about tactics, but what about those heads of departments who say they are quite capable of looking after their own affairs, thank you very much? For them the Report clearly has nothing to say of any consequence. Furthermore the feeling at this school meeting is right, that 'methodologies were at issue as much as language'—and, of course, the theories that lie behind methodologies. The Committee is indeed asking teachers to look at what they do in the teaching of their subjects. It is true that the context for this request is that of language: 'What uses do you, in fact, make of language in teaching? How much do you talk? What opportunities are there in your classes for your pupils to formulate their knowledge in their own words? Do they have opportunities to reflect on their work in talk or in writing?' These are indeed matters of methodology, and methodologies reflect hypotheses about learning, so the recalcitrants at this meeting were in effect saying, 'you are asking us to examine our professional expertise and we object to that. It is an infringement of our specialist autonomy'.

What was interesting and sensible on the part of this head of English was to try to get the head of another department (Mathematics) to chair a committee concerned with 'methodology and language'. Until heads of subject departments come to see that language is a major instrument of learning and is therefore their concern, initiating moves from a head of English inevitably appear as interference (or empire building) from a subject with a vested interest in language.

Finally, a thumbnail sketch by an English teacher of the situation in his school which expresses another common misapprehension.

> Since the Bullock Report the head, and heads of departments are committed to some kind of language policy. So far this has taken the form of the sort of support that the English department can give to the less able in other subjects: remedial staff support in mixed-ability lessons; encouragement to the less able to bring written work to remedial group English periods; the formulation of a common correction scheme. The school has a good academic tradition.

Because of the context in which the Bullock Committee was set up— problems of literacy, extended to include all aspects of the teaching of English—a great many teachers see the Report as concerned only with English teaching and remedial problems, and of course, these are its central concerns, but behind them lies its major theoretical contribution (Chapter 4 'Language and Learning') and its attempt to foster the implications of this by its recommendation that schools should establish language policies across the curriculum involving all teachers. This is not a matter of getting all teachers to cooperate in raising the literacy level of the backward, *but of increasing the quality of learning at all levels and in all subjects*. This is what it means.

One of the difficulties in implementing a language policy is that the operation has two directions—towards the classroom, and towards collective decisions by the staffs of schools. And both directions throw up very different kinds of problems. In the first, a teacher makes his own decisions about any changes he may want to make in his own teaching and classroom procedure; in the second he is initiating discussion, meetings, and decisions with his colleagues where the pressures and problems are different from those he encounters in the classroom.

The Report asks something new and difficult of teachers—not only that they should each as individuals foster children's reading and writing, but that they should *collectively* organize language policies for their schools. Here the pressures and problems are different from those of an individual teacher in the classroom. Here the problems arise from the organization of schools and the hierarchies within them. Who has the right to call a meeting? Whose responsibility is it? What are the attitudes engendered by this kind of activity? What unperceived questions lie behind the actual questions raised by the Report? No previous Government Report has required of teachers that they engage in collective action to achieve reforms, and there is therefore little tradition in schools to guide this kind of professional work.

Since the Bullock Report was published in February 1975 meetings and activity of all kinds have been taking place and a momentum slowly building up. What I shall try to do in this section is to document some attempts to implement the Report, choosing those which reveal both the nature of the problems and the kinds of changes which can (slowly) occur.

The very first steps have often proved the most difficult. Three teachers from different schools reported their initiating strategies as follows.

1. Get headmaster's support.
Work through some body already established? E.g., Academic Board of Studies.
If anything concrete agreed on, put to staff council (which consists of heads of departments, year heads+headmaster, deputy heads).
Some constructive ideas of what it *means* and what it has to offer as an educational policy would have to be formulated by somebody, and offered as a first step before any action taken.
Get interest and support of own department.
Get some evidence to present of value of such a policy.
Suggest books to read? Take Bullock Report as starting point?
Get advisers interested?
2. *School situation.* Small (500) rural grammar school, joining with local secondary modern school (at present $\frac{1}{2}$ mile away, 400 children) to form one-site comprehensive. Traditional approaches to most subjects. Little will to innovate (from top downwards).
Report back to Head, emphasizing Bullock. Ask permission to make following moves.
Begin arrangement of meeting (with two neighbouring schools) for Science or Maths teacher (outsider) to describe his attitudes.
Talk to own department (English) and describe course. Ask for follow-up suggestions since they have all been in the school longer than I have.
Approach those members of other departments with whom I have already had 'education' talk. Ask how they view English (ought I to say 'Language'?) in school.

Get them to complete questionnaire—possibly criticizing work of English department.

Collate these remarks and invite contributors to meet and discuss their views.

3. First approach the Head of Department who, happily is a member of the English Curriculum Study Group and in charge of Curriculum Development in the school.

Approach to Headmaster who is already interested in this area of thought.

Notice in staff bulletin—invitation to members of other departments to attend an informal 'talk-in' on this subject. (This approach may appear formal but as we are housed on a split site is necessary to ensure that all staff are aware of meeting.)

Informal approach to people known to be interested.

Meeting of interested staff (possibly 12–15) to discuss their requirements and how they would like to take action.

Possible (almost certain) visit from outside speaker to underline work started.

How difficult it all seems—and how time-consuming. How open to rebuff. It is not surprising that teachers who have not thought about the place of language in learning wonder what it is all about.

One of the recurrent tensions in trying to promote language discussion and action in a school is between short-term and long-term aims. Although there may be no end to the journey—in the sense that schools and their ways of working will continue to evolve—nevertheless it is encouraging to notice the odd milestone. Both long-term and short-term goals are needed and the relationship between them will need constantly to be reconsidered. So a short-term aim might be to get teachers to tape-record groups of pupils talking in their lessons so that there is material for group study and discussion on the role of talk in learning situations. But this really needs to be related to a long-term aim which might be something like this: 'teachers of all subjects should be aware of the importance of talk in making sense of new ideas and information—and find ways of facilitating it in their lessons'. A long-term aim like this might influence the discussion of the tapes so that the emphasis shifts from evaluation to a consideration of more fundamental issues. So the questions asked might be: 'What kind of relationship do you need to have with your pupils to enable situations like this to occur?' or 'could the teacher usefully intervene in this discussion?', rather than just 'how articulate are these children?' and 'how successfully have they answered the teacher's questions?'

A long-term aim should enable teachers to become 'active interrogators' of the material, rather than passive commentators, because they have a commitment to act.

Mike Torbe, writing from his experience as an advisory teacher says: 'The only profitable way in is action research. But the only profitable action research is the one that has visible consequences for tomorrow's teaching *as well as* long-term consequences. Thus, exploring teachers' questions ought to produce not just critical comments on existing questions which imply yes/no answers, but should also suggest questions which *don't*—such as, 'which part do you remember?' or 'did it make you think of something else?'. Exploring ways of structuring writing, rather than criticizing sterile ideas, should suggest alternative modes (log book, running records, branching diagrams) and how they might be introduced. Thus the working group can be seen as making a direct input to the major problem of teaching—*what do I do tomorrow?*

We have received many reports of school meetings to discuss action on the Bullock Report. We quote here one written by the deputy head (a science teacher) of a big high school—it is a personal report of the discussion on language across the curriculum in the Academic Committee.

1. The use of language—which means *learning*—is not the sole responsibility of the English faculty. They will lead, advise, inspire but each faculty must play an important part.
2. Pamela will produce a draft set of proposals for a common marking policy—spelling, punctuation etc. But more important, how do we encourage students to learn from their mistakes? Common errors fed back through the English faculty perhaps?
3. The 'thinking-it-out' which we adults tend to do with our 'inner-voice' (but not always) has to be rehearsed in a more overt way in childhood. And so we should be encouraging the exploration of ideas in oral work and expressive writing. How nice it was that nobody suggested it was the middle school's job and not also ours to do this, although I think we were all looking towards our third year.
4. How can we encourage more oral exploration of ideas? Is it more difficult in mixed-ability classes? (*a thought*: surely the oral mode of communication is less differentiating than the written). Are we too concerned with keeping our 30 quiet in our classroom so as to avoid upsetting our neighbours? All too passive?
5. Can we in our different subject areas look at ways of encouraging 'expressive' writing. How should we treat it when we get it, so as to avoid the teacher-as-examiner role?
6. So much gets committed to paper (exactly *why* is not always clear, but often in the name of examinations) that has not really been learned at all—copying from the board, out of a book, dictated notes. The process of mental digestion is not occurring. Surely, this is the rule of the whole business: receiving an idea, breaking it down and then reconstructing it individually in a way that fits within each person's mental framework. And this is precisely what language achieves.

7. Let us not forget how little we actually assimilated before which the person concerned cannot accept an idea; the experiences which construct the framework are simply not there.

8. Reading: we tend to think that (slow learners aside) there is no problem here, but there are important ways in which we can help: e.g., skimming and scanning.

9. I think we probably all went away with two feelings:
(a) I could devise more valuable language (and thus learning) experiences in my lessons, with all age groups and all abilities.
(b) I want to look in more detail at different examples of oral and written work: talk about it, criticize it and try it out.

10. To this end I propose that we ask Pamela, David and Trevor to organize, say, 3 workshop sessions in June/July (taking advantage of the slack period when many fifth-years will have left). One representative from each faculty to attend.

Let's hope that we shall see language in the curriculum appearing on more and more Faculty meeting agendas.

From the head of science (Drewry, 1976) in another school came the two questionnaires printed below, one for pupils in the second year and one for members of the science department.

A. *To pupils*
1. Write down the meaning of the following words.

Living	Distinguish
Animal	Similar
Dissolve	Concentrate
Refer	Temperature
Compare	

2. How do you know if you have learnt anything in Science?
3. Which of the following techniques help you to learn?

	Always	Often	Sometimes	Rarely	Never
Talking with teacher					
Talking in pairs					
Talking in groups					
Answering questions					
Asking questions					
Copying from board or books					
Writing up experiments					
Completing charts					
Writing in your own words					
Tests					

B. *To staff of science and maths faculty*
1. Please indicate how frequently you use the following techniques.

Discussion	Always	Often	Sometimes	Rarely	Never
Class—Teacher					
Group—Teacher					
Individual—Teacher					
Class—class					
Group—class					

Children only	*Always*	*Often*	*Sometimes*	*Rarely*	*Never*
Discussion in pairs					
in threes					
in three +					

Questioning
for factual recall
for development of ideas
for reasoning
to discover background
    knowledge

Writing
Completing charts and tables
Graphs
Traditional experimental
    write-ups
Free prose in paragraphs
Copying from the board
Copying from books
Notes
Dictation
Calculations
Sentences
Essays
One-word answers to
    exercises/tests

2. Do you assume that the children will understand the following words?

	Always	*Often*	*Sometimes*	*Rarely*	*Never*

Example
Living
Animal
Dissolve
Refer
Compare
Distinguish
Similar
Concentrate
Temperature

3. How would you introduce a new technical word, e.g., *Leibig condenser*?
4. How would you introduce work on a new concept, e.g., solution?
5. How do you know if learning has taken place?

Not only did this provoke considerable talk in the Science Department, but the idea was taken up by other teachers in the working party, and similar questionnaires were devised for other faculties.

About Westerhope Middle School, Durham, the Project team wrote:

We think it highly significant that we were not, in the first instance, given a language policy document. We were given a 'Statement of Intent' which set down in three pages the objectives of the school and the agencies by which those objectives might be carried out. This 'Statement of Intent' had been produced from many discussions between the headmaster and 'teaching staff who were to be the nucleus of a larger and changing staff of the new Middle School'. A rider was added that if any teacher having joined the staff was unable to accept the underlying educational philosophy of the 'statement', he should discuss this with the headmaster so that 'an acceptable solution to the dilemma' could be found. Accompanying this statement of intent was a document headed 'The Language Policy of a School'. With the language policy document went a video tape of a mixed ability class aged 12 to 13 recorded throughout three days and covering all the activities that this class engaged in, all lessons including drama, PE, craft and library, and occasions such as Assembly, school dinner and play time. The language policy document is meant to be seen as part of the school objectives and contains 11 aims. The first sentence reads: 'We hope that the video tape recording will illustrate as many of the following points as possible'. Clearly such a way of producing a visual language policy involved all the staff, disposed of the notorious gap between statements of objectives and actual practice, opened all that was going on to public discussion, and located the language policy firmly in the more general aims of the school.

The examples quoted above have focused on the practical—on what can we *do*? I now change the focus and quote from letters the Project has received which not only deal with the actual steps taken but which comment on the nature of the process involved—the changes in attitude and thinking which happen and the problems these give rise to. It also seems to me that these letters highlight the particular role of teachers of English. It is part of their professional expertise to know something about language, and to give a lead in implementing the Bullock Report, yet the paradox of the situation is that such a lead can be counter-productive.

Getting started on a language policy across the curriculum (by the Head of English in a Midlands Comprehensive School)

I suppose it was my own fault really. I first made notes about a whole-school policy on language in the Summer of 1972 sitting in the garden drinking bottled Whitbread and reading *Language and Learning*. I hadn't read *Language, the Learner and the School* at the time. Then the head retired unexpectedly, the Authority introduced a chaotic scheme of reorganization, and everything but survival got shelved for the best part of two years.

One of the first things the new head did was to ask for departmental reports, and when I sent in my first one about Christmas 1974 I spent some space on outlining the responsibilities of other departments for the language development of the children. He wrote on it something like 'I agree, but English Dept. must give a lead. Consult with DH'.

On 18 February the Bullock Committee Report was published amid such a blaze of publicity that within a week at least five of our 75 staff had heard of it. I'm not joking: an event in the educational world has to be pretty stunning to make that much impact on the staff of a Midlands Comprehensive School.

As a result, just before the end of the Easter term, the Deputy Head and I got notes from the Head asking us to look into this curious 'Language across the Curriculum' business. I lent the Deputy *Language, the Learner and the School* and my copy of Bullock; he hefted it thoughtfully in his hand. 'We ought to get one of these for the staff library', he said. 'How much is it? Christ, five pounds.'

A week or two passed. Then, ten minutes before a meeting of heads of department, the deputy asked if I would speak about LAC. Summoning up an ostentatious show of diffidence, stammering and mumbling like Brando ('I coulda been a contender, Charlie') I introduced the topic: it wouldn't do at all to let on that I or anybody else had *thought* about this beforehand, that it might actually affect their lives, or be dangerous in some other way. Even so, or maybe *because*, it stirred a surprising amount of interest. The Heads of maths and science got quite excited and started talking about spelling, always a good sign. Several others said, 'What? a spelling purge throughout the school? Good idea' and gave it their blessing. The craft, domestic science, art and geography departments ignored it completely. I lent out my last few copies of 'Gnat Killers'. A Committee was set up: deputy, head of maths, head of science, head of lower school history (we're hierarchy-mad in our place), head of craft and myself. We met in the autumn term (late September) to decide on action. In the meantime I had talked casually to perhaps 20 staff, including members of the Committee (I think this is probably the most important preparatory work). The most significant remark came from the head of science: 'They're not learning anything by the present methods we're using, so we might as well give this a go'. From a very Tory middle-aged Welshman, head of science since the school opened 16 years previously, this indicated the first stirrings of an almost overwhelmingly daring spirit of exploration and enlightenment.

First 'working' meeting of Committee: 26 September 1975. I had photocopied part of a tape transcript of a Nuffield Science lesson in a primary school. This was far enough from home, I felt, to allow fairly free criticism. I also had two sets of the pamphlets from the Schools Council project, a few quotations from Barnes, and a list of possible projects in areas of concern, something on the following lines:

1. The over-insistence of the use of technical vocabulary by insecure/ unaware teachers; this seemed fairly safe. Only the probationers would admit to being insecure and nobody would admit to being unaware.

2. The suitability or otherwise of textbooks in use in the school, paying particular attention to their language; similar consideration of duplicated material produced in school—was it better tailored to the children's needs, or did the teachers fall into the same traps?

3. A look at the extent to which talk is used as a means of learning in *our school*. They shied away from this a bit. We all know that teaching in many departments, from first year onwards, is by dictation of notes.

Open/closed questions came up here; I was censured for using jargon, and had to explain the term. Even the Deputy—three years in a College of Education before coming to us—had never heard of it.

There followed the usual teacherly talk about bad pupils/eccentric teachers I have known, spelling, whose fault it was if the kids write poorly, standards, and so on; but at the end of it we had agreed to produce examples of textbook work for a proposed seminar, plus a list of 'technical vocabulary' under the heading 'Things we all do wrong', most of the items to be unnecessary use of technical terms, undeclared use of technical terms and textbook phrases, unwillingness to abandon safe ground of subject language, and so on.

Next meeting, 6 November. The head of craft's list of 'Things we all do wrong' began:

1. incorrect stance before the class—we are inclined to loll about on the desk instead of standing up straight

2. incorrect dress: even in the workshop we should wear ties

3. use of colloquialisms and local slang.

In fact, he didn't produce this list at all in the meeting where he said practically nothing. He only showed it to me when I tackled him afterwards about his lack of involvement, explaining that when he heard what we were saying he realized he'd got it all wrong. In fact, many of his other ideas were very good.

An encouraging aspect of the business was the way teachers not on the Committee kept coming up and speaking about it. The head of biology borrowed the latest issue of 'English in Education' (9.2) to study the transcript of the biology lesson (pp. 52–58), and the head of history showed me two pieces on the way of life of North American Indians written by children of putative low ability. It's interesting that the teacher concerned commended them for their unusual length from these children, and for the extent to which the children were involved, although I'm not sure there's a lot of expressive language: I suspect that they had learned to 'do' Indians or 'do' days-in-the-life-of, or something, just as the top O-level streams are learning to 'do' the Stuarts.

Later the head of biology borrowed a cassette recorder and recorded some lessons; so did I and a history teacher. I already have a fair amount of teacher/pupil talk on tape, but for everyone else it's a whole new idea.

At the end of this meeting I was a bit worried when the deputy indicated that he saw the Committee's work as a sort of 'awareness purge', a bit like the periodic purges on smoking in the bogs. However, subsequent events lead me to think that he was just passing through a necessary transitional phase.

There was a long break before our next meeting, ostensibly to allow for the collection of material, and really because of work pressure and the mock exams, marking and reports; but we met again at the end of January to plan our first open event, a staff seminar, which took place on 12 February 1976, almost the anniversary of Bullock. The meeting was fixed for 4.20 p.m., ten minutes after school finished, and was, of course, voluntary. Twenty-two people attended, including the headmaster, but some of the younger staff stayed away because they had been offended by receiving notes from the head asking them to explain their absence from assembly that morning. However, it's difficult to see how such a seminar could have been much use if more had attended. The meeting lasted precisely one hour, and after a brief introduction from me (more Steve McQueen than Brando this time, but still not Cary Grant), discussion revolved around two of the three duplicated papers attached: we never got on to open/closed questions. Discussion was quite lively, at least a dozen people spoke, we even managed to avoid getting bogged down in spelling. I won't go into the discussion in detail, partly because it was a bit diffuse at times, but mainly because I recorded the whole thing, and hope to make an edited transcript soon. I outlined the next items in the programme: two more seminars between now and Whitsun, one on talk, and the other on writing. I hope that later meetings will enable us to look in more detail at textbooks, with the suggestion that certain criteria ought to be followed when making a choice; and, even more important, to consider the production of written and other source material by departments—some of the units currently being produced by, for instance, the history department, are depressingly like some of the less inspiring textbooks.

As part of the preparatory work for the talk and writing meetings, I announced that I would be badgering people individually to record their lessons (the PE dept looked amazed and horrified) and to provide examples of children's writing, with notes on the conditions under which it was produced.

A few comments.

You have to walk delicately. When I hinted that learning in science might be more effective if the excitement of discovery and exploration could be fostered and maintained, the head of science replied: 'But science *isn't* exciting; science is *dull*. It's our job to take any excitement out of it. We don't *want* people getting all excited about it.'

The head of craft said much the same thing about his own subject. His comment on 'Gnat Killers': 'This wouldn't do for us—it's just the sort of thing we *don't* want.' But he's mellowing a bit.

A great deal of work is likely to fall upon the head of the English department in implementing such a scheme—try at least to get secretarial assistance, which is something not normally available to staff in our school.

There was a strong feeling of the blind leading the blind to begin with. None of us knew where to start; and my main worry was that too definite a lead might produce a strong rejection by influential members of staff, hence the mumbling and shuffling.

Further along

A year later, the head of English in a Yorkshire Comprehensive School wrote this letter.

When I reread the piece I wrote last Easter about what I was trying to do in my school and what I felt were the guiding principles, I cringe at the easy way I accepted the long-term nature of the task of implementing a language policy. I still think that it is long-term; but having carried the process on I realize that I was, if anything, underestimating the complexity and difficulty of what I had undertaken.

I feel uncomfortable, too, when I compare my basic list of principles with what I have actually found myself doing. Again, I feel that if there is to be a real language policy, it is still imperative to begin with the present unacknowledged language 'policy'. Careful slowness still seems to me to be the appropriate strategy, small-scale discussion the basic tool. What I underestimated was the natural pressure to follow existing procedures and the extent to which this would become irresistible.

I was, if I was to prove my seriousness, compelled to go along with the ways serious business is conducted in this school—the matter is discussed with headmaster, aired at a meeting of heads of department, laid open at a staff meeting. These are unconvincing media for development here and I preferred to work in other ways. However, I was asked repeatedly to address the various groups; my reluctance had no meaning in this context and I certainly wanted to show my seriousness.

A sub-committee on curriculum development had seemed an appropriate place to begin. The discussions were exploratory, open, encouraging. The general attitude outside the committee was that our discussions would be futile and would change nothing. The pre-emptive attitude has turned out to be the crucial factor, covering as it does deep-seated ideas concerning the role of subject autonomy, and most particularly the fears of what change will bring. If we were to examine the reasons why most people teach a particular subject, the most frequent reason would be interest at school and college. As a basis for asking the question 'why?' about a subject, that is not a promising start.

As part of our discussions I have come to recognize more about the assumptions upon which schools operate through the teachers, teachers being the main element in the composition of the ethos or character of a particular school. There is no necessity for a subject teacher to question the subject. Thus, I have been unable to ask questions, however tactfully, and however pleasant the relations, without causing discomfort—and there is an easy place for any teacher to hide from discomfort—the teaching situation.

I don't know whether it is the same thing but I have sensed that the questions I have been asking are felt to be irrelevant to the process of teaching. For example: what is happening when a pupil takes notes? Does a pupil actually record what he sees and understands when he records an experiment in traditional form? Is discussion the same thing as question and answer?

What I have found most dispiriting has been the low level of thinking about education that has been typical of the discussions. Alongside that is an apparent lack of any feeling that anything is wrong.

What I am documenting of course are *exactly* the reasons why the pursuit is long-term. I foresaw them in general but I am nevertheless worn down by them in their endless particularity. I do not believe that I have the answer; I have not found that other subject teachers are often aware that there are any questions.

The last two terms have been dispiriting but in trying to evaluate what has happened, I have returned to firmer belief in my sense of the difficulty of the task and the possibility of change. Certainly a few people have indicated interest in talking about talk. When I am tempted to be depressed about the tiny step forward, I need to recall that I know of no convincing examples of actual cohesion on language in a school except those training children solely for examinations and employment. Now there's a thought.

Language policy across the curriculum

The head of English in a central London Comprehensive School wrote the following revised questions for departmental discussion.

1. Can you define the language skills demanded by your department in terms of demands on:

listening

talking

reading

writing?

2. To what extent can your department help in the teaching of reading?

3. Can your department make more use of work in small groups?

4. Can your department make more use of talk as part of the learning process?

5. How much does the work of your department encourage teachers to talk to individual pupils?

6. What attitude does your department have to non-standard English in speech and writing?

7. What sorts of questions should we ask, what sorts of assignments should we set to produce the most effective *learning* situations?

8. Are we asking for too narrow a range of writing from our pupils; in particular when we neglect poetic and expressive writing in favour of transactional writing are we asking children to run before they can walk, to parrot the language of teachers and textbooks, regardless of whether or not that language means anything to them? (I am sorry about the complexity of that question and its jargon—examples of expressive, poetic, and transactional writing are appended—see also Section 3 of the language group report.)

9. What can we do to get away from the banda worksheet and to improve the quality of our home-made resources? Do we need more sophisticated reprographic equipment?

10. What areas has the language group not yet discussed which it ought now to turn its attention to?

11. How has mixed-ability grouping affected your answers to the above questions?

These questions, which are an attempt to communicate with the rest of the staff, do not represent the thinking of any one member of the language group but were devised by those who attended the two meetings. In other words members of the group were happier with some of the questions than others. It should also be remembered that, as I stressed in the first article, each school situation is unique and what may be relevant in one school may actually be harmful in another. At our school members of the language group will be offering departments help with their discussions on the questionnaire—particularly with rather off-putting questions like 1 and 8.

Language across the curriculum talk has in fact featured very little since last summer (apart from one refreshing and encouraging report from a biology teacher who went on a one-day children's writing course) and we are now pinning our hopes on the response to the questionnaire and on the interest that is being aroused by ILEA's follow up to Bullock.

Dr Birchenough, in his letter to schools, quoted a piece from Bullock: 'each school should have an organised policy for language across the curriculum, establishing every teacher's involvement in language and reading development. . . .'

A policy document does not, in practice, 'establish every teacher's involvement' and I feel increasingly that, while it may encapsulate the views of a particular group at a particular time and even give guidance on specific issues which will outlive that group of teachers, it can never really be more than a point of reference. There is another quote from the same chapter which gets us, I feel, closer to the problem: 'The effort a child needs to apply in learning language must derive from the satisfaction of evolving from helplessness to self-possession.'

This is an enormous undertaking and the truth is that language is at the *centre* of any educational system we may devise. It is not just another aspect of school life like uniform, books, or afternoon games. When we make statements about language we often expose the heart of our educational philosophy. Language across the curriculum discussions aim to bring about increased understanding of how learning takes place and how best to bring it about. That is a *continuous* process and that is why it is so important that each school should appoint someone with specific responsibility for this language development with plenty of school time made available for working with staff (ensuring, amongst other things, that they come to meetings!)

Certainly, in our school, given the latent and manifest interest of many teachers (not all of them, by any means, in the English department) we *could* make much greater strides. As it is we are still dealing far too often with those already converted.

Teachers' ideas

We conclude this section by quoting from a teacher in the London comprehensive school referred to above who had worked hard in getting the language policy discussions started. He draws attention to the influence of the individual perceptions of the teachers involved in the discussions.

Bullock on language and learning makes the point that,

> It is a confusion of everyday thought that we tend to regard 'Knowledge' as something that exists independently of someone who knows. 'What is known' must in fact be brought to life afresh within every 'knower' by his own efforts.' (4.9)

If that is true for school students it is no less true for their teachers. As discussion progresses each teacher in his or her classroom or as a member of a team is going to have something of a personal approach to matters to do with language, whether writers of policy documents wish them to or not. Moreover, language is so much at the heart of the learning and knowing done by teachers as well as students that very strong feelings are generated whenever discussion of such matters moves beyond statements of personal opinions to a genuine questioning of our classroom practice. As we attempt to establish the sort of systematic policy that Bullock recommends we are immediately confronted with the same complex of knowledge, experience and relationships that we find in our classrooms.

The questions that Andrew (the head of department) has included in his article were originally circulated with a summary of the discussions from which they arose. As I read them in isolation now they appear a very mixed collection indeed. This again reveals something of the complex situation which produced them.

Each question relates to an area of concern amongst the teachers at our school, but they also reveal the different approaches to language that are current at the stage of development we have reached. The first question is a good example, particularly as we place it alongside question 8. Looked at together they may reveal a divergence of opinion within the school. The first question represents a behaviourist strand in the thought of some members of our group. Behind it lies I feel, an idea that, if only we can get it straight in our minds what we want the students to be able to do, we can go ahead and systematically teach them to do it. Along with that goes a notion that the performance of our students can be measured against these goals and the effectiveness of our teaching can be assessed. By contrast, the inclusion of question 8 in the list suggests a desire to change our attitude to the learning process so as to centre our approach on the students' experience of learning rather than on our perception of what they need to know.

If we are to be true to the spirit of the quotation from Bullock in my first paragraph, we must be prepared for both approaches to exist even if neither of them reflect our own opinions. It may well be that in the changing situations that exist in schools, particularly in inner-city schools, some sort of final policy document should not be our main aim; continuing discussion may well be more useful.

I conclude by a return to the practical—who can help?

The Bullock Report asks something new, and difficult of teachers— not only that they should each as individuals foster children's reading and writing, but that they should *collectively* organize language policies for their schools. This is proving particularly difficult in secondary schools because these are mostly organized in subject departments with nearly every teacher a subject specialist. Within this structure the English teachers are seen as being the language specialists. Many subject teachers feel that their concern is with teaching their subjects and do not see language as having much to do with this. Others, however, see children's language development as an important instrument in learning history, science, geography, etc.

Taking any kind of collective action is a very different kind of activity from taking individual action in one's own lessons. From our work with teachers over the last six months we have set out the following suggestions about who can help in getting some of the recommendations of the Report alive and active in schools.

Inside school

1. *The Head*. As the person directly responsible for the administration of the school, the head is very important to the success or failure of any innovation. The head can provide help—as we have seen in section 2—by:
   (a) supporting teachers in their purpose by taking an interest in it and by making it clear to all the staff that he/she is interested and supportive;
   (b) allowing opportunities for the teachers involved to inform their colleagues about what they want to do;
   (c) being willing to explain the work to parents, governors and other interested people.

2. *Senior Staff*. Deputy heads, heads of departments, curriculum co-ordinator—if there is one—are important in the same way as the head is, and without the help of the senior staff it will be difficult to get a language policy really working.

3. *The Pupils*. It will generally be desirable to talk to pupils about any experimental scheme. Ideally there ought to be some facility for children to participate in a critical way about the courses they are following and the problems as they see them. We have found the children acutely aware of language and its problems and transcripts of their discussions or their writing on these topics can provide good starting points for teachers' consideration of these same problems (the difficulty of 'book' language for instance).

Outside school

1. *Local education officers*. The chief education officer and other members of his staff, the secondary, primary and subject advisers or inspectors are responsible for all the administration of the area in the same way that heads are responsible for schools, and their influence is therefore very wide. They offer a great deal of support to teachers in the form of short-term or long-term in-service education and very varied kinds of assistance. In addition, some authorities have an officer specially responsible for in-service work with teachers, and he/she often works in conjunction with a University Institute of Education. Many of the starting points described in the previous sections of this Pack were initiated by or in conjunction with such local education officers. They will organize meetings or courses and, where possible, duplicate and circulate material for discussion. Some authorities, for instance, have made their own summaries of sections of the Bullock Report and have circulated these to schools; they have also, in co-operation with heads, arranged joint meetings between local schools and follow up activities of various kinds.

2. *Teachers' centres*. Teachers' centres are, of course, the responsibility of the local education authority, and the scope of their activities varies

a great deal. At best, the teachers' centre warden would be the first person to approach for help in getting a meeting outside school organized, and the facilities of the centre should be available for any discussion material needed.

3. *Parents.* Most parents are keenly interested in their children's progress but they often have very vague ideas about what goes on in schools. We had a report of a meeting in one school where the English department had approached the Parent Teacher Association and arranged for one of the education meetings to be on the Bullock Report. Most parents have little opportunity to discuss educational matters, and PTA meetings would seem to offer an opportunity of getting a wider understanding of the aims of the Report.

4. *Subject associations* (the Association for Science Education, the National Association for the Teaching of English, for instance). If subject associations could be persuaded to see the Bullock Report as one of their concerns they could be more influential than any small group within a school. It is perhaps worth noting that some twelve months ago *Mathematics Teaching*, the journal of the Association of Teachers of Mathematics, carried a long editorial on the importance of language in learning mathematics; and the National Association for the Teaching of English through its annual conference, branch meetings and publications has actively campaigned. These activities must have reached many more teachers than could be reached by a school or teachers' centre meeting.

5. *The Schools Council Project 'Writing Across the Curriculum'.* The Project team of three with a fair number of teachers working with it has provided discussion pamphlets, Packs of 'outcomes' from schools, lists of relevant books, speakers for meetings, and a book of outcomes from the Project (*Writing and learning :across the curriculum*, Ward Lock 1976, £1.35).

Ward Lock has now taken over the distribution of materials and enquiries should be addressed to them at: 116 Baker Street, London W1, or to the Schools Council at 160 Great Portland Street, London W1.

Implementing

Introduction

It is not the British way for 'authorities' outside the school to either prescribe, as in continental schools, or to demand, as is common in American school boards. In California, for instance, there is a state-wide Framework in Reading, a general syllabus, as it were. Within that each district draws up its programme, the San Francisco State Board, appointing a Director of Reading. He in his turn issues guidelines, which oblige the schools to draw up *their* programme. The demand is quite clear: 'The Director of Reading expects to have on file in his office as of September, 1973, a brief outline statement describing the Site Reading Programme in each school.'[1] Even the most energetic of LEAs have not done that, with the possible exception of ILEA, whose initiative was courteously but firmly sent to all schools, in the form of an invitation from the Chief Inspector:

> I believe it would be helpful if heads, in consultation with members of staff, would prepare written statements of the language and literacy policy of their schools, as has already been done in some cases. Such statements might contain descriptions of present approaches and ideas for further action. If these statements were given to the Inspectorate they would provide a source of ideas and information which could be made available to teachers throughout Inner London. I am therefore writing to offer the help of the Inspectorate in preparing such statements of policy and to invite you to send your statement to me at County Hall.

Elsewhere, however, the British tradition of school autonomy has left the initiative to the schools. The very way in which these two requests contrast reveals both the strength and the weakness of our system: the San Francisco demand is limiting compared with the broader request of ILEA. Yet the second could be dangerously vague for schools in doubt (as Nancy Martin shows some are) about what a policy concerns.[2]

In fact, surely no school could, or would want to, buy in a ready-made language policy. This book has tried to cover the possible ingredients of such a policy. Each school will want to make its own selection of ingredients, study them, and develop its own approach to each. Then it will want to put the ingredients together in the combination and with the relative strengths suited to its own situation. Each school will have to establish its priorities. It is not often realized how widely different these situations are. In some schools notions of appropriateness are a tense and complex issue, requiring careful analysis and really rather painful debate. In others there is an accepted approach that offers no serious problems. In some schools the initial stages of reading present difficulties to so many pupils that here will be a central issue. In others the crucial task will be to open up the range of writing tasks. Yet others may need to concentrate their debate on pupil talk. It is likely that the central issue of communication will be common to all, but the balance of drive will and should vary widely.

[1] San Francisco School Board, *Guideline for the District-Wide Programme*, undated, p. 4.

[2] In fairness it ought to be added that ILEA supported the request with a careful series of meetings, as have other LEAS such as Avon.

One difficulty has already been indicated in the previous section: some colleagues may over-simplify, and overlook the really important issues in favour of the relatively minor technical points. The opposite difficulty is to presume that it's all too difficult to understand, discuss, and put into practice. For instance, the underlying theory of Nancy Martin's account in Chapter 5 clearly is of considerable intellectual depth. Yet, on the one hand it is perfectly comprehensible to any teacher who gives it thought, and on the other hand the practical implications are not so daunting. At root it is a matter of considering the purposes of writing, and encouraging the use of the expressive category as the core out of which other forms will grow. It is also a plea for real communication purposes, which means a range of real audiences. Such considerations should not be presented as too complex.

The most hopeful first step is to set up a working party to look into Recommendations 138 and 139 in the light of the particular position in the school, and to make recommendations.

Running a working party

Presuming, following the kind of action Nancy Martin has described, that a decision has been made to set up a working party, is there any advice that can be given?

It is preferable if the very setting up of such a group can have wide backing, perhaps by submitting a general statement of the need, the purpose of a proposed working party, and its proposed composition to the heads of departments meeting or equivalent central planning group of the school. The composition of such a group needs care. Leaving it completely to volunteers is dangerous. In some schools this will mean a group who represent only some subject areas, or only one age-group of staff. In other schools it will mean a dominance by a few who are very knowledgeable in linguistics, frighten the other members into confused silence, and eventually draw up a statement which is far from acceptable to the staff as a whole.

Reports and observation, and general experience in staff decision-making, suggest that the group should include some volunteers, but should have a careful representation from a range of subjects, from a range of staff ages and seniority, and with a definite pastoral representative. It is vital that the pastoral role is not overlooked. I have tended to use and prefer the phrase 'whole-school language policy' precisely because it is not only the curriculum that needs attention. The quality of communication in tutor group, counselling, assembly, and interview needs attention equally.

The working party which we set up at Woodberry Down (actually to revise and expand the first attempt at a formal language policy) was chaired by the head of the remedial department, and included a pastoral representative, and the heads of the geography, craft, and maths (the last of whom was also Advisor for Recently Appointed

Teachers), a member of the English department (as a representative), one of the science department, the head of reading centre, and the curriculum co-ordinator. This is an example of a range of interests; there was also a range of teaching experience. For some schools the membership would have been too senior, and fewer responsibility holders would have suited better. However, as a general rule it is vital to have sufficient heads of departments to ensure that the voice of experience and seniority is heard early, and does not merely block later. Similarly, it is most important that the group is not ostentatiously composed as a progressive or any other clique, for if it is any recommendations will have less chance of being acceptable across the curriculum.

There are various possible starting activities for the working party:

1. *Basic reading.* All members would probably want to read a little in the subject—perhaps some of the books in the reading list, certainly substantial sections of the Report.

2. *Specialized reading.* Various members might volunteer to read one each of the more specialized books, maybe one each on writing, reading, and talking, and to report back to the group the main conclusions.

3. *Lesson observation.* Visits might be made by the working party members, initially maybe only to other members of the same departments, to observe what language difficulties and opportunities arise.

4. *Consideration of existing test data.* Any figures, e.g. from primary schools or the remedial department, already available might be collated and reviewed to establish what they show.

5. *Establish the staff's views.* A questionnaire has been used by a number of such working parties.

6. *Establish present practice.* Again, in some schools questionnaires have been used. Nancy Martin quotes one, and I have included that from Woodberry Down as an Appendix. Obviously no such questionnaire is ideal, and all drawn up without the benefit of professional planning, piloting, etc., suffer from technical flaws. Nevertheless, such a device has two advantages: on the one hand it informs the Working Party, and on the other it involves most of the teachers from the start.

7. *Visit other schools.* It is helpful if knowledge of progress towards a policy in other schools is acquired early on.

The Working Party will soon want to start reviewing possible recommendations: if it spends too long on the research stage the patience of the less committed, whose views are especially important, will run out. Where to start is a tricky tactical decision. If a school staff works in the order of my contents, from Chapter 4 onwards, for instance, it risks getting lost in disagreements about fundamentals; if, on the other hand, it opts for modest technicalities it risks getting stuck in simple agreements on the trivia—and, what is worse, believing the job has been completed. I recommend taking some sample areas, maybe writing (where the recommendations are likely to be fairly straightforward) and one technical aspect (such as study skills, which is obviously likely to be of real help in many subjects), and drafting initial observations and tentative recommendations fairly soon.

Experience suggests that these first drafts of preliminary sections should be discussed widely fairly early. This not only encourages the working party to be practical and consider its consumers from an early stage, but it allows an interchange of views at a formative stage. Without some such exchange, a working party can toil for a couple of years, getting farther away from the needs and atmosphere of its parent school, and eventually offering something which is entirely unacceptable and useless. As the working party should be trying to respond to the felt needs of the staff (even if it comes to consider that there are other needs not yet articulated), opportunities for the staff to express their problems are valuable. For instance, the Woodberry Down working party circulated the results of the questionnaire for full staff discussion at Departmental Meetings scheduled for the purpose, and devised the recommendations partly out of the staff's reactions.

The drafting of the report is a tricky business. Many I have seen are too long and off-putting. One school's was virtually a book, half an inch thick, an amorphous collection of material, which was difficult to find one's way about. Others are off-putting because of their pretentious and complex tone. Whilst it would be wrong to sacrifice accuracy or depth for the sake of an easy read, it is pointless to be obtruse. The most common danger is to include all the working party's *homework*, that is many items that were useful preparatory stages for the working party, but which are not necessary for the reader.

There is likely to be a statement about language, the result of the working party's consideration of the fundamental issues of language and learning. This is vital, but must be fairly brief, and drafted modestly and simply. It can consist of quotations if preferred, e.g. from Chapter XX of the Report, or from such writers as Professor Britton.

In my view the report must include actual recommendations for classroom practice, resources, and school organization. These recommendations must be suitable for all subjects, or suitably limited by qualification. They should not be too numerous or too strident in tone. Ideally, I consider that they should be grouped in such a way that the main points can be readily extracted. Finally, it is vital that the framework of the report is clearly tentative, and that the scope for later extension or amendment is made clear.

The draft report should then be submitted widely to those concerned—perhaps to heads of departments or maybe to the full staff in small groups. The governors should also be consulted at this stage, partly because this indicates the centrality of the issue to the curriculum, and partly to enlist their support. It must be remembered that they have a responsibility for oversight of the curriculum, and their views on some of the issues will be valuable. In some schools the parents will also be consulted.

Leadership

A policy needs institutionalizing. It must not die; it must not be overlooked. Its implementing therefore must be someone's responsibility, and it must be clear that the person whose responsibility it is has the personal knowledge, experience, and responsibility in the school to carry it out.

This leads to a strong recommendation that every secondary school should have a senior teacher, experienced and eventually suitably qualified, to lead and co-ordinate the school's overall language strategy, a Language Co-ordinator. His or her responsibilities should include: advising, leading, and co-ordinating teaching approaches across the curriculum; he should become expert in the grading of printed material (by the applying of age-level tests, for instance) and its suitability; he will have oversight of the diagnostic tests; he will lead the staff in-service training. In short, he would be the linguistic leader of the school. As the Report says, such a Language Co-ordinator should not always, or even usually, be an *addition* to the existing hierarchy. Our recommendation is that a senior member of staff should have those school-wide responsibilities. A number of organizational models seem to be open to schools. Each has its advantages and disadvantages, and only the individual school can decide which would be most likely to suit its particular conditions.

Nostalgic readers might with some justification say: 'But that is precisely what an old-fashioned Headteacher of a small secondary-modern school did.' Often, but not always, and usually without the wider qualifications now seen to be necessary. I concede though that I found the nearest to what I have in mind in two very small secondary modern schools. One head, in a small Midlands industrial town, had the reading ages of all his pupils in the top right-hand drawer of his desk. The other, on the outskirts of a small Derbyshire mining town had just finished hearing a boy read when I arrived. In the larger school, or in the grammar school, there is no equivalent. I am aware of the complications of a plea for yet another member of the senior staff, or of adding yet another task to those undertaken by the present holders of such posts. But if a language policy, including a reading policy, across the curriculum is to work it must be the stated task of someone, clearly described in his job specification, and well understood by his colleagues.

I could see that holders of each of the following posts could take on this responsibility:

Holder of role	Advantages	Disadvantages
The headmaster	Has overall vision and necessary status	In any but the smallest school is likely to be too busy to give detailed co-ordination
The deputy with curriculum responsibility, Curriculum Co-ordinator, or Director of Studies	As above. Additionally this responsibility would fit his overall curriculum concern	As above in the larger schools
The head of English	Likely to have necessary knowledge and interest	Difficult to have a HOD who is 'equal but extra' to the other HODs
	Likely to further broaden his experience into related and relevant fields	Already has a very heavy burden
Head of remedial department	As above, although not many have knowledge of the teaching of the higher reading skills	As above
A separate Language Co-ordinator	Given proper quali- fications and status could be best placed to look at all aspects of the school	Relationship with the rest of the hierarchy could be difficult
Teacher responsible for in-service training	Has experience and facilities to organize and lead teacher development	Is not actually responsible for departments' syllabuses

The logical choice for many schools might be the second person in the list, for this would clearly link the work with leadership of the curriculum and the departmental syllabuses.

As many schools have started consideration of their policy by setting up a working party, some have embodied the leadership responsibility in that committee. My view is that it is probably valu- able in some cases for such an initial working party to continue in existence to overhaul, and update, the original school statement, and to oversee in-service training in this field, but committees are not normally very successful at leadership.

Various combinations of responsibility are also possible. A working party can continue to develop the thinking behind the policy, whilst a deputy head in charge of curriculums is responsible for implementing it, and a teacher in charge of in-service training for promulgating it. What is certain is that it is not possible to establish a policy and leave it to work. Leadership is required.

The English department

The Report dealt in great detail with the teaching of English, and the staff and resources necessary for that. I have deliberately avoided straying into the wider field of the work of an English department,[1] wishing to focus only on the department's possible contribution to the whole-school policy. It must, however, be stressed that the whole-school policy requires from the English department both that the team do contribute in the agreed ways, and that there is a vigorous exploratory teaching of personal language, of literature, and of the imagination. Because I have needed to stress those uses of language that have not hitherto been extensively taught should not give the impression that I overlook the importance of good contextual teaching of English.

Those outside English departments should also know that the committee found disquieting deficiencies in the provision for English. Staffing was often random; a third of those teaching English having 'no discernible qualification' (15.14), many devoting only a part of the week to English. Timetable resources were frequently poor, that is, English was used to fit in round those subjects which more strenuously claimed blocking or had special period demands (singles *or* doubles). Rooms were frequently inadequate, again showing that English was fitted in to the rooms timetable after the needs of other subjects. This involved a large amount of split-room teaching, with not only teachers dispersed around a number of rooms, but the work of one class's English similarly distributed: 'A substantial proportion of English teachers—over one third—are itinerant within the school, carrying their books about with them from one lesson to another.' (15.21)

Storage space in general was shown to be poor, especially that vital equipment for English teachers—storage for books and papers in the classroom itself. Headquarters or Departmental bases, so vital for materials, were rare, and compared with other subjects the teachers were less likely to work *en suite*. The Committee also found internal deficiencies, of which weak planning, poorly worked out or nonexistent syllabus or 'instruments of policy', and unbalanced range of

---

[1] Apart from the Report, this is a very well written-up subject. A short list of books is included in the last part of the reading list, and my own views have been developed in *Towards the New Fifth*, Longman, 1969, with Graham Owens in *The Practice of English Teaching*, Blackie, 1970, and in Denys Thompson's *Directions in the Teaching of English*, CUP, 1969.

activities were the worst. Clearly an English department must be well set up and well run if it is to provide the core of language development and if it is to meet the school's expectations in specific tuition.

It is clear that some part of the work of the department, or any faculty in which it is placed, must deal with skills of broader application than merely required by that department. In other words, there must be confidence in the staff as a whole that the specific teaching required as an element in the whole language policy is actually firmly in some department's hands. It is also recommended in the report that the English department should 'lead, inspire, and encourage' the policy across the school.

There are many problems with leadership coming from the English department, and many ways in which the conscientious wish of teachers of English to help is made difficult. On the one hand teachers of English may know 'too much' linguistics, and find it difficult not to over-complicate and make too many learned references. They may also be simply too confident in language and adept in debate, thus failing to allow other language perspectives to be described. They can also be extremely ill-informed about how other subjects work, rather presuming that the language activities in a subject such as the crafts must be less good than that in English. They can, as I have implied in Chapter 5, also be very inexperienced in teaching reading outside literature, especially the initial stages and advanced non-fiction.

Even should they have none of these difficulties, the magnitude of the triple task is great: to give good learning experiences in their own subject, to teach specifically the agreed core of language knowledge and skills, to assist with the overall language policy. This last requires tact, as well as knowledge, experience, and energy.

In-service training

Like all curriculum development, a new language policy has in-service training implications. A school will no doubt want to encourage some teachers to attend courses; indeed it might manage to have a suitable teacher seconded for a year's course on language, such as that at the London Institute, or on some aspect of reading. It might persuade the LEA to send one or two more on 'short courses'. However, although such additional expertise would be valuable, especially if fed into the school as a whole by membership of the working party, such piecemeal action is insufficient for a whole-school policy clearly requires whole-school attention. This is clearly a subject for a coherent in-school in-service plan staged over a number of terms, at various levels, and on various aspects. The staff might well be asked which aspects they feel that they would like most help on. One might feel the need for basic phonics (say a recently reorganized Grammar School) and another might ask for work on language theory.

Schools have found the following activities useful. They have been initiated by working parties, the Headmaster, a senior teacher with responsibility for language, or the staff association:

1. *Listening to a visiting speaker*, on use of language (e.g. writing), an aspect (e.g. class and language), or teaching techniques (e.g. small group discussion).
2. *Study and discussion of existing documents*, e.g. the Writing Across the Curriculum's Discussion Pamphlets.
3. *Joint meeting with feeder schools.* There have been many useful discussions planning a continuous policy on such an aspect as reading.
4. *Departmental meetings*, to consider the detailed working-out of recommendations in the particular subject area.
5. *Seminars.* On the whole seminars are most successful when there is 'evidence' to consider, so that discussion is based firmly on actual material. Possible seminar topics:
(a) How do we talk in the classroom?
    Tapes of teacher talk can be analysed and conclusions drawn.
(b) How do the pupils write?
    A selection of writing from various subjects, or all the writing of one pupil a week can be studied.
(c) Accent and dialect in the area.
    A meeting with parents can be valuable.
(d) Reading demands of subject texts.
    Extracts from a variety of commonly used books can be duplicated to establish firstly whether they are suitable and secondly what help could be given to pupil readers.
(e) Punctuation explanations.
    How do we explain a point to a pupil who does not understand?
(f) Graphic demands of subject texts.
    Examples of the graphic support from a variety of commonly used books could be photo-copied for study: maps, diagrams, flow-charts, graphs, picture-charts, etc. Is adequate teaching given at the right point in school to prepare for these?

The major aim of the programme of in-school in-service training would be to share across the curriculum the present expertise, and to help especially those teachers with useful experience or expertise to offer it to others.

When the initiative for developing a language policy has come from the English department, there are serious dangers likely to come from the feeling of other teachers that the teachers of English are frankly getting above themselves and in the process knocking their colleagues. Indeed this is sometimes so. Those teachers of English who feel superior have used the movement for a whole-school language policy as a way of pushing themselves. Even modest and collaborative teachers of

English have sometimes found it difficult not to find themselves in unhappy postures of preaching to the unconverted. If they are the sole possessors of linguistic knowledge that has suddenly become prestigious this is not surprising—but it is usually counter-productive.

Many other parts of the school have relevant skills which they have developed to a greater extent, because of a greater need and more opportunity for practice:

1. *Exposition.* Mathematics and technical drawing teachers have had to face difficult problems of clarity and precision.

2. *Initial stage of reading.* Frequently only the remedial teachers have this knowledge and experience. Very few teachers of English have. It needs to be shared with all.

3. *Pupil discussion.* Craft and science teachers have often developed techniques for establishing and encouraging functional pupil communication.

4. *Reading for meaning.* Sometimes historians have battled with the use of expository print sources for so long that they have faced the problems of how to help.

5. *Bibliographical skills.* Chartered librarians or teacher librarians obviously are fully confident here.

6. *Discussion.* Social Studies and Religious Education teachers have frequently been obliged to develop modes of work helpful here.

7. *Reading aloud.* Drama teachers have useful skills and tips to pass on.

8. *Graphic aids.* Geographers have possibly the widest experience of using a range of illustrations in study texts, and of relating these to the total reading experience.

These skills and experiences can be shared. Members of each of the specialisms I have mentioned can be asked to prepare brief expositions of their experience for the benefit of others. Sometimes there is actually high-powered linguistic knowledge hidden in a teacher of another department, who, such is the tradition in departmentally run schools of specialists, has kept quiet about it. In one school I was in, by far the most knowledgeable student of linguistics was a mathematics teacher. Processes such as I have described break up the apparent linguistic monopoly of the English department, and work towards a true collaboration across the school.

Finally, it is, of course, helpful to have a staff library, and to make sure that a good selection of books on language is available. When all is said, however, the implementation of the policy depends on the head teacher and his or her senior colleagues. Only their continuing interest will ensure that an initial policy is actually carried out, that practice is thus altered, and that the policy itself is reviewed and revised. There can be fewer compelling tests of the leadership of a head teacher than the Committee's recommendation of a language policy across the curriculum.

9. Conclusion: A Language for Life

The title of the Report is no mere publishing expediency. It embodies the two overwhelming complementary truths, that the language we are working with in our schools is the language of the pupils' lives, and that we are helping them develop a language that must serve for life. Indeed, it has jokingly but truly been said that if we fail we have imposed by neglect a life sentence.

The task the adolescent faces in developing his language during the secondary years is immense. John Dixon has given one measure of the immensity when he says of the learning of the scope and range of writing is perhaps equal to the original learning of the mother tongue by the young child—except, of course, that the adolescent does not have the motivation or constant encouragement that the child has. Even in growth of vocabulary the challenge is huge, and the ways of moving from the childhood language to the full adult range complex and largely unknown. In reading, the growth from the average eleven-year-old's competence to that of the adult able to engage with most forms of print is a task so daunting that few manage it. Yet the ability to read at the highest level is essential to the individual's personal growth in the modern world, his ability to benefit from most of the learning opportunities at school and beyond, and his ability to take his place in society. Indeed, behaviour and adjustment problems not only lead to difficulties in learning to read, but, and this is less widely realized, the reverse is also true: reading difficulties are a frequent cause of difficulties of behaviour. And yet too often we are forced in schools into creating the paradox that those who need most help with this formidable adolescent task are given the least help, and are allowed to avoid language challenges and are given less language tuition.

There is no real question of those of us who teach content subjects arguing that we cannot afford the time, nor that the English department ought to get on with the job. In the first place a school language policy *saves* time in that it makes learning more effective, and the use of print sources more rapid and more thorough. In the second place, when we come into teaching we must recognize that we have entered a profession that is dependent on language: the curriculum depends on communication. We must therefore be willing to consider the language which is a large part of our professional task.

More than that, anyone looking to a single department, English, to tackle the whole of the task should be invited to look at the question from the other point of view: what right have teachers of English to teach the most fundamental aspects of the learning of each subject? Surely historians and scientists must want to keep for themselves the tasks of teaching the language of their subjects and the language skills special to their subjects. Such teaching is likely to both be more effective, because not dependent on the transfer of skills, and more realistic, because grounded in a real communication need. Certainly the new coherence of the comprehensive school, which I spoke about at the beginning of the book, depends on overall curriculum planning, which means relating the work of departments to each other, and this will be at its most important in language.

It is worth noting that mixed-ability grouping, favoured now by a large proportion of the schools for their younger classes, creates a special need for a language policy. The central thesis of a mixed-ability teaching organization is that pupils should work in the same setting and at broadly similar tasks. Further, it is presumed that where possible they should collaborate, learning more deeply through that collaboration. This thesis involves a careful consideration of the language aspects of the environment, for the ambitions of the pupil-grouping system can be made impossible by language demands that exclude some, or which cause further divergence. This is notably true of the teaching of reading. If no effort is made by specific individual or small-group tuition to improve basic skills, the least able will not be able even to begin to join in. They will be forced either to choose non-reading tasks (thus getting even less practice), or to choose considerably easier tasks. Similarly, but less often realized, if no tuition in the higher skills of reading are given, background, verbal reasoning levels, and earlier educational experience will push the pupils apart and defeat the grouping that is designed to bring them together. On the other hand, specific tuition in the reading requirements of the particular subject, and specific preparation for the reading tasks involved, help the full range of pupils to benefit from the reading tasks, thus allowing the mixed-ability group to be brought together rather than driven apart. Otherwise the naturally more literate pupils pull even farther ahead, leaving the less able reader to flounder more than they would in homogeneously organized groups.

We are at the early stages of applying the fundamental findings of linguists and psychologists of learning to actual schooling. Those of us who work regularly with a range of ordinary pupils in school urgently want more knowledge, and a more detailed working-out of the procedural implications of that knowledge. There is a real need for further study of the relationship between language and learning in the adolescent stages, and of actual classroom techniques. In particular, I should like to see three kinds of study:

1. Subject-based. Intensive study of the language requirements of each of the subject fields.

2. Micro-aspects. Intensive study of an aspect of language, such as the growth of vocabulary, to fill out the broader studies that we have had so far.

3. Institutional. Case-studies and action research on policies and their effects in particular schools.

In the meantime we must make workable what we do know. In this book I have tried to gather together sufficient of the components of a policy for a school to consider its own situation and to build its own policy to suit. It will be obvious to readers that such a policy cannot be drawn up and applied mechanically, and that the detailed technical aspects will be useless if there is not a constant attempt to create language contexts in which communication is genuinely stimulated and genuinely required.

There remains the importance of something rather more intangible, the need for the entire staff to value language in all its manifestations and in all its modes of communication. If a group of teachers care for communication and care for language, this will help the pupils through the intangible but powerful hidden curriculum as well as through the ostensible teaching curriculum that this book has covered. This concern has to be a positive one for enrichment and extension, not a negative one for fault-finding. It starts with a real concern for what pupils have to say, and goes on to suggestions for reading, listening, and writing. The older tradition of schools depended on a less specialized teaching staff. It seems that with the somewhat more professionally segregated specialist staff of today's subject departments, there may well be a tendency to equate language with English, and this with one department. Within this, there may be less sharing of each teacher's interest in books. In an ideal school, every teacher would be sharing his particular interests, formally and informally, in reading.

Edward Blishen puts the most important single simple point persuasively: 'I do believe that, within limits, with qualifications, a teacher with a passion for a book can touch those he teaches with that passion. . . . The main hope most children have of being turned into readers is that they will somewhere along the way meet an adult, and in the nature of things it probably has to be a teacher, who is himself or herself a true reader, and lays on a demonstration, naturally or with some aid from the teaching arts, or the theatrical arts if needs be, of what it is to be a reader, what you get from it, the excitement of it.'[1]

A language policy is a learning policy. To consider carefully the ways in which our pupils learn the subjects and skills we offer them, we find that language is at the heart of the learning process in each

[1] Reported in the *Times Educational Supplement*, 20 August, 1976.

of the subjects, even, or especially, in the apparently non-linguistic subjects like mathematics. To plan ways in which we can effectively improve our pupils' learning is inevitably to consider how we use language, the language environment of our school, the language expectations we have of our pupils, and the tuition and encouragement we give in language. If we do all this, that is if we establish a whole-school language policy, we shall not only have helped the subject learning of our pupils, we shall also have helped them with their own language for life.

Appendix One

Word Building: A Sample List

A school might wish to produce a list of roots, prefixes, and suffixes which would be available to all teachers. Such a list would be contributed to by all subjects, and would include the vocabulary for each. It would grow out of the school's subject needs. The following list is offered only as a sample to show what such a list might look like. Prefixes are easier to recognize than roots, but unfortunately some have more than one meaning. Also the final letter often changes into that of the first letter of the root. It is easier to consider them *not* in alphabetical order, but according to their sense.

Prefixes of direction

In, On, Upon, Into, Within
in- (or il- and im-)
en --
intro --
To, Toward
ad --
pros --
Away, From, Out, Of
ab --
e --
ex --
ec --, ex --
Below, Down
de --
cata --
infra --
sub --
hypo --
Above, Over, Beyond, Outside Of
extra --
ultra --
trans --
sur --, or super --
hyper --
Across, Through
per --
dia --
trans --
inter --

After, Behind, Back, Backward
re –
retro –
Before, In Front Of, Forward
ante –
pre –
pro –
Around, About
circum –
peri –
be –

Time and age

neo –	new
ante –	before
pre –	before
pro –	before
post –	after
re –	again

Size and number

bi –	twice, doubly, having two
maga –	great
micro –	small
multi –	many
poly –	many
omni –	all
semi –	half, or less than

Negative

a –	not
in –	not
non –	not
un –	not
anti –	against
contra –	against
with –	against
dis –	apart, away
ob –	against
mis –	badly, wrongly
hetero –	unlike
pseudo –	false

Positive

com –, con –, co –	together with
syn –, sym –	together
bene –	well
philo –, phil –	love
homo –	like

ortho –	straight, right, true
rect(s) –	straight, right

Suffixes

Suffixes do not control meaning as much as prefixes. They more usually alter the function of a word, e.g. change the way in which it can be used. Here are some useful ones.

showing capability or fitness	– able, – ible
a disease	– osis
a man who	– ary
	– ard
	– er
	– ent
	– ian
	– ist
	– or
dimunitives—that make smaller	– cle, – cule, – le
	– ette
	– kin
	– less
feminine	– ette
	– euse

To make into verbs:

– ate	ameliorate
– esce	effervesce
– en	lighten
– fy	rectify

To make into nouns:

– age	montage
– ance, – ence	appurtenance
– ion	reduction
	(thousands of others, all indicate the action)
– ism	communism
	('ism' means a doctrine or system)
– ity, – iety	sobriety
– ment	escarpment
– mony	parsimony
– ry	mimicry
– tude	turpitude
– ture	aperture

To make into adjectives:

– al (– cal)	inimical
– an, – ane	germane
– ar, – ary	sanguinary
– fic	soporific
– ic	didactic
– id	insipid

– ine	bovine
– ish	childish
– ive	furtive

There are also a number of combining forms, mainly from the Greek, which are frequently found at the end of a word, although not strictly suffixes:

– arch, – archy	ruler	anarchy, oligarch, monarchy
– cide	kill, killer	homicide, suicide, regicide, genocide
– cracy	form of government	plutocracy, meritocracy
– iatric, iatry	healing, treatment	psychiatry, geriatrics
– nomy	law, rule, system of knowledge	astronomy
– ology	study, science	entymology, biology, geology
– phile	loving	bibliophile, Anglophile
– phobe, phobia	fear	agrophobia, xenophobia
– tomy	slice, cut, cut away	tonsillectomy

Roots

The following roots are examples of those that might be taught:

Root	Meaning	Words using the root
fact	make	factor
		factory
		malefactor
		benefactor
flu, flux	flow	effluent
		confluence
		fluctuate
		flume
		flux
		influence
		influx
		superfluity
lat, from latus	to carry or bear	collate
		dilate
		dilatory
		elation
		relate
locut, loqu	speak	circumlocutory
		colloquial
		grandiloquence
		interlocution
		loquacious
		obloquy
		ventriloquist
mit, miss	send	remit
		remiss
		mission

reg –	rule	regulate
		regale
		regency
rog, rogat	ask, ask for	abrogate
		arrogance
		arrogate
		derogatory
		prerogative
		prorogue
		supererogatory
		surrogate
ven	come	venture
		adventure
vert	turn	vertebra
		vertex
		vertical
vol	wish, want	volition
		volunteer

Appendix Two

Example of a Working Party's Internal Questionnaire

Language in Education can be divided into four broad areas: Writing, Reading, Talking and Listening. The paper below is divided into these four sections, on which we request your views. At the end is a series of general questions which may require extensive discussion. There is no space on the sheets for answering general questions, so please number your answers carefully, e.g. A.2.i., etc.

A. WRITING

Below is a list of many of the forms of writing expected of our pupils. Please indicate which ones apply to your subject.

		APPLIES		
	A lot	*A certain amount*	*Not very much*	*Not at all*
I. i. Copying from the board or a book.				
ii. Making notes from a book or a film.				
iii. Making notes from teachers' talk.				
iv. Taking dictation.				
v. Transactional writing, i.e. typical language of science, intellectual enquiry, planning, reporting, instructing, informing, advising, persuading, etc.				
vi. Expressive writing, i.e. in which the personal views or opinions of the writer play an intrinsic part.				
vii. Poetic writing, i.e. writing to be shared rather than corrected. Creative writing.				
viii. Answering questions.				
ix. Answering questions in an expanded form.				
x. Filling in blanks in worksheets.				
xi. Collating information from different sources.				
xii. Lists and other condensed or 'incomplete' forms of writing.				
xiii. Translating or interpreting from one form to another, e.g. Languages or Maths.				

APPLIES

	A lot	A certain amount	Not very much	Not at all

   xiv. Map work, labelling, copying or creating diagrams, etc.

   xv. Other—please specify.

2.   i. Would you say that the types of writing demanded by your subject are mainly ones that can be marked as either right or wrong?

    ii. To what extent do you think the intended audience for a piece of written work can affect its quality?

    iii. Who, in the main, is the audience?

    iv. How important do you think it is to offer pupils a choice of kinds of writing or kinds of topics?

3. Assessing writing:

    i. Do you correct most of the errors of spelling, punctuation and grammar in a pupil's writing?

    ii. If so, do you write in a correct version or simply indicate that an error has been made?

    iii. How do you find out if your method is effective?

    iv. If you mark the pupils' writing, is your assessment numerical or a written or spoken comment?

    v. If you make a written comment, are you mainly concerned with content or presentation, and does the pupil know which your specific concern will be?

    vi. Are there times when your concern with content and presentation clash? How do you deal with this?

    vii. What do you see as the role of the English and remedial departments in relation to the written work of your pupils?

B. READING

Below is a list of the various types of reading activities that may go on in your lessons, please indicate which ones apply.

APPLIES

	A lot	A certain amount	Not very much	Not at all

1.   i. Reading aloud round the class.

    ii. Reading aloud to a teacher.

    iii. Reading aloud to a friend.

    iv. Following while the teacher reads.

    v. Individual self-chosen reading.

    vi. Individual reading for information.

    vii. Reading instructions from board.

    viii. Reading instructions from worksheet.

    ix. Reading instructions from textbook.

    x. Personal research in libraries—from a variety of sources, etc.

    xi. Reading one another's work.

    xii. Re-reading their own work.

    xiii. Other—please specify.

2. i. What methods do you have for finding out whether or not the pupils have understood what they have read?
    ii. How appropriate in general do you feel is the output from educational publishing for the needs of our children?
    iii. If you feel that the language of educational textbooks is not appropriate, how does your department deal with the problem?
    iv. Are you able to find a way by which non or reluctant readers are able to take part in the lesson?
    v. How much individual reading in and out of class do you expect from the least and most able? How hard are these expectations to realize?
    vi. How much help are you able to provide for the less able readers and of what kind?
    vii. How far do you find that backwardness in reading hampers your pupils' mastery of your subject?

C. PUPILS' TALK
Below is a list of the different kinds of pupils' talk that take place in the class room. Please indicate their frequency in your lessons.

	APPLIES			
	A lot	*A certain amount*	*Not very much*	*Not at all*
1. i. Silence				
ii. Answering teacher directed questions.				
iii. Asking questions.				
iv. Asking each other for help.				
v. Arguing/shouting.				
vi. Informal gossip/chatter.				
vii. Discussing work in hand informally.				
viii. Discussing work in hand in groups.				
ix. Discussing work in hand formally as a class.				
x. Expressing opinions and feelings.				
xi. Thinking aloud.				
xii. Memorizing.				
xiii. Taping.				
xiv. Asking teacher for help.				

2. i. To what extent do you find it useful for children to talk about what they are doing with (a) each other (b) you?
    ii. Which kinds of talk do you find most difficult to set up?
    iii. What constraints do you feel that you need to put on talk and why?
    iv. What improvements could your department suggest in order to make talk a more central or useful tool in learning?
    v. What are the most effective ways your department has found to utilize the value of talk?

D. TEACHERS' TALK
Below is a list of some of the different kinds of talk which we, as teachers, use. Please indicate the frequency in your lessons.

	A lot	*A certain amount*	*Not very much*	*Not at all*
APPLIES				

1.   i. Giving instructions to the whole class.
    ii. Demonstrating with explanation.
    iii. Talking to individuals.
    iv. Reading.
    v. Dictating.
    vi. Giving information.
    vii. Explaining.
    viii. Discussing.
    ix. Questioning.
    x. Encouraging.
    xi. Nagging / negotiating / reasoning / bribing.
    xii. Shouting/haranguing.
    xiii. Lecturing.
    xiv. Telling stories/anecdotes.
    xv. Praising.
    xvi. Criticizing.
    xvii. Joking.
    xviii. Chatting.

2.   i. What is the department's feeling about a class's attention span in listening to a teacher?
    ii. What ways have your department found to cope with the mixed ability reception of teachers' talk?
    iii. What are the problems in modifying your language to communicate the same ideas to different pupils?

E. LISTENING

    i. How do you help children to value one another's verbal contributions—and how do you help them to listen to you?
    ii. How important do you think is the role of teacher as listener?

F. GENERAL

    i. Does your subject have a specialized vocabulary which children need to master in order to do the work?
    ii. What problems arise in trying to teach this vocabulary?
    iii. How great an emphasis do you put on technical accuracy—written and spoken in your subject?
    iv. Do you think that your subject (or the speed at which a syllabus demands you should move through it) poses real comprehension problems for the pupils that you teach?
    v. What are the methods developed by your department to find out whether or not the children have understood the lesson?
    vi. Do you have any specific provision for coping with the range of ability in the classroom (i.e. for the least and most able)?
    vii. What are the problems in general, that the children bring to the lesson in terms of written and spoken language ability?

G.

    i. Bearing in mind the points raised in the language document,

'Reading to Learn . . .', what does your department feel can be done to support its work in relation to language and learning?

H.

   i. Corridors, Tannoy, Playgrounds, Assemblies, Tutorial and Pastoral care.

Any comments . . . in relation to language only please!

With thanks:
Language Policy Working Party
February 1976.

Appendix Three

Part of the Abbey Wood School
Report and Policy

Foreword

By the Headmaster, Mr C. E. Stuart-Jervis, Abbey Wood School, London.

This document came about because of the Chief Inspector's request for information from schools as to what they were doing, or about to do, as a result of the Bullock Report and its explicit concern for language development in schools.

Dare I say that Abbey Wood, largely due to the efforts of Brian Coxall and his colleagues in the English department, had been pioneering this work for some years, in a manner that predated Bullock. There had already been meetings in the school to discuss the problem. Heads of department and heads of faculty and house had all been involved to a greater or lesser extent and the topic was fresh in their minds.

Still, it is fair to say that we had not made as much progress as we would have liked. Bullock, perhaps, provided the spur needed to progress beyond discussion towards a more positive approach.

What emerges from the massive amount of work that Brian Coxall has undertaken is the need for all teachers to take responsibility for language development with their pupils, for all teachers to ask themselves somewhat searching questions regarding the type of language work involved in their discipline and for all teachers to return, to a greater or lesser extent, to the idea that we are all teachers of English.

This document, however, is also representative of the work of all at Abbey Wood, as it is also representative of their skill and thoughtfulness.

But though the report is an answer to the Chief Inspector's request, it is also a policy document and its content will occupy much time on agendas for many months to come, since one of its main aims is to develop a logical and coherent language policy within the school, one that will be fully understood and to which, eventually, all sections of the school community will be committed.

Introduction

1. These proposals are made in the light of the following:
(a) the work of the 'Language in the Curriculum' working party (Part 1) and Discussion Paper No 7 (Appendix 2);
(b) the Staff Meeting of 23 April, 1975, when it was agreed that 'all teachers

shared a responsibility for language teaching'. (Minutes Staff Meeting);
(c) Senior Staff Conference of 25 February, 1976 (Minutes quoted in Part 1);
(d) reports from Faculties on which Parts 2, 3 and 4 are based;
(e) our own conclusions.

Though writing, reading and talking are considered separately it should be remembered they are not discrete, but as the Bullock Report states: 'Language competence grows incrementally through the interaction of writing, talking, reading and experience, the body of the resulting work an organic whole'. (1.10)

2. There are several reasons for a language policy at Abbey Wood. When our children are in Junior Schools all the work is largely under the control of one teacher. At Abbey Wood they have a 'specialist' curriculum and they are under the direction of a number of teachers, all of whom make demands of language. One of the problems arising from a specialist curriculum is that teachers may make language demands which the pupils in any year are not equipped to meet, as subject teachers may assume their pupils are being prepared for these demands elsewhere. Development within each subject in the curriculum may well depend on coping with language in all its forms. The Language in the Curriculum working party came to the tentative conclusion that:

> For our young people, a week's experience in school may not be viewed as a time-table of various subjects, but as a mass of writing, talking and writing . . . development across the curriculum may depend on their skill in coping with a wide variety of language and finding language to make sense of new experience. (Appendix 2)

Whether this is true or not, we still have to ensure that our pupils can meet the language demands made on them. We should give particular attention to the part language plays in learning, for as the Bullock report says: 'The pupils' engagement with the subject may rely upon a linguistic process that his teaching procedures actually discourage' (12.1). It is hoped that such a language policy as the one proposed here will ensure that our pupils may be able to meet the language demands we make on them and, by directing our attention to the place of language in our teaching, ensure that pupils develop well in our particular subjects.

Proposed policy

Writing

1. Before presenting any written task to pupils an attempt must be made to establish the kind of writing expected. A check list is given below of the different kinds of writing which the Language in the Curriculum working party found were used in the school. There is considerable overlapping of these kinds within pupils' actual writing. But, once it has been established which kind is broadly required, it must be each teacher's responsibility to ensure that pupils are equipped to perform their task.

Check List

2. Expressive Function:
(a) Description of Personal Experiences
(b) Imaginative

The term 'expressive' means a personal language used in writing or speech. In writing expressive language closely approximates to speech at certain stages. It is permeated with the self and the pupil's relationships, and is the personal voice of the writer. Expressive language is a matrix from which other language functions develop. The implicit function of words assumes great importance as expressive language depends largely on a shared context.

Transactional Function

(a) Communication of Facts and Information
(b) Summary
(c) Objective Description
(d) Argument, Opinion and Persuasion
(e) Translation from other means of communication, e.g. statistics, maps, diagrams
(f) Note Making.

The term 'transactional' means language of direct instrumental intent, from one person to another. The referential function of words assumes importance in transactional language, i.e. that property of words which enables them to summon up ideas of an object or class of objects or relationship between objects.

Poetic Function

A highly ordered and patterned language used at an advanced stage in play, novel, short story, poetry and some other kinds of writing.

3. Pupils should be allowed to write about new material in their own words before attempting the adult, mature, perhaps objective language required by the subject. The personal language of the expressive function is a foundation upon which the more difficult transactional forms may be built.

4. No pupil should be required to copy out long extracts from books. Many projects are unsatisfactory because they are directly copied and the pupil has not understood the content nor, indeed, the language copied.

5. In correcting pupils' written work we need to be aware of the following:

(a) A writer needs to have an audience—no one will sustain written work if he feels he is writing for himself alone. The range of audience should be extended to include, besides the teacher, the class and others by displays, and publication in magazine form.

(b) As much reading of the pupils' work with the pupil present as can be managed is helpful.

(c) Pupils generally want to write correctly and marking should include attention about the mechanics of spelling, punctuation and grammar.

(d) Too much attention to some pupils' mechanical correctness may inhibit fluency, so it might be necessary to select one aspect for correction at a time.

(e) Comment by the teacher on the pupils' work should be specific and clear; repeating such phrases as 'excellent', 'could do better', etc., will not give the direction that a particular judgement will.

6. New vocabulary required by a subject should be carefully introduced and discussed with pupils.

Reading

7. In presenting reading material, teachers should ensure that the material is appropriate to the reading abilities of their pupils and, where anyone is unable to meet the demands, it is the responsibility of subject specialists to help improve reading skills. Where the problem is serious the Head of the Basic Studies Department should be consulted.

8. Staff should be aware of the three different levels of reading skills— Primary, Intermediate and Comprehension:

(a) Primary skill: where a child perceives individual letters, groups of letters and individual whole words.

(b) Intermediate skills: where a pupil needs the ability to handle sequences of letters, words and larger units of meaning.

(c) Comprehension skills: where a child who reads fluently is extracting meaning from the reading material.

In mixed-ability groups, particularly, there will be a wide range of these skills exhibited. The teacher needs to be aware of the level of skill in each pupil and be responsible for the development of reading required by the subject. (The background to reading process is described in the Bullock Report, Chapter 6.)

9. The reports from departments and faculties generally indicate that in the curriculum the main purpose of reading is to gather information—using the word 'information' in its broadest sense. Of all the ways of gathering information—by observation, personal experience, listening and discussion—reading is by far the most difficult. Where pupils need to read to gather information, they must be helped in how they are to record the information they need, to avoid copying whole sections from books.

10. Where subjects need pupils to consult various books in the Library or within the classroom, pupils need to be taught how to find the relevant material.

11. The Bullock Report sums up the responsibility of subject teachers when it states that they need

> to be aware of the processes involved, able to provide the variety of reading material that is appropriate, and willing to see it as their responsibility to help their pupils meet the reading demands of the subject. (12.7)

Oral

12. The reports which formed the basis of Parts 2 and 3 indicate that the importance of spoken language in the educational process is clearly recognized. Learning requires making knowledge personal and understanding requires the pupil to come to grips with new material, firstly in his own language.

13. Staff should give time in a variety of ways for pupils to talk about the material that is presented to them, whether in small groups in a structured situation or the more formal class discussion. In preparing lessons, situations need to be devised to give pupils the opportunity to explore the particular topic in their own expressive language.

14. The reports from faculties indicate that there is considerable difficulty experienced in enabling children to use the specific technical language of a certain subject. This needs to be introduced carefully. The bridge between personal language and impersonal language should be given serious consideration in lesson planning.

15. Teachers need to structure situations, both formal and informal, in small groups and whole classes where demands are made on pupils spoken language, and to encourage pupils to speak appropriately in these different situations.

16. The kinds of questions staff ask are also important. There is a difference between the question for which there is but one answer and the open question which invites exploration and speculation.

17. More than any other aspect of language, oral work needs further investigation and faculties should document useful practice for the benefit of other subject areas.

Worksheets

18. Worksheets are now in fairly common use, particularly since the school changed to mixed-ability teaching. Subject areas are aware of their advantages and disadvantages and time is ripe for a thorough evaluation of their use.

19. Presentation of worksheets needs careful consideration. They may be constructed in some haste and there are indications that many are poorly presented. With the developing Media Resource Centre and secretarial assistance, there should be no excuse for inferior presentation. Worksheets need to be prepared well in advance of the lesson.

20. Modern methods of examining in several subject areas reduce the demands made on pupils language skills. Many worksheets, perhaps reflecting this, may also carry the danger that these demands are being avoided in the classroom. Consequently the content of worksheets needs to be evaluated, so that we make similar demands of reading and writing on pupils as the former teaching methods did.

21. The levels of ability to which the worksheets are devised needs to be seriously considered. One worksheet for a whole mixed-ability class may mean that only the average pupil is being provided for.

22. The place of worksheets in subject teaching; the need for them to make language demands on pupils; and their presentation needs to be considered further.

Conclusion

This Language in the Curriculum policy has these purposes:

1. to make teaching staff more aware of the language demands made on pupils;

2. to propose that staff should teach the particular language skills required by their subject;

3. to suggest the place of language in the curriculum;

4. to improve the quality of teaching;

5. to implement the recommendation in the Bullock Report that: 'Each school should have an organized policy for language across the curriculum, establishing every teacher's involvement in language and reading development throughout the years of schooling.' (Principal Recommendation 4)

Appendix Four

Part of the Woodberry Down School Working Party Report

Headmaster's preface

There is no doubt that a great deal of learning is by means of language, both because access to ideas, knowledge, and explanations is largely through words, but also because we come to understand ideas by putting them into our own words, by thinking them through. Equally, there is no doubt that a great deal of an individual's growth in language is by means of his or her learning in the subject areas.

Thus the heart of a whole-school language policy is a simple common-sense pulling together: success in the subjects is likely to be greater if we all consider the use of language and our teaching of it; conversely, the pupils' own language, which they so desperately need for their personal, intellectual, and occupational lives, is likely to develop better if we all consider the part language plays in our particular subject area.

At Woodberry Down the number of issues discussed across the whole school and the tradition of inter-departmental meetings (e.g. the house staff Meetings and the lower school pastoral conferences) have put us in a good position to explore how we can, whilst very much retaining our individual subject variety, work together in a concerted approach to and through language. Nevertheless it is a daunting challenge, and I and all the members of the Woodberry Down staff are grateful for the thorough care which our working party have put into considering the question, sounding out our views, researching, and drafting their recommendations. I am personally grateful to each of the members of the working party, and especially to its Chairman, Chris King.

Throughout their deliberations, I know that the working party were anxious to consider each one of us teaching in our classrooms; to respect our personal approaches; to help our subject specialisms; and to reflect our individual wishes. They therefore set up a dialogue with us all, as the Chairman outlines in his introduction. Their final recommendations were then open to scrutiny both individually and at a full heads of departments' meeting. The full staff voted unanimously that the recommendations as amended by this heads of departments' meeting would be accepted. Here, in a revised and polished form, then, are the results of this careful exercise in collaboration and consultation.

Like me, each member of staff probably feels that he or she would have altered a point here, or added an idea there. Opportunities for continuing collaboration will continue, for in accepting the recommendations of the working party, I am also accepting the recommendation that there should be further staff study and in-service work, and that the working party should continue to stand. I have therefore asked Mr King to chair a continuing working party, whose terms of reference are: to assist the Senior Teacher: In-Service (Mr Phipson) with the dissemination of the working party's recommendations; to monitor the implementation of the recommendations; to further consider the language policy, with special regard to language in the mixed-ability group and the use and teaching of reading across the curriculum.

Individual teachers, departmental teams, or house staff groups are invited to contact Mr King with ideas or suggestions.

In the meantime I ask every member of staff to study the recommendations carefully and to implement them vigorously. This document is as near as one can get to a policy for all the teachers actually by all the teachers. Let us make it work, and contribute to its further growth.

Michael Marland
August, 1976

Membership of the Working Party, June 1975–July 1976

Simon Clements, Second Deputy and Curriculum Co-ordinator
Margaret Despicht, Careers Advisor
Ann Dubs, in charge of Reading Centre
Judith Hemming, representing the English Department
Chris King (Chairman), Head of Remedial Department
Joan Nicholson, Science Department
Tom Peryer, Deputy Head of Keller House, representing the House Staff
Christine Peters, Head of Geography
George Phipson, Head of Maths and Advisor to Recently Appointed Teachers
    (from 1 September, 1976, Senior Teacher: In-Service)
Keri Priddle, Head of Technical Studies

Terms of reference

The working party was set up by the Headmaster in June 1975, with the following terms of reference:

'To consider the present whole-school language policy, "Learning to Read and Reading to Learn", and to recommend any additions or modifications, including any strategies for in-service-training.'

Chairman's Introduction

The following report is the product of three terms' work towards a whole-school language policy.

Following initial staff discussion in 1972–73, Mr Marland wrote 'Learning to Read and Reading to Learn' in January 1974, which was open for discussion and was finally agreed in June 1974. Bullock reported in February 1975 and it was apparent that the whole-school language policy document was not being operated by the whole school.

A working party was set up late in the summer term of 1975 to consider the whole-school language policy as embodied in 'Learning to Read and Reading to Learn', and to make recommendations. The original intention was to draw on as wide a range of curricular experience as possible by involving subject teachers from all major departments. Pressure caused by other commitments made this impossible, and the majority of working party discussions have involved only a small 'nucleus' of members.

The first term was spent largely in defining our own terms and attitude to the matter and the second was spent in producing, distributing and collecting responses to a wide-ranging questionnaire, which was discussed by all departments. From our discussions and the collation of departmental opinion a draft report was circulated for further departmental discussion in May. Observations from these discussions have been included where appropriate, including those from the heads of departments' meeting in July, and the whole process forms the following report.

This report is not a consensus view; neither does it contain only majority views and common ground. It is democratic in the sense that it embodies views held by many staff, which have arisen in response to local necessity and practice. For this reason the working party thought it important to retain the main body of the survey because it is rooted in the reality of our experience and will be valuable for all future discussions. It is therefore included as Part Two.

It is the working party's view that this paper is not definitive. We consider that it is not possible to be absolute in this situation, and we recommend strongly that the paper should be further discussed by departments in relation to their work. The process should be on-going and to this end we further recommend that the working party should be made into a standing committee with the aim of stimulating staff thought in this area as an aspect of internal in-service training. A standing committee of this kind would produce occasional papers, organize seminars, and invite outside speakers.

The first part of the report lists our recommendations in four sections: Writing, Reading, Talking and Listening, and General. The second part is the report on our survey. We also include a short list of books for further reading.

Chris King, Chairman
August, 1976

The Recommendations

SECTION A: WRITING

A.0 GENERAL

A.0.0 Writing is very important in our education of our pupils, both as a mode of learning and as a useful skill for adult life.

A.0.1 All teachers of subjects in which writing can be encouraged and all tutors should devote time to the development of the pupils' writing, even if at first sight this seems to be at the expense of the subject. However, the total educational diet of the pupil must be watched, and teachers should also ensure that there are times when there is no pressure on the pupils for writing.

A.0.2 The less able should not be encouraged to avoid writing, and, equally important, the more able must have the stimulus to write at length whenever possible.

A.1   TYPES OF ASSIGNMENT

A.1.0   Writing skills grow by a gradual differentiation into the specific requirements of special forms. Pupils are less likely to master these forms both if they are forced into them prematurely or if they later have inadequate practice in them.

Writing can be classified into three function categories: expressive, transactional, and poetic. (These are defined at the end of this Section.) Of these 'the expressive is a kind of matrix from which differentiated forms of mature writing are developed'. (Britton, *et al.*, 'The Development of Writing Abilities', Schools Council, 1975, p. 83.)

A.1.1   In the first two years, the expressive function should be the most frequent one, and pupils should not be forced to develop transactional writing out of it too rapidly. However, they should be encouraged to move towards other modes whenever possible.

A.1.2   Teachers should provide considerable variety and choice in the writing assignments that they set.

A.1.3   As pupils get older, especially in year 3, all subjects in which writing is used should provide opportunity for expressive, transactional, and even poetic writing.

A.1.4   Departments should discuss how writing in each of these categories can be encouraged in their subjects.

A.1.5   Questions for writing should be formulated as far as possible so that they invite the pupil to give an expanded answer. There should therefore be very little reliance on worksheets which require one- or two-word answers or which use multiple-choice questions unless there is a special reason.

A.2   AUDIENCE AND RESPONSE

A.2.0   Writing develops better if the writer is aware of a real audience, whose response he values and can know.

A.2.1   Teachers should talk about the pupils' written work with them, or give a written comment. Despite the time involved, teachers should not rely on only a grade or figure to provide the pupil with an assessment.

A.2.2   Opportunities for creating pupil audiences should be developed as much as possible, and all departments should explore ways of doing this, such as displaying written work in subject rooms, having regular displays in tutor rooms, reading work to the class, duplicating pupils' work, producing small magazines.

A.2.3   House heads and heads of schools should also be utilized as readers by sending pupils with good work to see them as a supplement to the entering of commendations in the diary.

A.2.4   Parents should be encouraged to look regularly at their child's books, and this idea should be put across at parents' meetings, etc. Opportunities should be devised for parents to see and discuss their children's work at school.

A.3   TECHNICAL HELP

A.3.0   Responsibility for technical accuracy should be carried by all teachers who expect pupils to write. Although one of the main aims of the English department is to help children achieve technical accuracy, and therefore tuition is given in English time, this must be reinforced by the attention and support of all other teachers.

A.3.1 The most important aspect of technical accuracy is the division of sequences of words into 'sense groups', especially sentences, which should be marked off by full-stop followed by upper-case letter. All teachers correcting written work should make this a top priority amongst technical aspects.

A.3.2 When a pupil makes a great number of technical errors only one or two points should be corrected in each piece, and pupils should be made aware which particular errors are being corrected.

A.3.3 When errors are picked out, the corrected version should be written out by the teacher.

A.3.4 In commenting on technical aspects of writing, teachers should use the technical language: word, space, letter, syllable, stress, vowel (short and long), consonant, upper case, lower case, paragraph, indentation.

A.4 HANDWRITING AND NEATNESS

A.4.0 Neatness, careful presentation, and good handwriting should be encouraged by all teachers.

A.4.1 First-year pupils should have some provision for practice in hand-writing and other aspects of work in the curriculum.

A.4.2 All departments which include writing in their work should spend some time encouraging a legible style of handwriting, and should give priority to the correct use of upper- and lower-case letters (especially the letters 's', 'w', 'b', 'd', and 'r').

A.4.3 Parents should be encouraged to help their children with hand-writing and spelling, although this should not in any way diminish the school's responsibility. The Head of Lower School should find appropriate times to raise this matter.

DEFINITIONS—(A.1.0 ABOVE)

Expressive This kind of writing may be very close to talking—but on to paper—with the same sense of being a conversation, free to offer personal thoughts and feelings whatever the topic happens to be. The 'writer' is very much part of what he is saying/writing and feels confident that his opinions will be of interest to the reader. His sense of audience is likely to be himself, a friend or 'trusted adult'.

The Writing across the curriculum project sees expressive writing as the core, or central function out of which the following two functions emerge:

Transactional The job being done here is one of communication, the task of the writing is either to pass on information of one kind or another, or to persuade or command. There is always an audience other than the writer involved—and the more clearly the writer can envisage his audience the more effective the writing is likely to be. The writer's personal views are only a feature of transactional writing in so far as they are strictly relevant to his main topic.

Poetic   The job here is to handle language creatively—to *make* something with it. It presents an experience, rather than passing on information, and shapes it into some kind of verbal art form (poem, story, play, film script, etc.).

SECTION B: READING

B.0   GENERAL

B.0.0   The ability to read at the highest level is essential to the individual's personal growth in the modern world, his ability to benefit from most of the learning opportunities at school, and his ability to take his place in society. The school must therefore take the teaching of reading very seriously. Behaviour and adjustment problems not only lead to difficulties in learning to read, but, and this is less widely realized, the reverse is also true: reading difficulties are a frequent cause of difficulties of behaviour.

By 'reading' we do not mean simply the basic decoding of print, but reading for understanding.

B.0.1   There is no point at which a pupil can be said to have 'mastered' reading. Reading development is a continuous development, which is always incomplete. It must therefore be taught at all levels.

B.0.2   Reading involves language, and therefore is best taught in an overall context of talking and writing.

B.0.3   Subject teachers have reported that their pupils are set back in their work in their fourth and fifth years by their reading. It is therefore important that the first three years are used to prepare pupils for the kinds of reading in subject areas demanded by the middle-school curriculum.

B.1.   OCCASIONS FOR READING

B.1.0   Pupils should encounter and be helped through print, and they should not be encouraged to circumvent it. Teachers should ensure that there are a variety of suitable occasions for reading as an integral part of the study of most subjects. (We were worried at the interim report of the Schools Council's Effective Use of Reading project, which indicated that reading took very little time in many subject areas with younger classes.) Our survey suggested also that it is possible for a pupil to avoid reading for a large proportion of the week. The problem is partly related to the efficiency of mixed-ability teaching, and also to the view taken by some departments that subject knowledge rather than literacy level should be the main forms of assessment. Obviously there is a need for a balanced approach.

B.1.1   Where possible subject teachers should set aside time for individual reading as a class for reasonable lengths of time. (The same School's Council project also found reading tended to be in brief intermittent bursts.) We feel that as a school we should be aiming for more continuous reading. This reading must be relevant to the subject concerned. The least able can be given individual assistance in this context. The problem of the more able marking time in this situation should be considered by departments.

B.1.2   In certain circumstances and for specific purposes, it is good for pupils to read aloud to the rest of the class, or in groups, or to the teacher. (Teachers should be aware of the problems involved in asking poor readers to read aloud; reading to a larger audience will require the support of proper and adequate teacher intervention.)

B.1.3 Subject departments should, where appropriate, consider with the senior-sixth-form tutor the use of sixth-formers to help lower down the school. A formal structure should be devised to enable this to proceed with the greatest effect and continuity.

B.2 TEACHING READING

B.2.0 Every teacher should be a teacher of reading in his or her own subject.

B.2.1 Staff should use the phonic method of reading words, and should teach syllabification. 'Difficult' words should be approached in this way even with older and more competent pupils. (See Appendix.)

B.2.2 It is good practice to ensure that a child sees a word as well as hears it.

B.2.3 When worksheets are distributed to a whole class, they should be read aloud in advance, either by the teacher or pupils or both, at least at the beginning.

B.2.4 When a passage is given to a pupil to read, the teacher should normally prepare for the reading by (i) putting it into context, (ii) explaining the general aim of the passage, so that the pupil knows what to expect, and (iii) new ideas or concepts should be explained.

B.2.5 Pupils should have their attention drawn to the logical structure of the argument, and to the key words which carry the argument forward, e.g. 'since', 'as a result of', etc.

B.2.6 Pupils should be encouraged to isolate and read out key sentences which convey a particular idea or piece of information.

B.2.7 Whenever possible pupils should be encouraged to comment on the clarity or effectiveness with which a point is made in whatever is being read.

B.2.8 Pupils should be taught the study skills and their way round a book (e.g. use of table of contents, index, etc.).

B.3 RESOURCES FOR READING

B.3.0 Departments must ensure that reading resources are sufficient, suitable, and varied.

B.3.1 The librarian, Education Library at County Hall (via the librarian), and local libraries will provide special collections, given adequate notice, on short-term loan.

B.3.2 The practice of providing book boxes for classes, containing suitable books at a variety of levels, may be exploited by many departments.

B.3.3 When departments produce their own material, they should consider how appropriate it is, not only in the level and kind of reading, but in the quality of reprography. We recommend, therefore:

(i) The use of improved versions of past worksheets, rather than producing afresh on each occasion.

(ii) The maintenance of departmental libraries of worksheets for reference and improvement.

(iii) A central library of individual sheets. (*Note:* The Headmaster has acted on this recommendation: cf. Standing Instructions on Reprographic Facilities.)

B.3.4 The ILEA through its inspectorate should capitalize on and co-ordinate the individual production of materials in schools with a view to producing more of its own material. The SMILE, MRC, English Centre and others have begun to work in this direction, but we consider that the great amount of time spent in production of materials would be reduced if this could be continued in all subject areas. Further liaison between LEAs on a reciprocal basis might increase the materials available at a cost less than that of purchasing through educational publishers.

B.3.5 Representations should be made to the ILEA to discuss with educational publishers the suitability of their products specifically in maths, science, history, geography and humanities, and reading books for the older less able pupil.

B.3.6 Subject departments should encourage the use and understanding of the library in conjunction with the librarian.

SECTION C: TALKING AND LISTENING

C.o. GENERAL

C.o.o Spoken language is the basis of all other language activities, yet creating an appropriate situation for constructive language development is one of the most difficult objectives to achieve, since small group work is not always possible.

C.o.1 In the long term, each department should consider its criteria regarding both spoken language and vocabulary development. Material and references to encourage discussion should be made available.

C.o.2 The drama and English departments should consider their joint and linked role in this area.

C.o.3 Informal contact by children with staff in school should be encouraged. Individual contact with pupils offers a better, more natural chance for development in the pupils' spoken language. It is often difficult to achieve in a full classroom.

C.1 PRACTICE

C.1.o A child has to 'use' a word, concept or idea in speech many times to make it his or her own.

C.1.1 Although children should be helped to extend their vocabularies and develop their spoken language structure we should be careful not to make any of them scared to open their mouths in formal situations.

C.1.2 A pupil's contributions should be valued—he or she should be listened to in the same way that we as teachers expect to have our contributions listened to. Pupils should also be encouraged to value each others contributions.

C.1.3   If a teacher has an 'expected' answer in mind to a verbal question in class, he/she should nevertheless be prepared to find unexpected answers interesting in their own right.

C.1.4   Practical work provides a shared focus for talk to take place. Where practical work is involved pupils should be given every opportunity to use talk to organize themselves, to talk through what they are discovering and to reflect on and explain what they have done and learned.

C.2   LISTENING

C.2.0   Listening is a difficult and important skill.

C.2.1   Positive training in listening should be the responsibility of all departments. Comprehension passages, formulated within subject framework should be given regularly in the lower school. These should be oral but require written answers whenever possible so that progress may be monitored. A standardized oral comprehension test should also be given in the first year.

SECTION D: GENERAL

D.0   VOCABULARY

D.0.1   We recommend that each subject department should draw up a list of its specialist words. It may be helpful to consider two categories—words specific to the subject (*pipette, contour, glaze*) and words given a specific meaning within the subject (*power, divide, perspective*).

D.0.2   Departments should set aside time to teach this vocabulary in whatever way each department finds appropriate. Lists should be distributed to staff for information.

D.0.3   The teaching of new words and their subsequent retention could be helped by a subject 'dictionary' drawn from the list mentioned above. Where the word describes a physical object this could be in the form of a 'picture dictionary'; however, this idea does help to highlight a problem in teaching vocabulary.

D.0.4   Some works are the names for objects or activities and can be learnt as such (*countersink, test tube, vertex*); but for some words their full meaning is only assimilated from contextual use and may differ from subject to subject (*revolution, improvise, moral*). Thus a word should be taught and its meaning understood in a context.

D.0.5   Even when the pressures of time and pupils make the ideas outlined above seem idealistic we would suggest that an ordinary dictionary is available and its method of use known.

D.1   MIXED-ABILITY TEACHING

D.1.1   The benefit to pupils with poor reading and writing of being in a group with all abilities will only accrue if the provision for coping with such a group is equal to that which they would receive under a different grouping principle. A particularly worrying element is when the main body of work or text is the same for all pupils and only at the end is there extra work or more difficult questions.

D.1.2   We would recommend that where a department has not got suitable material for mixed-ability teaching either the lack should be remedied or the mode reconsidered.

D.1.3   As a subject for further debate that of the efficacy of mixed-ability teaching in extending the most able and helping the least able while ensuring that those in the middle are suitably taught should be considered. The view has been expressed that in this situation everybody suffers, and it seems appropriate that the question should be raised at some future date.

D.2   RESOURCES

D.2.1   A selection of books dealing with the study of language across the curriculum should be held in school for staff reference.

D.2.2   The library should include more taped resources.

D.3   GENERAL PRACTICE

D.3.1   Teachers should consider starting some of their lessons by recapitulating what was learned in the last lesson.

D.3.2   In the giving of formal tests, the questions could be read out loud so that poor readers are tested on what they have learned of the subject matter and not penalized by their reading difficulties.

D.3.3   It is valuable not to rely on any one method of evaluating what the pupils have learned from a given body of work. There are times when this can be equally well done orally as in a written form.

D.4   REPORTS

D.4.1   More consideration should be given to the language used in the writing of reports.

D.5   SOCIAL COMPETENCE

D.5.1   All pupils should, by the time they leave school, be able to cope with being presented with all types of forms—e.g. application, tax, union registration, insurance, holiday booking, HP, cheques, and documents such as vaccination certificates, pension statements, mortgages, tenancy agreements, bills, record-player instructions, etc.

D.6   ENGLISH DEPARTMENT

D.6.1   It would be helpful if the aims, policies and practices of the English department could be disseminated so other departments could see their own requirements and efforts in a context.

D.7   TOPICS FOR FURTHER STUDY ARISING OUT OF THE REPORT

D.7.1   Woodberry Down departmental analysis of reading and writing as in the Schools Council report, Writing across the curriculum 11–18.

D.7.2   Staff familiarization with phonic method of teaching reading.

D.7.3   Whether sixth-formers, parents, etc. could and should be involved in helping backward readers, etc.

D.7.4   The efficacy of mixed-ability teaching in ensuring that all abilities are properly catered for.

D.7.5   The structuring of talk as a medium of learning and its 'dual', listening.

D.7.6   Also listed in document—seminars on Study Skills, Vocabulary Development, Comprehension Technique, Spelling Rules, Oracy, Social Literacy.

D.7.7 A collection should be made of the range of Woodberry Down pupils' writing.

D.7.8 A series of lectures and/or seminars should be arranged on pupils' and teachers' language as an important aspect of in-service training.

Appendix Five

A Policy Document for the School-wide Reading Programme in Culverhay School

Section 1

(a) *Why read?*

Some of the ways in which reading is an aid to children's development.

(1) information from reading used in work with other people.
(2) understanding of oneself and others through reading.
(3) behavioural changes as a result of reading.
(4) therapeutic effect of reading.
(5) building a personal philosophy from reading.
(6) understanding of the natural world and society.
(7) meeting personal needs with reading.

(b) *What is reading?*

Reading encompasses so many aspects of human behaviour that one definition could only be inadequate. The list below, of some of the many activities associated with reading, gives an indication of the complexity of the subject with which this booklet deals.

At secondary level reading involves:

1. *Vocabulary development.*
   (a) learning new words through wide reading.
   (b) learning of key words and concepts in various subjects.
   (c) learning of technical abbreviations, symbols and formulae.
   (d) consulting of dictionaries.
   (e) studying of words in various contexts.
   (f) study of word origins.
   (g) recognition of common words.
2. *Word recognition.*
   (a) division of words into syllables for pronunciation.
   (b) use of phonetic approach.
   (c) knowledge of prefixes and suffixes.
   (d) recognition of 'overtones' of words.
3. *Comprehension and organization.*
   (a) comprehension of sentences accurately.
   (b) comprehension of main idea of a paragraph.
   (c) recognition of the organization of an article or chapter.
   (d) summarizing what has been read.
   (e) learning to read critically—distinguishing essential from non-essential.

    (f) examining truth or correctness of statements.
    (g) recognition of fact and opinion.
    (h) bringing own experience to bear on what has been read.
    (i) noting of cause and effect.
    (j) drawing of inferences and conclusions.
    (k) reading 'between the lines'.
    (l) integration and organization of information from various reading sources.

4. *Reading interests.*
    (a) voluntary reading.
    (b) use of school and public library.
    (c) finding a personal value in reading.
    (d) appraisal of quality of reading material.
    (e) development of particular reading interests.
    (f) reads more for personal information.

5. *Study skills.*
    (a) sitting still long enough to read.
    (b) using 'skimming' for different purposes (to obtain facts, a general impression, main ideas, possible questions, plot).
    (c) reading of maps, charts, graphs, diagrams, formulae.
    (d) reading of 'out-of-school' material.
    (e) location and selection of information on a particular topic.
    (f) familiarity with a variety of sources of information.
    (g) learning how to take notes.
    (h) use of Survey Q3R method when appropriate.
    (i) reading more rapidly with adequate comprehension.
    (j) application in other forms of ideas gained from reading (making pictorial or graphic records, solving problems, entertaining, teaching).
    (k) forming of a study habit.

6. *Approaches to reading.*
    (a) setting of specific objectives before beginning to read.
    (b) use of a variety of methods and rates of reading.
    (c) reading with intent to organize, remember and use ideas.
    (d) relating of reading to circumstances outside the immediate environment.

7. *Personal development.*
    (a) meeting personal needs with reading.
    (b) use of information from reading in work with other people.
    (c) understanding of oneself and others through reading.
    (d) understanding of the natural world and society.
    (e) behavioural changes as a result of reading.
    (f) using of reading to build a personal philosophy.
    (g) therapeutic effect of reading.

From this greatly abridged selection it is apparent that reading can never be one teacher's task or even one department's responsibility.

(c) *Why a school-wide reading programme?*

Reading permeates the curriculum and is a major source of knowledge in every subject field. Therefore, instruction in reading should be an integral part of every reading experience in every subject. ('The Reading Curriculum', Amelia Melnick and John Merritt (eds), The Open University Press, 1972.

The level of reading skill required for participation in the affairs of modern society is far above that implied in earlier definitions of literacy. ('A Language For Life', The Bullock Report, HMSO 1975.)

Every school should devise a systematic policy for the development of reading competence in pupils of all ages and ability levels. (Bullock Report.)

All subjects are involved with reading and this booklet is intended for information and as a check-list for all teachers concerned about the reading abilities of the students they teach.

Section 2 : The school reading programme

(a) *Area of concern*
All students. (Those with special difficulties in literacy will be dealt with on a withdrawal basis.)
The problem areas are principally:
(a) students with ability but lack of interest in reading.
(b) students with low ability and too difficult books.
(c) students with no recognized need for reading in their lives.
(b) *General procedure*
All students will have reading tests on entering the school: at the beginning of the third year, and again at the beginning of the fifth year.
The tests used are:
(c) *Reading tests*

Test	Ability measured	Age range	Score
1. Daniels and Diack Graded Test of Reading Experience	Reading Experience Age	6–14+	In years and months
2. Holborn Reading Scale	Individual test of reading age	5 years 9 months – 13 years 9 months	In years and months
3. Neales Analysis Test	Reading Age Comprehension Level Speed and Fluency		In years and months

Non-verbal tests

1. Compound Series Test by Morrisby	I.Q.	
2. Standard Progressive Matrices by J. C. Raven	Intellectual Capacity (5 grades from 'intellectually superior' to 'intellectually defective'—with a very broad band for the 'intellectually average'	

The results of these tests are avilable to all staff so that the appropriate ability levels of any teaching group can be readily determined.

This information is *an indication of a particular situation* and must not be regarded as an end in itself.

The English and Remedial teachers provide all students with tuition in the basic reading skills (word recognition, basic vocabulary, spelling, appreciation of literature, comprehension, etc.) and through a School

Reading Programme it is hoped to develop these skills more fully and to encourage all aspects of reading.

(d) *Reading procedures across the curriculum*

Set out below is a check-list of some of the procedures which will help the teacher play an effective part in the successful practice of reading within his or her own subject group. This is not intended as a prescriptive list, nor as a 'reading course'; it is merely a list of activities which singularly or collectively aid reading development.

Text testing techniques

(1) *Initial informal testing* in each subject using a passage from a textbook or reference book. By timing the reading and testing the comprehension the teacher will learn much about how well students read the text used in his or her subject.

(Through the standardized tests the teacher already knows the range of reading ability represented in his teaching group.)

Tests of this kind can show:

(a) the accuracy with which the student comprehends main points.
(b) how well the student can draw inferences.
(c) the student's ability to organize his own ideas as he reads.

Questions need to be designed to elicit such responses.

One basic test to check the suitability of a particular text for a group of students is the cloze procedure.

This requires the student to fill in a gap, usually a whole word, which has been left in the text. To do this the child must employ a number of criteria:

(i) select a word which is grammatically correct.
(ii) select a word with the correct meaning.
(iii) choose a word which best fits the vocabulary employed by the author.

Implications for the teacher are

(a) the production of the text with appropriate gaps from subject text books.
(b) to choose texts from the beginning, middle and end of book so as to obtain passages that are representative of the whole book.
(c) to decide whether to employ 'structural deletion' (omission of not more than one word in ten), or 'lexical deletion' (where certain parts of speech are omitted, e.g. nouns or verbs). Each method appears to have some merit and provides a quick and easy way of checking a student's comprehension of a chosen text.
(d) the use of standardized 'cloze' tests would simplify the testing of reading suitability.

(The 'cloze' method has however, several aspects which still need answers:

(1) What is the appropriate number deletions for each age group?
(2) Will *deletion*, or *omission* affect the results?
(3) Will results vary according to whether the test is given orally or in writing?
(4) What variables affect performance (e.g. vocabulary, sentence completion, level of comprehension, intelligence and socio-economic status)?
(5) What is the particular effect of heavy 'lexical deletion' (parts of speech)?
(6) Can a percentage score be equated with individual reading levels or even individual frustration level?

(2) Make a careful choice of textbooks, taking into account the vocabulary; illustrations; simplicity of reference sections (glossaries, indexes, footnotes, marginal headings); general layout of main ideas; clear expression of author's

ideas; level of previous knowledge required by student in order to find the book useful; the aims and achievements of the book (are they consistent with those of the teacher)? Even the most well-produced books need teacher assistance. The generous provisions of suitable reading material is essential. *Implication*—Familiarity with the variety of material available in any one subject may involve time spent by the teacher in bookshops, at publishers' exhibitions and at teachers' centres. Further capitation may also be necessary.

Pupil testing techniques

Before the teacher decides to use a particular textbook it is necessary to determine a student's level of ability *informally* using the following technique:

(i) at the *independent level* the student is able to read aloud fluently with 99 per cent accuracy. If he has more than one error in every 100 words then he has not reached this level with this material.

(ii) If however, the student has not more than 5 errors in every 100 words and can answer 75 per cent of the questions asked by the teacher, then the student is at the *instructional level* with the material.

(iii) With 10 or more errors in every 190 words and a literal comprehension rate of 50 per cent or below, the student is at *frustration level* and the material is too difficult for him.

This is an indication of a student's reading level on a particular text and the value of such information lies in the future action that the teacher takes. *Implications*—Thoughtful recording of this information is necessary if it is to be turned to advantage, and provision of a large selection of texts for differing abilities may be costly in time and money.

It must be stressed that there is no *one* reading activity which is the panacea for all. Any of the methods suggested above may be modified and used according to the purpose and material. To attempt the use of a single technique may well be a handicap to the teacher and the student.

Study techniques

1. Use of Survey Q3R

This system can be used on textbooks in many subjects, but is most successful with material which is organized in a logical pattern. Five stages are involved:

Step 1—Survey:

Look through the passage to be read and note the information given by

(a) Boldface type, (b) Pictures and captions, (c) Charts,
(d) Drawings with captions, (e) Maps and diagrams, (f) Summaries,
(g) Questions.

Step 2—Question:

Questions can be formulated by the student and teachers. (Use of the information contained in the headings and sub-headings is adequate for this purpose). Questions can be written down one side of a piece of exercise paper folded lengthwise and answers need only be represented by *key words*; there is no need to write long answers.

Step 3—Read:

Read to find the main points which answer the questions. (There are usually one or two main points in each section). Remember the important facts.

Step 4—Recite:

Return to the question sheet. Cover the answers and attempt to answer the questions—reference to the key words and/or the book may be necessary. The immediate recall and then the practice until all the questions can be answered without reference to key words or text is very *important*.

Step 5—Review:
Answer the questions again some three or four weeks later and before examinations. Re-read only the passages which have been forgotten.

This is a basic study method and as such is capable of being modified.

2. The teacher may cater for the various reading abilities within a group by setting different reading tasks—simple factual work for those of low ability; more critical comments on content and style from those of high ability. Simpler texts or pamphlets on the same topic, ready-made or written by the teacher, will meet the requirements.

Critical reading (reading and selecting for the purpose of reaching a conclusion).

The demands on adults to read critically are greater today than ever before. A single textbook, considered to be authoritative, is not conducive to critical reading. Comparative reading should be attempted whenever possible. Questions that require only verbatim reproduction of the ideas read, deter critical reading.

Students must read slowly; subject teachers should plan to allocate adequate time for instruction in critical reading. (Less material may be covered but the quality of the student's reading should improve).

Implications—Considerable time needed to prepare the various pamphlets required by the group and the arrangement of the different tasks for students of varying abilities will demand careful organization and grouping in the classroom. The teacher may need help when determining the reading levels of the material he intends to create. Collection over the years of work from able students provides material which has been simplified from more difficult sources.

3. In every subject the student should be taught to glance through the book, chapter or passage to obtain an idea of the content, and then see how this fits in with his previous knowledge and his present and future needs.

Implications—Discussion of possible goals is important for greater understanding.

Anticipation of what the author is going to say sharpens interpretation and usually increases concentration.

'Literal' comprehension is important and though the techniques might be learned in an English lesson, the amount of practice required must come in other subjects 'Inferential' comprehension is time consuming but must be made part of the reading in every subject if a student is ever to realize the implications of what he reads.

4. The use of various reading aids must be pointed out; pictures, charts diagrams; indexes; tables of contents; footnotes; glossaries; chapter, section and marginal headings; and questions.

In each subject an understanding of these aids makes the difference between true comprehension and a vague impression.

Implications—Subject time must include this specific instruction.

5. Speed of reading and what to look for in any reading task needs to be indicated by the teacher.

If a student is to study a basic book in a subject then he needs to read slowly in order to grasp the main points and supporting details of each paragraph.

Implications—When to 'skim' and when to 'study' a text must be pointed out to the student so that he may make more economic use of his time.

6. For the student in any subject the setting of specific goals which are

attainable through reading is an important factor in creating a 'felt need' for the reading experience. Having a purpose creates 'readiness to read'.

Implications—Use of blackboards and/or printed sheets for thought-provoking questions related to the reading, greatly aid the student's success with these goals. The problem-solving method with answers attainable through reading needs similar preparation.

7. Time to discuss what they have been reading is necessary for the students' understanding; this sharing of reading-experiences with others is important.

Implications—The Bullock Report stresses the need for group discussion on any reading material attempted by a class and time must be available for this activity within any 'reading' allocation. The ability to read critically can be fostered by discussion.

8. The student can be helped to acquire the particular vocabulary for each subject—through lists and glossaries specially prepared by subject teachers and students, and kept for reference in exercise books or notebooks. The preparation of a glossary by the student is a much more impressive experience than the ready-made glossary in front of him (there is a glossary of 300 words peculiar to Music).

Implications—Initial tests are useful to determine the extent of a student's identification of particular terms used in a subject. (In mathematics where sentences are stripped to the minimum there is little to suggest the meaning of a particular term in a sentence; its meaning must be taught.)

9. Students should be encouraged to use school and public libraries to supplement subject reading—specific references must be given to the student.

Implications—The teacher's own knowledge of specific references available in public and school library is essential for any real encouragement of pupils' reading.

10. The student's previous familiarization, through some other medium (television, film-strips, tape-recording, records, illustrations, models, etc.), with the material to be taught is often an aid to more effective reading in a subject. The poor reader who turns to the 'practical' subjects because he does not have to read about them, is disillusioned when he is confronted *immediately* by a textbook or a worksheet that teems with technical vocabulary.

Implications—Pre-viewing of films, television programmes and similar material, and the organization of its presentation to the students must become part of the 'reading' preparation. The Resources Department has facilities for aiding all subject teachers with this preparation.

11. The student should be able to plan his reading, to know what can be read for a general impression and what needs detailed consideration. He must be taught to read selectively and to summarize the information gained not merely by copying but by making his own notes. Students who are too dependent upon the teacher for what they should read become unable to make their own choice.

Implications—When the student plans his own reading activity, he becomes an 'active' not a 'passive' reader and can take advantage of the opportunities that the printed words offers for exercising reason, existing knowledge and imagination:

(a) the time, extent and place of the reading can often be chosen.

(b) past or future parts of the text are always readily available for re-reading.

(c) the reading process can be halted at any moment to allow time for thought, for note-making or for making comparisons with other sources of information.

Section 3: Criteria for success

(a) *The success of reading improvement* across the curriculum depends on the following conditions:

    (a) All the teachers must recognize the need for such a scheme.

    (b) That this is not seen as a short-term measure.

    (c) Financial support for materials must be available where necessary (it may be necessary to limit the scheme to a small area of the curriculum at first).

    (d) Senior staff need to be sincerely and actively in favour of reading improvement on a school-wide basis.

(b) *Some questions and answers*

    Q. Does a teacher need specialized training to help any student to improve his reading?

    A. It depends upon the level of proficiency, but a good teacher who is friendly and understanding will help students improve their reading.

    Q. Do I neglect the teaching of my subject for the teaching of reading?

    A. Unless the teaching of reading is part of a student's instruction it is probable that he is a less-efficient student. The teacher becomes frustrated; he/she feels teaching is impossible if the students are unable to read satisfactorily. The check list provided will help subject teachers with a basic approach.

    Q. With a limited number of periods for my subject how am I expected to do the job of an English teacher as well as my own?

    A. The teaching and practice of reading is the concern of all staff and all students. Helping students to read more efficiently is part of the teacher's professional role and is not an extra task or the task of one department.

    Q. What can I do in a class of varied ability when I have to use the same textbook for all students?

    A. By recognizing that there is no one method which will improve all students' reading and by employing a variety of the activities suggested, some satisfaction should arise for students and teacher.

(c) Conclusion

If reading is understood to encompass all those activities listed in the opening part of this booklet, and more, then any evaluation of a reading scheme across the curriculum is extremely complex. Subject teachers need (a) to state their specific objectives when attempting any reading with their students, (b) to consider carefully the material they are using and, (c) to choose the reading activities most likely to achieve the stated objectives and which are most appropriate for the subject content.

When the reading has been completed the evaluation takes place and must include an assessment of the success or failure of the objectives. In the light of this knowledge there must then be a review of (a) *materials* and (b) *reading activities* in order to determine whether future policy will be a repetition or a modification of these.

The aim of evaluation is to promote growth not merely measure it.

The evaluation of the total reading programme across the curriculum is a major undertaking requiring separate consideration with representatives from many areas of the school.

The aim of all developmental reading programmes is to produce effective readers who like to read, who are not deterred by any reasonable difficulty, who are independent and analytical in their reading, who are capable of literary appreciation, and who are interested in the possibility of a better life and a better world. (*The Improvement of Reading*, Strang, McCullough and Traxler, McGraw-Hill, p. 198.)

Appendix Six

Television

Any policy concerned with language must include reference to the dominant entertainment and art form of our age, television. The Report firmly recommends:

> Television is now part of our culture and therefore a legitimate study for schools. The school has an important part to play in promoting a discriminating approach to it, but it is equally important that children should learn to appreciate the positive values and variety of experiences the medium can provide. (C & R 295)

I hope that once and for all its recommendations have pulled television out of 'audio-visual aids' and placed it in a position analogous to the books, cassettes, or pictures in a resource centre. Unfortunately even the report, for all its strong encouragement, places its discussion towards the end of the chapter (22) entitled 'Technological Aids and Broadcasting'. It should rather be in the continuum of talk, reading, and literature, for the approach to television is analogous to the approach to literature.

The emphasis should not be on criticism—that is, the 'Reading and Discrimination' approach. It should certainly not be heavily influenced by concentration on what is judged to be meretricious in an attempt to drive that enjoyment out of the pupils' lives. The sharing of enthusiasm is more educative than the attempt to persuade to see the faults in certain programmes.

There is now a substantial body of books and articles to help the teacher, but it is sad that although the Committee declared: 'We were impressed by such work as we did see' it also registered concern at how little was being done.

The question the school has to ask is where the approach will come. Some schools respond to this recognition of a curriculum need by putting a new subject on the timetable, and thus there are schools that study separately the mass media or screen appreciation. I do not regard this as an adequate approach to television, which is, after all, a medium not a subject. The Report recommends three complementary approaches:

> (a) the group study of television programmes, extracts, and scripts alongside other media dealing with the same theme;
> (b) the study of a full-length television work in its own right, with associated discussion and writing;
> (c) the study of television as a medium, with some exploration of production methods, comparison with other media, and analysis of the output of programmes.

Although (c) is necessarily limited to one subject area, (a) certainly and (b) to some extent should be used in a number of subject areas. (a) could be thought of as a thematic approach, and simply suggests that the material, normally on video recordings, should be used as often as possible as one of the elements in the range of learning resources assembled for a particular topic. This is fairly easy in the Humanities. (b) is likely to be especially the province of English, when plays or dramatizations will be the main fare. However, documentaries and other presentations must not be overlooked across the curriculum.

All work is probably helped if there is a specialist option in the fourth and fifth years, for not only does that permit the pupil whose interests have been aroused to pursue the topic more methodically, but also it helps a teacher develop a special expertise, skill in presentation, and range of material which will help the work elsewhere. Obviously, however, this special patch is not so important to the school as a whole as the contextualized approach. This will be helped by a proper stocking of books in the library, displays in various parts of the school, references in assemblies, and the like, but above all the use of television material in a variety of subject areas, with teacher-led responses and questions about the nature of the medium as evidence.

A Language Policy Reading List

This is a short list of further reading that Heads and Deputies would find useful and realistic. All quotations and references have been credited in footnotes with full bibliographical details.

Titles have been grouped into sections, and alphabetically by author within sections. The key books, such as might make the initial reading for a Working Party, are asterisked.

Bullock Committee

*A Language for Life, HMSO, 1975.

Language

*Britton, James, *Language and Learning*, Penguin, 1970.

Bruner, J. S., *Language as an Instrument of Thought*, in Davies, A. (ed.), *Problems in Language and Learning*, Heinemann Educational Books, 1975.

Creber, Patrick, *Lost for Words: Language and Educational Failure*, Penguin, 1972.

Gleason, H. A. Jr, *Linguistics and English Grammar*, Holt, Rinehart and Winston, 1965.

Lawton, Dennis, *Social Class, Language and Education*, Routledge and Kegan Paul, 1968.

Open University, *Language in Education*, Routledge and Kegan Paul, 1972.

Pride, J. B., and Holmes, Janet (eds.), *Sociolinguistics, selected readings*, Penguin, 1972.

Quirk, R., *The Use of English*, Longman, 1962.

Trudgill, P., *Sociolinguistics*, Penguin, 1975.

*Wilkinson, Andrew, *The Foundations of Language*, OUP, 1971.

Wilkinson, Andrew, *Language and Education*, OUP, 1975.

Williamson, A. (ed.), *The State of Language*, University of Birmingham, 1969.

Talking

*Barnes, Douglas, *From Communication to Curriculum*, Penguin, 1976.

*Barnes, D., Britton, J. and Rosen, H., *Language, the Learner, and the School*, Penguin, revised edition, 1971.

Reading

Cashdan, A. (ed.), *The Content of Reading*, Proceedings of the 1975 UKRA Conference, Ward Lock Educational, 1976.

Fader, Daniel N. and McNeil, Elton B., *Hooked on Books*, Pergamon Press, 1969.

Harris, Albert J. and Sipay, Edward R. (eds), *Readings on Reading Instruction*, David McKay and Company, N.Y., second edition, 1972.

*Herber, Harold L., *Teaching Reading in Content Areas*, Prentice-Hall, 1970.

Mackay, D., Thompson, B., and Schaub, P., *Breakthrough to Literacy*, Teachers' Manual, Longman, 1970.

Melnik, Amelia and Merritt, John (eds), *The Reading Curriculum*, University of London Press, 1972.

*Melnik, Amelia and Merritt, John (eds), *Reading: Today and Tomorrow*, University of London Press, 1972.

Moyle, D. (ed.), *Reading/What of the Future?* Proceedings of the 1974 UKRA Conference, Ward Lock Educational, 1975.

Pumfrey, Peter, *Reading: Tests and Assessment Techniques*, Hodder and Stoughton, for UKRA, 1976.

Robinson, H. Alan, *Teaching Reading and Study Strategies*, Allyn and Bacon, Robinson, H. Alan and Thomas, Ellen Lamar (eds), *Fusing Reading Skills and Content*, International Reading Association, 1969.

Schools Council, *Children's Reading Interests*, being Working Paper 52, Evans/Methuen Educational, 1974.

Smith, Frank, *Understanding Reading*, Holt, Rinehart and Winston, 1971.

Turner, J., *The Assessment of Reading Skills*, UKRA bibliography No 2, UKRA, 1972.

Walker, C., *Reading Development and Extension*, Ward Lock Educational, 1974.

Writing

*Britton, James *et al.*, *The Development of Writing Abilities (11–18)*, Macmillan for the Schools Council, 1975.

Martin, Nancy; D'Arcy, Pat; Newton, Bryan; and Parker, Robert, *Writing and Learning Across the Curriculum, 11–16*, Ward Lock Educational, 1976.

Peters, Margaret L., *Diagnostic and Remedial Spelling Manual*, Macmillan, 1975.

Peters, Margaret L., *Spelling: Caught or Taught?* Routledge and Kegan Paul, 1967.

Writing Across the Curriculum Project, *From Information to Understanding, Why Write?, From Talking to Writing, Keeping Options Open, Writing in Science, Language and Learning in the Humanities, Language Policies in Schools*, Ward Lock Educational, 1976–7.

Listening

Wilkinson, Andrew; Stratta, Leslie; and Dudley, Peter, *The Quality of Listening*, a Schools Council Research Study, Macmillan Education, 1974.

Organizing Resources

Beswick, Norman, *Organizing Resources*, Heinemann Organization in Schools Series, 1975.

Beswick, Norman, *Resource-Based Learning*, Heinemann Organization in Schools Series, 1977.

Library Association, *School Library Resource Centre: Recommended Standards for Policy and Provision*, The Library Association, 1975.

Schools Council, *School Resource Centres*, being Working Paper 43, Evans, 1972.

Teaching of English (General)

Central Committee on English (Scottish), *The Teaching of English Language* (Bulletin No. 5), HMSO, 1972.

Currie, W. B., *New Directions in Teaching English Language*, Longman, 1973.

Dixon, John, *Growth Through English*, OUP, revised edition, 1975.

Dixon, John and Stratta, Leslie, *Patterns of Language*, Heinemann Educational Books, 1975.

Owens, Graham and Marland, Michael, *The Practice of English Teaching*, Blackie, 1970.

Whitehead, Frank, *The Disappearing Dais*, Chatto and Windus, 1966.

Index

lunar phases and have speculated that a slight deformation of the Earth's crust could trigger an earthquake. Geoff Chester, an astronomer and public affairs officer with the U.S. Naval Observatory, has stated, "The same force that raises the 'tides' in the ocean, also raises tides in the [Earth's] crust." The earthquake in Greece and Turkey in fall 1999 occurred after a total solar eclipse, and the tsunami of 2004 happened at the time of the full moon. Although the ability of the Earth's gravity to trigger moonquakes has been confirmed, the converse phenomenon is still doubted by others.

Aboriginal Australians in coastal areas have noted a correlation between the moon's phases and the tides. They once believed that the high tide runs into the moon as it sets into the sea, making it fat and round. Conversely, when the tide is low, water pours from the full moon into the sea below and the moon consequently becomes thin. For travelers from ancient Rome and Greece who were used to the relatively motionless Mediterranean Sea, it was a surprise to experience the strength of Atlantic tides. When, in 55 B.C., Julius Caesar arrived on the coast of Kent, a few miles northeast of the chalky cliffs of Dover, to invade England for the first time, he inadvertently anchored his ships at the peak of a twenty-foot spring tide with a strong following wind. After a short foray inland, he returned to find his fleet stranded high and dry on the tidal flats. His army had to face a brutal attack from the Britons.

Myths and tales abound of correlations between the moon and the natural life on Earth. Farmers once thought that the harvest should occur when the moon was full. Rural folk bundled wheat on the threshing floor "during the moon's age" so that it would dry better; grain brought in during the waxing moon would be moist and soft and might burst during threshing. *The Old Farmer's Almanac,* published annually since 1792, recommends that its readers, "plant flowers and vegetables which bear crops above the ground . . . during the *light* of the Moon; that is, between the day the Moon is new and the day it is full." However, the authors continue, "flowers which bear crops below ground should be planted during the *dark* of the moon; that is, from the day after it is full to the day before it is new again." Some gardeners in France, in contrast, according to Camille Flammarion, attributed to the light of the moon in April and May "an injurious action on the young shoots of plants." "They are confident of having observed that on the nights when the sky is clear the leaves and buds exposed to this light are blighted—that is to say are frozen, although the thermometer in the atmosphere stands at several degrees above zero." Farmers plagued by fleas were told to look for them in the moon's glow, because the tiny creatures were assumed to be drawn by the shimmer and would emerge from the clothes where they were hiding.

Do trees feel the moon's influence? In some areas of central Europe, lumber was cut exclusively in the winter—for example, between

Christmas and Epiphany—and then only according to certain rules. This makes sense because
wood is generally dryer during this time of the
year, but some people believe that lunar phases
influence the quality of wood as well. Sometimes
"moon wood"—lumber cut during a certain lunar
phase, typically before the new moon—is said
to have unique properties, such as resistance to
weathering or suitability for use as cookware in a
fireplace.

Preference for moon wood may sound like
superstition, but in fact there is some evidence
that it exhibits unique stability. By means of exact
measurements, the engineer Ernst Zürcher from
the Bern University of Applied Sciences has discovered that wood felled just before a new moon
has more water in its cells—and thus is heavier
and more stable—than wood felled immediately
preceding a full moon. Zürcher attributes this
phenomenon not to the moon's attraction but to
the tree's water economy, hypothesizing that in
its new phase the moon shields the Earth from
solar winds that directly influence how water
binds in the tree's cells. Zürcher also believes that
the light of the full moon is capable of bleach-

Farming in tune
with the moon

ing wood. Experiments by the Swiss Federation of Technology's Department of Forestry Science, in turn, have shown that trees planted during a full moon grow better, giving weight to a piece of folk wisdom, though not all scientists agree with these conclusions.

The daily rise and fall of the oceans' water level governs much of the life on their fringes, where habitats are exposed and submerged every day. Anthony Aveni reminds us in *Empires of Time* that "many feeding and reproductive cycles of organisms that inhabit the ocean receive their input signals from the tidal period, which, in turn, is regulated by the position and appearance of the moon." He speculates that "moon-based cycles may be a reflection of our original ascent from the sea." Aveni looks at a coral from the Devonian period (350 million years ago) and recognizes it as "an archaic month calendar frozen in time": horizontal growth ridges have been deposited during full-moon phases. This detail proves that when the moon was closer to the Earth, our months were shorter, and there were thirteen of them in a year rather than just twelve. More important, we have evidence that organisms have been attuned to the evolving lunar period over hundreds of millions of years.

How, then, do these cycles and rhythms manifest themselves in living beings? When Aristotle was observing sea urchins, he could discern that their ovaries tended to swell during the time of the full moon—a phenomenon that modern scientists studying sea urchins at Santa Catalina Island in California have confirmed. When Columbus reached the New World on October 12, 1492, he and his companions saw a candlelike flicker in the water off the coast of the Bahamas about an hour before the moon rose. The naturalist Lionel Ruttledge Crawshay (1868–1943) conjectured that this light may have been caused by the luminescent Atlantic fireworm. During the moon's last quarter, the female fireworm rises to the water's surface after sundown and emits flashes of light. This spectacle attracts the male worm, which in turn speeds through the water flashing like a firefly. Using range maps for the worm, Crawshay postulated that Columbus's first landfall would have been at Cat Island and not San Salvador, as is commonly assumed.

Today, chronobiologists can demonstrate links between biological processes in animals and daily, seasonal, lunar, and tidal cycles. A number of organisms living in tidal zones are attuned to the pulse of the ocean. The Pacific palolo worm (*Eunice viridis*) found in the coral reefs around both the Samoan and Fiji Islands offers a particularly interesting example. This creature breeds in October and November (spring in the Southern Hemisphere), when the moon is in its last quarter. The Samoans have developed an intricate system of monitoring this process, which includes close observation of the scarlet blossoms of the flame tree (*Erythrina indica*), whose flowers' emergence appears to be synchronized with the palolo's

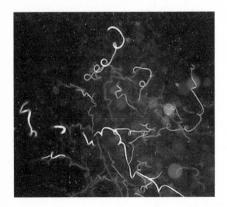

Palolo worms

approach. Shortly after the full moon, when it is low on the western horizon at daybreak, the Samoans know that the palolo feast is just seven days away. At sunrise, the ends of the swarming worms, engorged with sperm or eggs, break off, rise to the surface, and thrash about. Most islanders scoop up a mass and eat the palolo raw as they catch them; others later devour this caviar as a fried delicacy.

Can we establish with certainty what is going on? "Moonlight penetrates quite well in tropical water and these organisms are much more sensitive to light levels than we are," says Paul Brown of the National Park Service, who has repeatedly dived at eighty feet in the middle of the night, with moonlight as his sole illumination. "Palolo tend to be fairly shallow in Samoa and are most common on the reef crest and slightly below, and so are typically in depths of twenty to forty feet." Brown adds: "Even though the breeding process is clearly connected to the lunar cycle, the exact trigger mechanism is still unknown."

The grunion, a small member of the silversides family that lives off the sandy coast of southern California, provides a fascinating example of the attunement of spawning rhythms with the tides. Twice a year, around the end of February and the end of August, thousands of these tiny fish approach the coast. Grunion spawn on the beach chiefly during spring tide—more precisely, during the first three to four nights after a new or full moon in the first two hours following the high tide. Swimming against the flow of the outrushing water until they reach the beach, these small fish transform the beach into a glimmering silver surface.

Accompanied by one or several males, the somewhat larger females dig themselves into the sand with their tails and fins. There they are encircled by the ejaculating males. After they lay their eggs a few centimeters deep in the sand, the females work their way up, loosening themselves to be carried back into the ocean by the retreating waves. It will be two weeks until the next spring tide, but under cover of the sand the young fish develop as if following a secret plan. When the water level again reaches the point of the beach where the eggs were buried, the pounding of the waves lays bare the young grunion, which emerge to be carried with the waves to the ocean.

Other examples from the natural world show a closer affinity of other species with moonlight itself. In the early 1980s scientists discovered a mysterious rite of procreation among corals. Just after the full moon, the researchers observed, corals dissolve in an orgy of reproduction, sowing the waters with trillions of swirling eggs and sperm that merge to form new life. The details long remained obscure—it was unusual, for example, that the breeding phenomenon occurs during different months, though usually in the

Life in the oceans is strongly affected by tides and the moon. Painting by Ernst Haeckel (*Arabian Corals,* 1876)

summer. In 2007 a group of seven scientists from Australia, Israel, and the United States was able to substantiate that the moon regulates this synchronized mass spawning. This synchronicity is possible because the corals contain ancient photosensitive molecules that trigger this unique response to moonlight. "It is an amazing molecule really, triggering the largest spawning event on the planet," says Ove Hoegh-Guldberg from the University of Queensland in Australia. More research is needed to fully understand the secrets of how the corals keep an "eye" on the moon.

How about the moon's influence on land animals? Pliny (A.D. 23–79), known for his interest in the natural world, believed that monkeys are depressed while the moon is waning and happy during the New Moon—but only the ones with tails. According to the Roman writer Aulus Gellius (ca. 130–180), a cat's eyes grow and shrink just as the moon does. The even more imaginative Claudius Aelianus (ca. 170–222) wrote that elephants grazing in the forest when the sickle of the new moon appears gaze at the crescent and wave torn twigs as if begging for the grace, mercy, and protection of the goddess. Aelianus further claimed that the ibis closes its eyes during lunar eclipses until the moon shines again. Pliny, Gellius, and Aelianus, of course, were relating fanciful tales, and later folk wisdom about lunar effects on animal behavior is likewise unreliable. Still, moonlight has been shown to influence the activity of many animals. Contrary to intuition and folklore, some smaller rodents—and even bats, such as the South American fruit bat—are less active than usual on moonlit nights, exhibiting "lunarphobia" or moonlight shyness. The three-striped night monkey or northern owl monkey (*Aotus trivirgatus*), indigenous to Venezuela and parts of Brazil, has the opposite activity pattern: "lunarphilic," during the full moon it is energetic all night.

A key element of lunarphobia or lunarphilia is an animal's place on the food chain. An animal more likely to be caught by predators under bright moonlight, or a hunter that has evolved excellent night vision, is apt to be more active on dark nights. Lunar cycles may also affect the haul-out patterns of harbor seals, when they leave the water for a variety of reasons, including to breed. When the moon approaches its full phase, the seals of Snake Island, British Columbia, stay at sea more often, possibly because the bright moonlight could facilitate the hunting of prey that migrates vertically and thus yield a higher catch. In England, the lapwing usually searches for food during the day and rests at night, but during the winter months the bird looks for food even at night if the moon is full, unless it is covered by dense clouds. And cranes tend to return to their roosts later after sunset if the moon is bright. The skylark (*Alauda arvensis*), a small bird that can be found in Europe and Asia, provides another fascinating example for the influence of moonlight: As a comprehensive study undertaken by French scientists during four fall seasons has shown, the onset of migration of this bird is determined by the phase of the waxing gibbous moon, which facilitates the navigation conditions during their flight to southern Europe and North Africa, from where they return at winter's end. This timing allows the bird to benefit from optimal conditions of illumination for almost a week.

Eric Warrant of the University of Lund in Sweden, who specializes in the vision of animals, has observed the behavior of African dung beetles (*Scarabaeus zambesianus*). He found that this animal is, by virtue of particularly sensitive eyes, able to navigate by the dim polarization pattern of moonlight. The beetle maintains a straighter rolling path under these conditions, the moon acting as an orientation cue.

Dung beetle

As crepuscular animals, active predominantly at dawn or at dusk, wolves are naturally associated with the moon: their wavering howls were once thought to be a message intended for that heavenly body. But the facts provided by canine experts are sobering. Even if a wolf points its faces toward the moon and stars, it is not really howling at the moon. The reason for its posture is purely acoustic — projecting its call upward simply allows the sound to carry farther.

The iconic wolf

Thanks to the high pitch and drawn-out notes, a wolf's howl can travel as far as six miles in the forest and ten miles across unobstructed tundra.

This final example may disappoint some romantic nature lovers, but the fact remains that some rhythms in the animal world are related to the tides and the phases of the moon — that some animals possess internal clocks able to "predict" the lunar cycle. Their lives are indeed affected by the moon — but what about our own?

CHAPTER TEN

Esoteric Practices

If we could travel back in time, we'd arrive at a point when only a fine line separated alchemy, folk medicine, and medical science. The moon figured in all of them. Premodern healing systems are attracting renewed interest today, but two centuries ago or so they were the first choice of much of the populace. The Berlin of the late eighteenth century, for example, was a major center of the Enlightenment, but also the home of a moon doctor, a Mr. Weisleder. By the early 1780s Marcus Herz, one of the most renowned Jewish physicians in the city, decided that this dubious (but irksomely popular) character must be a traitor. Herz scrutinized his rival's practices and recorded his observations in a long piece entitled "Pilgrimage to the moon doctor in Berlin." At five o'clock one summer afternoon, when Herz arrived at the "temple of Asclepius" in a "wretched beer house of the lowest class" in the Jakobstrasse, he found crowds of people from all around the city climbing the back staircase to the building's second floor. There, it was said, Weisleder would soon begin to treat patients of all religions and from every stratum of society. He would cure any and all ailments, mutilations, and physical handicaps, from fractures and hernias to

The Hungarian astronomer Maximilian Hell (1720–1792) thought that he could heal certain ailments by redirecting the forces of the moon.

inflamed eyes and impaired hearing. "When we got inside," Herz wrote, "all I could see was the empty, dirty, and low-ceilinged parlor of a common craftsman. A big table in the middle and a few stools at the wall represented the whole furniture." A rumor spread that the moon doctor had been called away by a princess and was also awaited by the prince, so the crowd would have to wait for five hours. Doctor Herz bit his lip, fearful that laughter might cost him injury, or even his life. He decided to return in a month.

On his next visit, he again found a large crowd waiting for Weisleder; he counted a few hundred, many of them in elegant garb. He pushed his way through the crowd to reach the front of the line, waiting for the second-floor "divinity." Twelve patients were admitted at a time. An assistant named the ailments, one after the other, and the sick came when beckoned. Doctor Herz pretended to have an attack of gout. The moment Weisleder made his appearance, the sick paused for a moment in veneration. Herz was less impressed. This tall, gaunt man, indifferently walking about the room holding a pipe, was a former hosiery knitter with "combed out hair" clinging to his coarse blue frock. "His physiognomy was one of the vilest kind," Herz wrote. The moon doctor used the southern window, while his wife treated female patients at the north window. These charlatans distributed no pills, drops, or bandages; each simply instructed the patient to extend the afflicted body part out the window toward the moon. Weisleder stood still, folded his

hands, and murmured a secret text in a voice too soft for the visitor to discern. To be effective—such was the claim—this procedure had to be repeated on three successive days during the first quarter of the moon; Herz later learned that some patients had to subject themselves to this routine for three or four months. Weisleder claimed to have healed conditions that had been declared incurable by regular doctors, including the fracture of a renowned "Madame N.," who would not have to use a bandage anymore. He demanded no

The curious moon doctor drew large crowds to his house. He made his patients point their ailing body parts in the direction of the moon.

fee for his service, but some patients gave his wife money.

Herz commented sarcastically on the astral medicine and supernatural cures of the moon doctor. He even traced the doctor's ministrations to the digestive problems of the former hosiery knitter, who he assumed had been weakened by his constant sedentary position and suffered from inhibited circulation. The man's condition, Herz speculated, "formidably" produced wind "of a hippocratic kind," which, instead of going its natural course, rose up and transformed the man into a fortuneteller, a visionary, a doctor. Herz would have been thrilled to have observed an actual cure, he wrote, but all he saw was an "abominable appearance," the "sight of debased and befouled humanity." Some in the general public acknowledged that Weisleder's cures were often ineffective; some even ridiculed him. Inexplicably, though, many defended and even patronized him. Herz called the moon doctor's practice a common form of "mischief," the mere thought of which was capable of producing in him "a surging heat" for years to come. Later he learned with satisfaction that Weisleder had to close in the face of undeniable evidence that his procedures produced no effective results.

The moon doctor is representative of a whole tradition of practicing medical astrologers, who believed that there are celestial and lunar influences on particular diseases. Some drew an analogy between the human body and the atmosphere, both of which were thought to be influenced by the moon and sun. The German doctor and healer Franz Anton Mesmer (1734–1815), for example (whence the term *mesmerized*), was convinced that certain psychological symptoms in his patients changed with the moon's phases. Believing that there are "invisible fluids" within the body, he attached magnets to patients' bodies to induce "artificial tides" that would influence those fluids. His practice was somewhat akin to widespread bloodletting, which was meant to lessen the viscosity of bodily fluids.

The belief persisted among many astrological doctors that women were particularly liable to lunar influences. Men were assumed to be reasonable and resistant to natural conditions, but women were closely associated with nature and considered more susceptible to celestial influences. The Dutch anatomist Theodor Kerckring (1640–1693) described a French matron as beautiful and round-faced at the full moon, but when the moon was waning, her face became deformed, her eyes and nose turned on one side. During this phase, she would stay inside, waiting until the moon changed again.

Richard Mead (1673–1754), Physician in Ordinary to Saint Thomas's Hospital, Southwark, seems to have been particularly aware of the conflicting theories of his age. On the one hand, Mead embraced the mechanistic philosophy of Newton, but on the other hand, he clung to belief in celestial influences on particular diseases and to traditional lore linking meteorological conditions to phases of the moon. Reflecting on the loss of reverence for the moon, he lamented that "we can scarcely believe that the present Moon is the same which shone in the days of old. —It is true her aid is sought in navigation; otherwise, she now wanders round the earth for little more than the amusement of juvenile astronomers."

And even Erasmus Darwin, for all his scientific bona fides, wrote in *Zoonomia; or, The Laws of Organic Life* (1794–1796) about the lunar and solar influences on a number of diseases, declaring that over longer periods of time the heavenly bodies could "produce phrensy, canine madness, epilepsy, hysteric pains or cold fits of fever." At the same time, Darwin stressed that "the periods of quiescence and exacerbation in diseases do not always commence at the times of the syzygies [straight-line configurations, producing full and new moons] or quadratures [right-angle alignments, at the first and last quarter] of the moon and sun, or at the times of their passing the zenith or nadir; but as it is probable, that the stimulus of the particles of the circumfluent blood is gradually diminished from the time of the quadratures to that of the syzygies, the quiescence may commence at any hour, when co-operating with other causes of quiescence, it becomes great enough to produce a disease." In his painstaking attempt to prevent returns of diseases, Darwin then asked,

> Do not the cold periods of lunar diseases commence a few hours before the southing of the moon during the vernal and summer months, and before the north-

ing of the moon during the autumnal and winter months? Do not palsies and apoplexies, which occur about the equinoxes, happen a few days before the vernal equinoctial lunation, and after the autumnal one? Are not the periods of those diurnal diseases more obstinate, that commence many hours before the southing or northing of the moon, than of those which commence at those times? Are not those palsies and apoplexies more dangerous which commence many days before the syzygies of the moon, than those which happen at those times?

One can't accuse Darwin of giving the matter too little thought.

Often doctors saw a connection between fever and the moon if only to the extent that fever was thought to "ebb and flow" in parallel with the lunar cycle. In the tropics, where the heat, so medical men tended to believe, relaxed the fibers of the body, this link became particularly evident. Many physicians who worked in British military hospitals in India or the West Indies maintained that the moon has "a more corruptive power" there, in a process comparable to the rapid putrescence of exposed meat and fish. Benjamin Moseley (1742–1819), a member of the Royal College of Physicians of London, recalled that "all the soldiers in the military hospitals in Jamaica, under my care, in dysenteries and intermittents, almost certainly relapsed at the lunar syzygies." Moseley's friend Sir Richard Worseley observed similarly that "afflictions of the skin and eye" in the eastern Mediterranean "fluctuated according to the waxing and waning of the Moon." These doctors often modified their procedures according to the medicine practiced by indigenous peoples, who also tended to believe in the power of the sun and moon.

Times of crisis inspired particularly strong beliefs in lunar influences on human health. When, in 1817, cholera first became endemic in India, it was linked to the influence of the moon. The surgeon Reginald Orton associated the new and full moons and the periods when the moon is closest to the Earth with disease. According to his experience, it was at such times that the majority of cholera deaths occurred. Such explanations, incidentally, were not generally held by the medical profession in Orton's British homeland.

Although the nineteenth century was marked by many industrial and intellectual advancements, as more people moved into cities and lost touch with nature and the nighttime sky, some popular interest in occult knowledge nonetheless survived. Often the moon was associated with evil forces. In fact, a number of secret societies embraced magic rites that celebrated supposed supernatural powers. In various cultures the moon is accorded a place in regional medicinal systems unrelated to the tradition of Western medicine. Acupuncture treatment, for example, is particularly recommended during the day before and after the full moon. The Chinese principle of yin represents both the Earth and the moon, while yang stands for the sun. During the time of the full moon the human body is believed to be in a state of excess; doctors, as a general therapeutic principle, aim at restoring a state of balance. Ayurveda, an indigenous tradition in use for several thousand years in India, is recognized by the World Health Organization as a sophisticated medicinal system. According to its precepts, certain seasons and lunar phases enhance the efficacy of plants used to treat certain ailments. Some traditional medical systems in Africa and South America also recognize the influence of heavenly bodies; one common belief is that herbs collected during the full moon have maximum curative power.

A French tarot card

Belief in occult wisdom and rituals involving the moon thrive in some Western societies as well, as if the Enlightenment had merely pulled a thin veil over millennia of traditional lore about the moon and its alleged involvement with terrestrial life. In fact, an underside of irrational thought has always counterbalanced rationalism, resisting the dramatic change wrought by science and technology in humanity's worldview. Full-moon rituals are common among New Age, New Pagan, and occult groups in some Western countries. Adherents to the Wicca movement, for example, perform rituals to "truly bathe in the touch of the Divine." Typically, the moon of each month is associated with a characteristic image:

the storm moon of March, the wind moon of April, or May's flower moon. The specific ritual homage to the current "magic moon" varies, but the rite is usually performed on an altar, with candles, burning incense, and spiritual music.

In spring, when "the frigid cold of winter has been replaced by the promise of new life and growth," offering, "a chance at fertility and abundance, rebirth and regrowth," the altar will be decorated with fresh cuttings from the garden or packets of seeds to celebrate the arrival of the full moon. Harking back to a time when the well or spring was seen as a sacred and holy place, participants bring small bowls of water and form a circle. The ritual is led by a priest or a priestess. Facing the moon, the celebrant holds the bowl of water to the sky and says, "The moon is high above us, giving us light in the dark. She illuminates our world, our souls, our minds. Like the ever-moving tides, she is constant yet changing. She moves the water with her cycles, and it nourishes us and brings us life. With the divine energy of this sacred element, we create this sacred space." The priestess then dips the cut flower into the water and, while walking a circle, sprinkles water on the ground with the petals of the flower. Next she approaches the participants, carrying a large bowl; each person pours water from a small bowl into the larger one, telling where the water was collected—for example, "This is water from the creek behind my grandmother's farm." When everyone has contributed water, the priestess uses the cut flower once more, stirring and blending the water in the large bowl with the stem of the flower, then listening to "the voice of the moon" above. As the forehead of each participant is anointed with the blended water and an individual symbol, such as a pentagram or a triple moon image, the priestess adds, "May the light and wisdom of the moon guide you through the coming cycle." Finally, she reminds them to "consider how our bodies and spirits ebb with the tide, and how we connect to the cycles of water and of the Moon."

What should we make of such a full-moon ritual? Obviously, the followers of these cults are not satisfied with the worldview contemporary science has to offer. Wiccan practice is at odds with the doctrines of the scientific establishment, but it should make us consider how we deal with the natural world. Even if we may not share such beliefs, they merit respect as attempts to restore a lost magic and faith in nature and the moon—efforts that may ultimately address the deeply felt human need to feel more connected with a world whose forces and mechanisms often evade our grasp.

While a critical examination of the premises of folk wisdom is always worthwhile, the significance of premodern belief systems in Western societies should never be underestimated. It is worth asking ourselves why, for example, in contemporary Denmark fewer nurses and medical doctors graduate from universities than the number of graduates from schools of alternative healing.

CHAPTER ELEVEN

Spurious Correspondences

No one can deny the existence of certain atmospheric influences on the human body. Sunlight, the duration and intensity of which directly correspond with the seasons, not only burns us if we get too much of it, it can affect our moods. Thunder and lightning can arouse strong fears and end lives. Extreme changes in atmospheric pressure can cause lungs to collapse. Correlations have been shown between the seasons and the birth and death rates. But does the moon actually influence the physiology of humans? If so, how much?

A whole body of folklore explores the moon's influence on various phenomena of human life. According to German folk wisdom from the 1800s, for example, those who worked by moonlight risked being slapped in the face by an invisible hand or even going blind. Use of the spinning wheel at night, a common household activity among poor people before the Industrial Revolution, was thought to yield worthless cloth—or, in an extreme version of the superstition, the threads that would be woven into a rope that would end up around a relative's neck. Hanging laundry to dry in the moonlight was believed to wear out the fabric or cause it to absorb poisonous night dew. Moonlight was not to fall

The legendary gambler in Gold Rush California Eléonore Alphonsine Dumant (also known as Madame Moustache) was not shy about using a gun at night onto the marital bed, and children were not to be conceived by its glow, because such pregnancies were fated to end in miscarriage or the birth of an insane child. People who urinated in the moon's direction were at risk of developing a badly swollen eye, and those vomiting in its direction could expect a rash of the mouth. Dancing under the moonlight, especially in a tight embrace, was considered dangerous because Earth's crust was thought to be particularly thin under moonlit conditions. Moreover, the dancers' tapping feet could create vibrations that would lure ghosts from beneath the surface.

Of course such superstitions are hardly limited to European traditions. They can be found in various cultures around the globe. According to a saying from the Philippines, bathing during a full moon results in insanity—and during a new moon, death. In some yogic traditions, both full- and new-moon days are observed as holidays because of the exceptional energy released by the relative positions of sun and Earth. The energy of the full moon is believed to correspond to the end of inhalation. An expansive and upward-moving force, it makes practitioners feel emotional rather than well grounded and makes exercise difficult. In contemporary India some doctors avoid performing surgery under a waxing moon, which they believe encourages scarring. And there are other lunar taboos. For example, some cultures consider the time of the new moon unfavorable for undertaking business.

In Western cultures, some have claimed that

humans can shift shapes to become werewolves, wolves, or other animals, demonstrating violently aggressive behavior and sometimes devouring raw meat. Supposedly, a summer night with a full moon shining directly on the subject's face is particularly conducive to this so-called lycanthropy. In 1977 Harvey A. Rosenstock and Kevin R. Vincent described in the *American Journal of Psychiatry* the case of a forty-nine-year-old married woman who, at the full moon, imagined herself to be a wolf. She became sexually aroused, experienced homosexual impulses, suffered "irresistible zoophilic urges and masturbatory compulsions." This profile may sound odd to contemporary ears and even make us smirk. Today, the extremely rare medical condition of lycanthropy is believed to be caused chiefly by schizophrenia or bipolar disorder—serious scholars rarely if ever link it with the moon. Yet one residual lunar mythology may be connected with a medieval belief in werewolves. Some people claim it's a good idea to trim or cut hair during the full moon if you want your hair to grow thicker and fuller; if cut at the new moon, it might grow back too fast.

We have seen several examples of medical astrology positing dubious correlations between human ailments and lunar phases. While modern science has found little evidence of any causal relationship, some people persist in seeing connections, particularly in the realm of mental health. At the notorious Bethlehem or Bedlam Hospital in London, until 1808 inmates were chained and flogged during certain lunar phases to prevent

"Lunacy"

violence. Joseph Daquin (1732–1815), a Swiss doctor considered a pioneer of psychiatry, believed that the inmates of a mental asylum suffered literal lunacy: the condition was particularly evident during nights with a full moon. Although these observations were not "scientific" in the modern sense, they reflect some strong impression the writer must have experienced. The so-called Lunacy Act of 1842 characterized a lunatic as a person "afflicted with a period of fatuity in the period following after the full moon." The impetus to define a medical condition and pass a law based on it speaks to the conviction that must have reigned regarding the moon's effects on man.

Even our language reflects such superstitions. The erroneous connection of the moon with insanity is reflected in various Indo-Germanic languages—*lunatic, loony,* or *moonstruck* in English, *lunatico* in Italian, *lunático* in Spanish. The Italian idiom *avere la luna,* like the French *avoir des lunes,* means to experience a nervous breakdown. The German *Laune* for *mood* is etymologically connected to *luna,* implying that a changing state is influenced by a phase of the moon. Shakespeare was pointing toward something along these lines in *As You Like It* when Rosalind calls a tormented, confused, hopelessly enamored boy, "but a moonish youth." And while we may recognize today that these connections are false, surely the lines between fact and fiction are not so neat.

In some cases these erroneous ideas have become so enmeshed with our thinking that we hold them to be common sense. For example, conventional wisdom accepts that a full moon interferes with the phases of deep sleep. But is this really the case? Someone who has spent a sleepless night may realize afterward that there was a full moon, and some people have claimed that the light of the

full moon wakes them. Light has been proven to influence activity patterns during sleep, but the first reaction to excessive light would probably be to turn over or cover one's face.

Many people think that they are influenced by the moon.

Many such cases may simply be instances of self-fulfilling prophecy. To a certain extent, something can become reality if we expect it to happen. Applying this logic to case at hand, one predisposed to believe that the full moon interferes with sleep might selectively discard the memories of sleepless nights with no full moon. Could it be *awareness* of a full moon might interfere with the sleep of someone who believes in the connection? Sleeplessness may have many causes, and could even indicate a serious health condition, so for some, blaming the full moon might provide a welcome and comforting explanation—though possibly a dangerous one.

Various studies across nationalities and cultures have sought a correlation between the moon's phases and sleep patterns. In the vast majority, no such correlation could be established, but a few exceptions suggest that a minor connection may exist. A study in 2006 in a suburban area of Switzerland involved thirty-one volunteers over six weeks, including two full moons. Researchers found that subjective sleep duration varied somewhat with the lunar cycle, from six hours, forty-one minutes at the full moon to seven hours at the new moon. There was also evidence that the amount of fatigue the subjects felt in the morning was associated with the moon's phase, with participants feeling more tired when the moon was full. Then again, those taking part in the study had to be aware of the moon's current phase, which may have led them to expect that they would sleep better or worse. The results can scarcely be called conclusive.

Another phenomenon often associated with the full moon—lack of evidence notwithstanding—is sleepwalking, or lunatism. Sleepwalking is a complication in which the sufferer "gets stuck" in his sleep, usually during the first third of the sleep cycle. He may not be fully conscious but is able to perform routine tasks, such as getting up and dressing. Usually, the sleepwalker has no memory of his nocturnal activities the next day. In extreme manifestations of lunatism, the affected person arises and, in an attempt to get oriented, is attracted to sources of light. In the stereotypical manifestation of this phenomenon, the somnambulist may venture onto a rooftop in a futile attempt to get closer to light—often, the light of the moon.

In an earlier time, before artificial light became ubiquitous, moonlight may indeed have been responsible for the disruption of sleep—just as today sleep can be disturbed if someone leaves a lamp on at night or has a streetlight outside an unshaded window. And longer spells of continuous sleep disturbance can certainly have detrimental effects on psychological well-being. But because the moon is a relatively minor source of illumination for most people today, it seems unlikely to have much effect on sleep. Charles Raison at Emory University goes so far as to hypothesize that the persistence of the belief in "the moon's power over the mind" may in fact be a "cultural fossil"—"a memory of an actual effect that no longer obtains." He adds: "Perhaps the moon once had a power over the functioning of the brain that it has since lost."

Might other forces emanating from the moon influence us? Could the moon's gravitational pull—which moves the vast oceans—have a direct effect on humans, whose bodies, after all, are more than three-quarters water? The psychologist Arnold L. Lieber made this claim in his popular book *The Lunar Effect* (1978), but it has been disputed by contemporary scientists. The astronomer and cosmologist George O. Abell once formulated the curious calculation that a mosquito exerts a greater gravitational pull on your shoulder than does the moon. Others have said that the pull would be less than that of a wall of a building at a distance of six inches. One critical consideration is that the water of the oceans is unbounded, whereas the water in the human body is contained within our tissues, which shield it from the moon's tidal forces.

Another common belief, widespread since the eighteenth century, is that a correlation exists between certain lunar phases, the menstrual cycle, and fertility. The word *menstruation,* of course, derives from *mensis,* meaning *month,* which is again related to *moon.* But etymology and causality are two different things. A decisive disparity exists. Although the average menstrual cycle is 28 days, with some variation from woman to woman and month to month, the length of the lunar month is consistently 29.53 days and therefore out of step by a significant margin. The similarity of the time spans appears to be mere coincidence. Nor has a convincing argument been posited why natural selection should favor a method of reproduction based on the lunar month. Furthermore, while the estrous cycle of most primates approximates the human menstrual cycle—ranging between 25 and 35 days—one would expect that other mammals would synchronize with the lunar cycle as well. On the contrary, rats have an estrous cycle of 4 or 5 days, elephants of 16 weeks. Some research, though, has shown some interesting correlations. A study of 826 women, for example, found that menstruation was relatively common around the new moon. Another study of 140,000 births in New York City showed a slight increase of fertility during the last quarter of the moon. On the other hand, a study of a Dogon village in Mali, West Africa, showed no lunar influence on menstruation, even though these villagers don't have electric lightning and spend most nights outside, exposed to moonlight.

Do women living together menstruate in sync with one another? This phenomenon has been observed among some women in some settings, but serious scientific investigations have not established any connection with lunar influences. Is there really a rhythmic, cosmic, or lunar cycle that has been disturbed or masked by civilization, specifically by indoor electric lighting? Such a theory might seem plausible until one considers that many other mammals on the planet have reproductive cycles that are not synchronized with the lunar cycle.

Perhaps the clearest conclusion to be drawn from this range of evidence is that, by virtue of its sheer ubiquity, conventional wisdom dies hard. We see, for instance, that a single study that finds a minor correlation between phases of the moon and this or that phenomenon usually receives a tremendous amount of attention, as if validating our hunch. For instance, Italian biomathematicians and gynecologists have studied the relationship between lunar position and the day of delivery. Trying to avoid what they perceived as pitfalls in earlier research, they focused on all the spontaneous, full-term deliveries occurring at one hospital in Fano in the Marche region during a three-year period in the early 1990s. Although they observed a "statistically significant" connection, they found it "too weak to allow for predictions regarding the days with the highest frequency of deliveries." But exactly such a predictive power would be necessary to optimize the resources of a hospital. In short, even the most rigorous scientific studies, able to lay bare phenomena imperceptible to a single individual, have revealed only effects of the moon so weak as to be irrelevant to our daily lives. Still, belief lives on. Another study with a similar focus on births and birth complications, carried out in North Carolina, considered more than 564,000 births between 1997 and 2001 and could establish no connection between frequency of deliveries and lunar phase.

Maybe we can simply blame the media. Any entertaining anecdote about some alleged lunar effect, no matter whether it stands up to scrutiny or not, is sure to arouse public interest and, therefore, to be widely reported. Studies that yield unspectacular results attract no media coverage. In 1997, for example, J. M. Gutiérrez-García and F. Tusell studied almost nine hundred suicides reported by the Anatomical Forensic Institute of Madrid but could establish no significant relationship between the lunar cycle and the suicide rate. Ivan Kelly, James Rotton, and Roger Culver, psychologists from the United States and Canada, examined more than one hundred studies on lunar effects to get a clearer picture. They analyzed numerous phenomena—homicides, traffic accidents, crisis calls to police or fire stations, suicides, kidnappings, stabbings, and various other examples of unusual behavior—and concluded that the studies failed to show a reliable correlation with the full moon or any other of the moon's phases. Remember the splash those reports made in the papers and on the Internet? It's just not much of a story if during the full moon . . . nothing unusual happens!

The belief in lunar influence on our lives has a long pedigree. Given its constant perpetuation in books of pseudoscience and other popular publications, it's not likely to give way anytime soon to a more realistic sense of the moon's power. But scientists and intelligent laypeople alike must continue to question assumptions passed down the generations. Russell G. Foster and Leon Kreitzman have provided many insights into the ways biological rhythms affect life on Earth, but they remind us: "Despite the persistent belief that our mental health and a multitude of other behaviours can be modulated by the moon's phase, there is no solid evidence that the moon can influence our biology." Or, to put it the other way around, as the writer of popular astronomy Bob Berman does, "If all the Moon's alleged powers were real, we'd be a species of lunatics controlled from outer space."

CHAPTER TWELVE

Visions of the Moon

We have seen that mapping the moon began as soon as telescopes provided a closer look. By the middle of the nineteenth century, at the time when Jules Verne was writing his books about lunar travels, moon maps had become more sophisticated and detailed than early ones, but they still failed to create a truly vivid image of the moon. How could anyone give an impression of what it would be like to actually stand on the moon rather than merely to look at it from afar? How could the still invisible and unknown be made visible? Paradoxically, to achieve such verisimilitude required creation of an illusion.

Driven by such an ambition was the Scotsman James Nasmyth (1808–1890). After making a fortune as a manufacturer, Nasmyth had retired to devote himself to astronomy. In the early 1840s he constructed highly detailed and three-dimensional models of the moon's surface out of plaster, then photographed the models. Much of Nasmyth's fascination with the moon came not from hours spent behind a telescope, as one might conjecture, but rather from a trip to Mount Vesuvius near Naples, where he "became in a manner familiar with the vast variety of those distinct manifestations of volcanic action, which at some inconceivably remote period had

The pogo stick was supposed to be powered by a small rocket and designed to permit a hopping range of twenty-four miles.

This illustration by James Nasmyth shows the lunar Appenines with the crater Archimedes—a region visited by the Soviet *Lunik 2* probe in 1959 and the U.S. *Apollo 15* mission in 1971.

produced these wonderful features and details of the Moon's surface." From the top of the volcano he could look down into the "pit from which the clouds of steam are vomited forth." To produce the models for his photos, he also drew on the expertise of his father, the landscape painter Alexander Nasmyth.

After thirty years of research and work, James Nasmyth published *The Moon: Considered as a Planet, a World, and a Satellite.* Printed in 1874 by John Murray, who was also Charles Darwin's publisher, Nasmyth's impressionistic photographs presented an artificial moon, the lunar landscape strangely illuminated. Prepared using the arduous Woodburytype process, with a relief mold made of the image in lead, these images were regarded as technically superior to daguerreotypes of the real moon. They provided the viewer with a thrill typical of special effects. Even *Nature* magazine, with its reputation as a guardian of rational science, endorsed the book, a reviewer writing that "no more striking or truthful representations of natural objects have ever been laid before his readers by any students of science."

Shortly before Nasmyth's publication, the American clergyman Edward Everett Hale (1822–1909), in two short stories published in the *Atlantic Monthly* in 1869 and 1870, formulated the idea of an artificial moon. The Brick Moon, measuring about two hundred feet in diameter, was to be hurled into space by flywheels rotating in opposite directions. In Hale's fictional account, the satellite failed to assume its scheduled orbit when launched and seemed lost, but it later reappeared five thousand miles above the Earth, populated by self-sufficient lunarians equipped with a gigantic telescopic lens made of ice. Of course, there is hardly a material less appropriate than brick for a spacecraft, but all peculiarities of this fantastic story aside, we may still consider it an early precursor to the idea of a space station.

In the early 1850s, while Nasmyth was at work on his models, a hemispherical model of the moon's visible side was constructed by Thomas Dickert in Bonn, Germany. The model was about twenty feet in diameter, with fea-

This moon globe produced by Thomas Dickert was presented to the German public in the 1850s and later shown in the United States.

tures partly based on Johann Heinrich von Mädler's map. Dickert's model was transported to the United States and ended up in Chicago's Field Museum, where it helped to popularize interest in the moon's makeup and astronomy in general. In 1898 Elias Colbert, former director of the Dearborn Observatory in Evanston, wrote in the *Chicago Tribune:* "A close examination of the mammoth Moon enables me to testify to its accuracy as a real model and not mere show. It reproduces with marvelous fidelity the features of the lunar surface and really exhibits a great deal more detail than would be noticed by the unpracticed eye in viewing her through a first-class telescope." The mammoth moon can be seen as an early precursor to the space shows at the world's fairs of the twentieth century. This specimen can still be seen at the Science Central Museum in Fort Wayne, Indiana.

In 1925 the optician and geophysicist Frederick Eugene Wright (1877–1953) established in Washington, D.C., a "Committee on the Study of Physical Features of the Surface of the Moon." Wright's interests in both the moon and the underwater seascape of the Caribbean led him to be dubbed "Moon-Man" and "the modern Jules Verne." To facilitate the mapping and interpretation of the lunar features, he projected negatives of the moon onto one-foot chemical flasks that had been coated with photographic emulsion. They extended over a distance of 130 feet in an underground tunnel. This so-called Wright's Sphere was an early example of rectified photography, designed to restore the original spherical spacing of features, with each assuming its relative position on the globe.

Later in the century, visual portrayals of the moon fell more to artists in the popular, even the pulp, press. A subgenre of space art begins with the rise in the 1920s of science fiction magazines and popular science magazines such as *Science Wonder Stories* and extends through the 1950s and 1960s, the golden age of space enthusiasm, with certain techniques typical of each period. Some consider the American Chesley Bonestell (1887–1987), whose space art was first featured in *Life* in 1944, as the father of astronomical art, but artists such as Lucien Rudaux of France, Klaus Bürgle of Germany, Ralph A. Smith of England, and the American Fred Freeman also made seminal imaginative contributions. All science fiction artists creatively confronted the challenge of

Technological
moon fantasy as
envisioned by
the space artist
Klaus Bürgle in
the late 1950s

making unseen worlds seem real to a demanding
public—believability being more important than
accuracy, which, until late in the era, was a mystery anyway.

These artists, of course, worked in tandem
with writers who were painting word pictures
of the unknown worlds. And despite the strong
competition from the planets—above all Mars—
the lunar theme continued to be eagerly taken
up by writers of science fiction, with the turn of
the twentieth century marking a general shift to
the dark and dystopic. The moon imagined by the
British author Herbert George Wells in *The First
Men in the Moon* (1901) is an aging, porous body
interspersed with large interconnected caves. The
inner moon is full of air and contains the habitats of the lunar beings, who dare to come to the
surface only occasionally. Two British travelers

launched from Earth with the help of a substance with antigravity properties experience a dramatic landing. After their space vehicle rolls into the depth of a lunar crater, they are captured by ant-like moon creatures. These savants have balloon-like heads, but their limbs have completely degenerated; their race is doomed. In keeping with its elegiac tone throughout, the book ends with only one traveler managing to escape and return to Earth.

Attack by insectlike lunarians in Wells's *The First Men in the Moon*

Inevitably, science fiction was soon matched with the new technical possibilities of moving

The rough
landing of the
rocket in the
right eye of
the moon—a
memorable scene
from Georges
Méliès's *A Trip
to the Moon*

images. Georges Méliès, a French actor, theater owner, producer, and director, became the leader of the newly emerging French film industry. His fourteen-minute-long film *A Trip to the Moon,* based on the novels of Jules Verne and H. G. Wells (1866–1946), premiered in 1902 and was soon acclaimed worldwide. The rocket piloted by the long-bearded Professor Barbenfouillis (played by Méliès himself) and five professors lands in the lunar eye, and the visitors are flung to the ground by a violent explosion of unknown cause. They then settle in for the night, only to be awakened by snowfall. Retreating into an area under the surface, they discover strange rock formations, waterfalls, and gigantic mushrooms. But the lunar subsurface turns out to be a kingdom, whose strange beings attack the intruders. When the besieged Barbenfouillis strikes a creature with the tip of his umbrella, it explodes in a cloud of smoke. When more Selenites appear, the visitors are caught, bound, and presented to the king, but they manage to break free and defeat the king.

Ultimately, the travelers flee to their rocket, now perched on the edge of a cliff. It is launched when the passengers' weight tips it over the edge and, with one Selenite on the tail of the rocket, it falls toward Earth. After a brief journey through space, they manage to alight safely on the surface of one of Earth's oceans. Today, this film seems a simple and laughable series of static tableaux, but it remains a real pleasure.

About the same time in New York, the idea of a flight to the moon was combined with panoramas and movement to produce an "electroscenic mechanical illusion." A Trip to the Moon in the Luna Park of Coney Island in Brooklyn was a collectively experienced spectacle. The visitors entered the cigar-shaped and brightly illuminated ship *Luna IV,* which soon began to move back and forth, its large batlike wings moving up and down. Facades whizzing past in the background created the illusion that the ship was moving: from Manhattan to Niagara Falls and then up into space; wind-producing fans enhanced the impression of flight. When the goal of the trip came into view, the wings began to move faster. The ship had encountered a thunderstorm above the lunar surface and just missed landing in an extinct volcano. The visitors were asked to disembark and inspect the rocky landscape with its strange mushroomlike vegetation. Here they encountered midget moon men with rows of long spikes on their backs, who escorted them through stalactite-laden caverns and across chasms spanned by spiderweb-like bridges, all the while singing "My Sweetheart's the Man in the Moon." When they arrived at the fantastic palace of the Selenite king, moon maidens performed a dance for the visitors and offered them bites of green cheese. The guests left the attraction through the mouth of a moon calf and exited into the daylight.

The moon was no source of innocent amusement for Karl-August von Laffert, who foresaw a catastrophe of biblical dimensions in his novel *The Demise of the Luna* (1921): the moon falls to Earth, creating an enormous tidal wave and nearly eradicating humanity. In a time when Europe and Asia have combined to form a kingdom with Eurasiapolis as its capital, the inhabitants have for years awaited a lunar cataclysm, a calamity understood as an inevitable fate. Because of previous volcanic activity, houses on Earth are constructed on a system of clips or are hung from elastic straps for safety. A system of astronomical and meteorological stations has been established across the planet, including one in the Himalayas. After shedding its "icy shell," the moon becomes red, because its metallic core has been exposed, and a second Great Flood ensues. Attempts to transport part of humanity to another planet fail because Earth is the only hospitable place in the solar system to receive "the divine fruit of human seed." All the resources of science and technology are devoted to saving humans, plants, and animals, "to bring them over to the imminent alluvium." A rescue ship has been built—a sort of Noah's Ark where animals have been selected by natu-

ralists—as well as a num-
ber of large airplanes for
selected humans. Earth
and moon finally come
together at "flat, almost
tangential paths."

Compared with Laf-
fert's apocalyptic vision,
many of the lunar fic-
tions of the era's movies seem restrained. Most
focused on the idea of humans traveling to the
moon rather than the moon coming to Earth. The
short film *A Trip to the Moon* almost has the feel
of a fairy tale. The German filmmaker Fritz Lang,
though, who had set the standard for dystopic
futurism with *Metropolis* (1927), was not known
for restraint. His 1929 film *Frau im Mond* (translated literally in Britain as *Woman in the Moon,* but
released in the United States as *By Rocket to the
Moon*), based on a novel by his wife at the time,
Thea von Harbou (1888–1954), combined melo-
drama with scientific speculation.

Wolf Helius and his assistant Hans Windegger,
enthusiasts of space travel, team up with a Professor Manfeldt, who expects to find water, oxygen,
and gold on the moon's far side. Another member of the group is Friede, the former girlfriend
of Wolf and now the fiancée of Hans. Helius has
been blackmailed into including on the trip the
gangster Walt Turner, who has stolen precious
material from the scientist. If Helius doesn't cooperate, Turner will sabotage his rocket. Finally,
after taking off from a water basin, they discover

Scene from Fritz
Lang's film *By
Rocket to the Moon*

a sixth passenger, a young stowaway named Gustav. When gold is discovered on the moon, Manfeldt and Turner fight, and the professor falls to his death in a crevice. Turner is fatally shot trying to hijack the rocket. When it is time to leave, a shortage of oxygen on the ship condemns one passenger to stay behind. Windegger draws the short straw, but Helius chooses to let him return in the company of his beloved Friede. Upon takeoff, with the boy assuming the role of the pilot, it becomes clear that both Friede and Windegger have remained on the moon, where, in the final shot, they embrace passionately.

Albert Einstein was among the guests at the premiere of Lang's last silent film, which the press celebrated as "a miracle becoming reality." That judgment may not have held up over the years, but in one important respect, Lang's science fiction inspired reality. The film popularized the launch "countdown." One title card read, "20 Seconds to go—lie still—Take a deep breath," and ever-smaller numbers built suspense toward the climactic liftoff.

Forty wagons of sand were brought in from the shore of the Baltic Sea for the artificial moonscape created for the film in the legendary film studios of Babelsberg near Potsdam and Berlin. It's still surprising to see the similarities between Lang's portrayal and the first flight to the moon forty years later: the appearance of the rocket, its flight, even the separation of the capsule. One of the most fascinating similarities is evident when the spaceship orbits the moon. As the lunar surface passes by, we are aware of the extreme curvature of the lunar globe, so much smaller that Earth. The film transcended its medium, popularizing the idea of rocket science, an effect that cannot be underestimated. Indeed, the shots of the rocket in Lang's film were so convincing that shortly after the Nazis came to power in Germany, they banned the film as a security threat to the rocket development program.

Encouraged by the inspiration of science fiction novels and movies, the fantastic idea of colonizing the moon began to be taken seriously. The New York World's Fair of 1939 and 1940, attended by forty-five million people, presented human flight as the culmination of human technological achievement. The fair's World of Tomorrow simulated an American landscape of 1960 as seen from an airplane, and the Rocketport of the Future imagined a trip from New York to London that would take only one hour. Only civilian applications of future flight technology were heralded at the fair, but its military uses were soon to become reality. Spatial and temporal limits were suspended as never before in the exhibits, but the moon was still a distant vision. By 1962, though, at the Seattle World's Fair, NASA had an exhibit on rocketry, satellites, and human space flight.

The picture began to change soon after World War II, when space began to be promoted as a strategically important issue in the new nuclear age. As early as September 1946, a year after the explosion of the atomic bombs in Japan, *Collier's* magazine ran a story by George Edward Pendray about the moon's colonization. Without providing supporting evidence, he warned that

> rockets only a little faster than the German V-2s could bombard the Earth from the moon. With the aid of a suitable guiding device, such rockets could hit any city on the globe with devastating effect. A return attack from the Earth would require rockets many times more powerful to carry the same pay load of destruction; and they would, moreover, have to be launched under much more adverse conditions for hitting a small target, such as the moon colony. So far as sovereign power is concerned, therefore, the moon's control in the interplanetary world of the atomic future could mean military control of our whole portion of the solar system.

Even though the factual basis for Pendray's thesis was feeble, many readers probably accepted that the moon would be an ideal fortress and military base in a nuclear war. In an October 1948 article, *Collier's* carried the argument further by propagating a "Rocket Blitz from the Moon." Rockets were shown rising out of lunar craters and resulting nuclear fireballs spreading across New York City. Many doomsday scenarios devised during the postwar era had their origins in the sky; UFOs sightings became common at about the same time.

We have to remember that a flight to the moon was not yet an event the general public could envision clearly. The interplanetary epic *Destination Moon* by George Pal (1950) was the first American science fiction movie to take up the topic seriously, staging the trip to the moon as "the greatest challenge ever heard in American industry." The film is not particularly suspenseful, and its dialogue sounds rather stiff to contemporary ears, but its rational treatment of its subject matter was revolutionary. *Destination Moon* includes an inventive comic-strip sequence dismissing the "comic book stuff" characteristic of most depictions of moon flight, and provides a primer on the science of rockets, the effects of gravitational pull between Earth and moon, and the logistics of getting back "from this piece of cheese." During the flight the astronauts experience severe pain, their

First men
on the moon
in the movie
Destination Moon

faces are distorted, and they suffer from space sickness. The film also serves as a document of the Cold War. Spurred by fear that other forces plan to go to the moon, and that the United States might be attacked from outer space, the "race" is on. The spacecraft is financed and manufactured by U.S. private industry, before the U. S. government buys the technology.

A few years after its more hysterical ventures into lunar speculation, *Collier's* began in 1952 a series of articles to advocate for space travel as an inevitable human endeavor. The spectacular and influential series predicted a successful manned expedition to the moon within twenty-five years. Wernher von Braun (1912–1977) theorized that three rockets would be necessary to reach space, and that the crew would number no fewer than fifty. The designated vehicle would be clumsy, weigh fourteen million pounds, and include various tanks and such elements as thermal protection shields and a base frame for the landing, but a streamlined rocket wouldn't be required for moving through airless space. The *Collier's* articles also made a passionate plea for active science once we arrived on the moon. Artificially triggered moonquakes, for example, might reveal insights about its inner composition.

When lunarians made it into fiction at all at this time, the portrayal was even more ironic than in the nineteenth century, when at least some serious scientists still clung to the possibility of lunar life. A character in Kenneth Heuer's *Men of Other Planets* (1950) posits that "the really fash-

Collier's

OCTOBER 18, 1952 • FIFTEEN CENTS

MAN ON THE **MOON**

Scientists Tell How
We Can Land There
In Our Lifetime

ionable people on the Moon . . . probably live on the side where they can see the Earth." Heuer's explorers make a daring expedition and find that the far side is covered by forest, whose floor comprises a deep layer of ash. The inhabitants are vile. They appear to be digging holes to plant a garden, but

> they set nothing in the holes; and having covered them with sticks and ashes, they go a little distance off and hide behind the bushes, watching; and we notice that as each walks he puts his foot down care-

A series of articles published in *Collier's* magazine between March 1952 and April 1954 introduced Americans to the possibilities of a trip to the moon.

fully, constantly on the alert as to where he treads. We should like to go and work with these men; and not looking where we step, we sink suddenly into one of the holes. And eyes gleam from behind the bushes, and we hear the men laugh in the dark.

An equally wry portrayal of lunar life is Arthur Hilton's *Cat-Women of the Moon* (1953), a low-budget science fiction film originally released in 3D, about an expedition to the moon that finds a group of women in a lunar cave, where some breathable atmosphere remains. They are the last remnants of a human civilization on the moon, and they haven't seen any men for centuries. Grandiosely promoted as "the most startling picture of the century," the film revolves around the ultimately failed attempt of the cat-women to steal the spaceship and return to Earth.

Even our amusement parks took a more sensible stance by this time. Disneyland enlisted the cooperation of Trans World Airlines for its Rocket to the Moon experience of 1955. Von Braun himself contributed advice the attraction's design. After entering the Disneyland "space port" and viewing a fifteen-minute briefing film on rocket flight, 102 "space travelers" entered a theater with seats arranged in circular tiers. The ship's progress toward the moon could be tracked through "space scanners" in the floor and on the ceiling, while "sound effects and vibrations gave the 'feel' of an actual aerial journey." First the southern California coastline passed by, and then, after the ship

Kenneth Heuer imagined a "black and mystical forest" on the far side of the moon, with a vile race of beings lurking in the bushes.

Brainse ríomhghlaise
Finglas Library
Tel: (01) 834 4906

Camp on the moon

breached the sound barrier, the noise abated. TWA's Captain Collins informed the passengers that they were now 119,000 miles above the Earth. "To prevent the rocket from hurtling indefinitely into space, the captain must brake the tremendous speed, accomplishing this maneuver by an end-over-end flip so that the rockets are in a position for use as brakes." A sister spaceship, already returning to Earth, could be observed performing the exact same maneuver. The highlight of the trip was a close-up view of the lunar surface, where the captain pointed out the major mountains and valleys. To illuminate the "dark" side, some flares were dropped. Following another end-over-end flip, the "rocket" returned to Earth—"a half-million mile journey accomplished in a quarter of an hour, a preview of the world of the future in Tomorrowland." Walt Disney also produced two related television programs: *Man in Space* and *Man and the Moon.* Both aired in 1955, and Disney himself introduced the programs.

In the 1950s the idea of permanent human settlements on the moon was still an implausible fantasy, but writers devoted considerable imaginative effort to creating such fictive colonies based on the latest scientific developments. In *The Exploration of the Moon* (1954) Arthur C. Clarke made the establishment of a permanent and eventually self-supporting lunar settlement contingent upon the development of nuclear power sources. Just a year before the first actual moon landing took place, Clarke, eccentric as usual, speculated in *The Promise of Space* that the entire moon could become habitable. Because "the most daring prophecies turn out to be laughably conservative," he was convinced that "what the human race will do with the moon during the centuries to come may be as far beyond imagination as the future of the American continent would have been to Columbus." He saw in the moon "a virtual Rosetta Stone that, if properly read, may permit us to learn how the solar system, the Earth and the continents on which we live were formed." Clarke also conjectured that by the year 2020 a trip to the moon would be a distinct possibility for anyone who so desired — "perhaps to see grandchildren who, having been born under lunar gravity, can never come to Earth and have no particular desire to do so. To them it may seem a noisy, crowded, dangerous, and, above all, dirty place." He also imagined "committees of earnest citizens" 150 years later "fighting tooth and nail to save the last unspoiled vestiges of the lunar wilderness." He may yet be proven right on this last point.

Plans to build structures on the moon were by no means limited to the United States. In the second half of the 1950s, besides pursuing their project of interplanetary flights, the Russians declared their intent to build lunar cities with artificial atmospheres to make human life possible. The first so-called Red City was to be erected in a lunar crater under a dome of glass. Sophisticated aluminum doors would form an airlock, and the interior of the city would be further compartmentalized by glass walls with double doors to minimize the possible damage resulting from falling meteorites. These Soviet lunar settlements were expected to be self-sufficient, including locally grown vegetation — which, due to the difference in gravity, could be expected to look dramatically different: "A radish will be as tall as a date palm grows on Earth," declared Nikolai A. Varvarov, the head of the astronautics section of the Russian civil defense organization. "Onions

will send forth sprouts 33 feet long." Since Mars and Venus were envisaged as the next targets, and the moon considered a mere way station, the settlements were pictured as bases for the manufacture of spaceships and the production of fuel. Anatoly Blagonravov from the Soviet Academy of Sciences maintained that "the hoisting of any national flag on the Moon should not be the basic goal." The projected time frame was fifty years, when the moon would have effectively become the seventh continent of planet Earth.

The science journalist Fred Warshofsky chronicled some bizarre speculation about lunar possibilities in *The 21st Century: The New Age of Exploration* (1969). One was the cultivation of algae on human fecal waste under intense sunlight; the algae would be fed to chickens, which would then be consumed by the humans. The moon's reduced gravity would allow fruits and vegetables to grow faster and larger than on Earth. But in a near-perfect vacuum, how could

An ironic illustration of a Kremlin-like structure built under a glass dome on the moon accompanied a 1958 article in *Science Digest*.

a structure filled with breathable air be insulated against leaks? Engineers at the General Electric Company envisioned gouging out a chamber under the surface with nuclear explosives. After decontamination, the cavity would be lined with a strong plastic membrane, then filled with air. This plan was based on blast experiments with pumice rock, which was assumed to resemble the rock of the lunar surface. The astrophysicist Fritz Zwicky (1898–1974) of the California Institute of Technology suggested a way of creating an atmosphere on the moon's surface. First, the moon's mass would have to be increased, either by supplemental waste materials from Earth or by nuclear explosions that would fuse parts of the moon, flatten the lunar mountains, and cut its diameter in half with no lost matter. The gravity would subsequently increase with the smaller surface area so that carbon dioxide could be held in the atmosphere. At another point, Zwicky counseled "to shoot the moon" and to nudge the sun to then observe the consequences from Earth—this was all part of what he called "experimental astronomy," a revolutionary response to humans' history as passive observers.

But secret military scenarios were also being hatched. At the height of the Cold War, in 1958, the U.S. Air Force explored a plan to trigger a nuclear explosion on the moon at least as strong as the one used at Hiroshima. The point of the proposal, made public only decades later, was simple: to display military strength with an atomic mushroom large enough to be visible from the Earth. Leonard Reiffel, the physicist in charge, explained that the bomb was to explode on the lunar edge, so that the ascending cloud would be illuminated by the sun. Reiffel hired the American scientist and astronomer Carl Sagan to calculate how the cloud would expand in space around the moon. The possibility that the explosion might interfere with future scientific research on the moon didn't concern the air force; the public relations triumph justified the means. Although some details of this project ("A119" or "A Study of Lunar Research Flights") remain obscure, with some documents still classified, Reiffel confirmed in 2000 that it was "certainly technically feasible" and could have been accomplished with an intercontinental ballistic missile. Upon the disclosure of the secret plan, David Lowry, a British consultant and nuclear historian, commented, "It is obscene. To think that the first contact human beings would have had with another world would have been to explode a nuclear bomb. Had they gone ahead, we would probably never have had the romantic image of Neil Armstrong taking 'one giant leap for mankind.'"

Scenarios such as this one remind us how thin the line between fact and fiction has always been in the realm of space—thin enough that science fiction could sometimes turn into science fact. But fictional life on the moon often drifted into the realm of the grotesque. The American novelist Robert A. Heinlein wrote *The Moon Is a Harsh Mistress* (1966) when the Apollo program was in full swing, but his moon is not a peaceful one. Instead, it is a forced-labor penal colony in revolt against Earth in a fight for independence. Lunar society of 2075 is characterized in ways that would surprise the most experienced cultural anthropologist: by a form of group marriage and extended families; murder is not prosecuted; insults to women are punishable by death. The dialect of the "Loonies" is a sort of abbreviated English with a strong influence of Russian grammar. Earth rapaciously extracts resources from the moon, where famine and collapse of the colony are imminent. The rebellious Loonies ultimately force Earth to provide recompense for their grain production.

Any attempt to chart the moon in recent imagination has to include Stanley Kubrick's *2001: A Space Odyssey* (1968), which was based on a screenplay by Arthur C. Clarke and premiered in movie theaters just fifteen months before the first moon landing. The film is a speculative reflection on extraterrestrial life and the revolutionary influence of space technology on the future of mankind. It incorporates groundbreaking special effects that continue to amaze viewers. Some critics interpreted it as a critique of modern technology, while others read it as a vehicle for providing a spiritual dimension to the space euphoria of the time. The story revolves around the discovery of a monolith buried below the moon's surface, which, it is assumed, was created by an unknown civilization at some distant point in history.

The late twentieth century brought no return to the lunar utopias of a more optimistic age. In the novel *In Peace on Earth* (1987), by the Polish author and science skeptic Stanisław Lem (1921–2006), the moon fulfills the role of a strange repository. All weapons are transported to the moon to prevent human self-destruction on Earth. But these weapons have the ability to multiply by using the energy of the sun and exploiting the lunar soil.

A lunar mining station is the setting of *Moon* (2008), by the British filmmaker Duncan Jones. The protagonist is an astronaut near the end of a three-year contract supervising the extraction of helium-3, which has become the primary source of energy on Earth. His mission is purely commercial, with no scientific considerations—the astronaut's task is to launch rocket-propelled canisters filled with helium-3 to Earth. The film gives a sense of the alienation experienced in a hostile environment, and the eerie dislocation that might accompany long periods of isolation on the moon. The film's echoes of *2001* and its 1960s-influenced concept of the future give it a retro feel, but its grimy surfaces suggest an even darker flip side to Kubrick's vision.

CHAPTER THIRTEEN

Before and After Apollo

History often rewards great breakthroughs but ignores the preparatory steps that made those achievements possible. The Apollo program, for instance, has been documented in great detail and still receives ample attention, but what of the extraordinary labors that led to that summit? How was flight to the moon realized in practical terms after Jules Verne & Co. designed the blueprints?

When President John F. Kennedy approved the program soon after taking office, he proclaimed that humans would be on the moon before the close of the 1960s. As a result the technicians of the NASA saw their budget suddenly increased tenfold and were able to draw on a tremendous body of technological knowledge. The American public had already embraced the supposedly infinite possibilities of the Space Age. Of course, technological progress doesn't just happen. In an earlier era, the railroad had promised to diminish travel time like no other means of transport had done before. It became a metaphor for the changing relationship of humans to the material world—for the domination of the landscape by technology. In fact, at a time when fast movement was uncommon, travelers found rail travel to be a mind-altering experience, and often a con-

Chandra, the Indian goddess of the moon

229

fusing one. The railroad became an essential asset in the American Civil War, and its influence was still critical to rapid delivery of weapons and matériel during the First World War. Airplanes were quickly put to military use, too, and soon proved highly effective as weapons. In fact, military concerns drove much technological progress in the nineteenth and twentieth centuries.

In the human imagination the way to reach the moon is simply to fly there, but from a purely technical perspective, moon "flight" converged with the history of rocketry at the end of the nineteenth century, when it became clear that rocket propulsion would be needed to transport humans into space. Technology then evolved more or less independently in Russia, the United States, and Germany. Given the backwardness of czarist Russia—an atmosphere that hindered experimentation—it is all the more surprising that Konstantin Eduardovich Tsiolkovsky (1857–1935), a teacher of mathematics, developed what is now seen as the basis of theoretical astronautics. The American Robert Goddard (1882–1945), the Frenchman Robert Esnault-Pelterie (1881–1951), and the German Romanian Hermann Oberth (1894–1989) also independently worked on effective rocket systems.

Tomorrow's geniuses are often today's lunatics, and initially, such endeavors were regarded as esoteric. When Oberth wrote a dissertation on rockets in space and submitted it to the University of Heidelberg, the professors initially rejected it. But by 1923 his thesis, "By Rocket into Planetary Space," inspired wide interest and encouraged further research on the topic of space flight. Oberth was vindicated.

The space program was a pastiche of new and established technologies. As Walter A. McDougall has asserted, it combined "four great inventions: Britain's radar, Germany's ballistic rocket, and the United States' electronic computer and atomic bomb," each "the product of humankind's most destructive conflict—World War II." Immediately after the war, before the start of the space race, these discoveries were applied to the development of intercontinental missiles. The German V-2 rocket program had already established that flight outside the atmosphere was possible. Given the American advance in bombers, the Soviet Union felt particularly compelled to compensate for its lag in warhead technology with a rocket force.

Research on the moon and the idea of flying there attracted people who combined visionary and technical inclinations. Wernher von Braun was a key figure in this effort. From his background as the chief rocket engineer of the Third Reich and a member of the SS, he ultimately became head of launch-vehicle development for the Apollo moon program—an odyssey requiring not only exceptional skills and intelligence but also an opportunistic grasp of diplomacy. As his biographer Michael J. Neufeld has put it, few individuals have "shaken the hand of Eisenhower, Kennedy, Johnson and Nixon—but also Hitler, Himmler, Göring and Goebbels." It

Wernher von
Braun, right,
on the shooting
range of the
German Rocket
Society in 1930

is known that the German underground rocket
factory that produced the V-2 rockets—called
Vengeance Weapon 2 by the Nazi Propaganda
Ministry—used slave laborers from a nearby
concentration camp. It is also known that von
Braun was involved in the facility's planning and
operation. He clearly benefited from war crimes
and bears at least moral if not legal guilt. Even
under the most adverse circumstances at the
end of a lost war, von Braun managed to control the outcome of events. Realizing that rocket
research could best be continued in the United
States rather than in Great Britain or Russia, he
allowed himself to be captured by the Americans
and brought more than one hundred key members of his team along. He was the world's most
experienced rocket engineer and soon developed
a broader vision of man's venture into space. He

suffered a setback in 1954, while working on the advanced rockets program in Huntsville, Alabama, when he was denied the necessary financial support to launch a satellite to ensure that the United States would be the first country in space, but the success of the Soviet Union's Sputnik program boosted his promotion of space exploration.

Humans would not have set foot on the moon so early without the fierce antagonism and competition between the two superpowers during the Cold War. Without the political will and, as Roger D. Launius has put it, "the desire to demonstrate the technological superiority of one form of government over another," the Apollo program would not have been possible. After the USSR launched the *Sputnik 1* satellite in October 1957 and, a month later, *Sputnik 2* with its dog passenger, Laika, the space race gained momentum. Suddenly, Americans found themselves in second place. Because the Soviet Union was known to be working to develop nuclear weapons, its small orbiting orb was seen as a potential military threat to targets inside the United States. The extreme secrecy surrounding the Soviet space program—neither the launch site (the Baikonur Cosmodrome, in the desert of what is now the Republic of Kazakhstan) nor the name of the chief designer of the rockets (Sergei Korolev) was known at that time—certainly contributed to the sense of urgency of the American efforts. This perception translated to a massive infusion of public money into the space race. The Apollo program

A reunion of pioneers. Wernher von Braun is second from right and Hermann Oberth in the foreground.

cost approximately $25 billion and was driven by Cold War politics rather than important scientific goals. It became, as Launius reminds us, "the largest nonmilitary technological endeavor ever undertaken by the United States."

It is easy to forget that President Kennedy not only failed to define a clear purpose for the program, but also, recognizing the vast costs of the effort, repeatedly tried to persuade Soviet Premier Nikita Krushchev to pursue a joint expedition to the moon — to no avail. The Apollo program had critics from the beginning. According to polls taken in the year leading up the Apollo 11 launch, there was never a clear majority in favor of it. Some objections were based on the belief that "God never intended us to go into space," in the words of one response offered. This is a minority opinion in the population as a whole, but it reminds us of the metaphysical feelings the moon has always evoked. Should this eternal symbol in the sky be touched, its secrets unveiled? In fact, the Apollo program assumed various quasi-religious reverberations. Norman Mailer, for example, characterized the space capsule itself as a sacred object. He also speculated about the extent to which Nazi ideology might have been enmeshed in the space program, mediated through Wernher von Braun and other former German officers.

Так может выглядеть межпланетный скафандр.

Most objections, though, had nothing to do with metaphysics and everything to do with economics. As early as 1964, the sociologist Amitai Etzioni characterized the race to the moon as a "monumental misdecision" in his book *The Moondoggle*. The space program, Etzioni argued, produced neither major economic development nor a better understanding of the universe. "Some of the claims are safely projected into a remote and dateless future, others should have never been made; still others exaggerated out of proportion to their real value." All of scientific manpower devoted to space, he wrote, should instead be put into health care or education. "Above all the space race is used as an escape. By focusing on the Moon, we delay facing ourselves, as Americans and as citizens of the Earth." Etzioni later served as senior adviser to the White House under Jimmy Carter. For the eminent science historian and independent humanist thinker Lewis Mumford (1895–1990), the Apollo program was simply a waste of money, "an extravagant feat of technological exhibitionism." He likened the manned space capsule "to the innermost chambers of the great pyramids, where the mummified body of the Pharaoh, surrounded by the miniaturized equipment necessary for magical travel to Heaven, was placed." Still, the Apollo program

Russian cosmonaut, 1950s

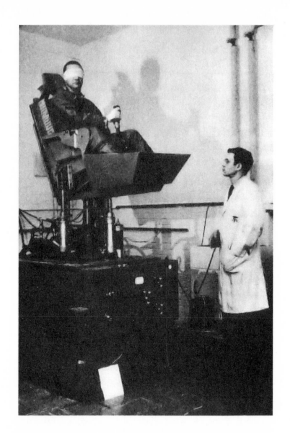

A stage of the aptitude test for astronauts of the Mercury program, the first spaceflight program of the United States, 1959–1963

received consistently favorable media coverage. In the wake of the 1968 assassinations of Martin Luther King, Jr., and Robert F. Kennedy, as well as ever more distressing reports from Vietnam, the push toward the moon acted as a counterbalance for the national psyche.

The technological wonders of yesterday rapidly become the banalities of today. The late 1950s and the 1960s held the promise of a world in which nuclear power would be the key to solving problems, replacing scarce energy resources, sharply reducing pollution, and even relieving poverty. The mere threat of nuclear annihilation would put an end to war. Hypersonic air travel would shrink our world while settlements on the moon would expand its capacity. Half a century later, the dreams that became reality have shown a nightmarish side. Implementation of nuclear power has been hamstrung by grave doubts about safety. The supersonic Concorde, a plane developed in 1962 with more than twice the speed of conventional aircraft, was eventually abandoned after a spectacular crash in 2000.

The live transmission of the shaky images of the astronauts' ghostlike steps and their metallic voices from the moon, which triggered a global wave of excitement at the time, has become part of our cultural memory, a relic of a century past. Perhaps the most significant lasting image was that captured when the cameras were turned toward home. For the first time, the entirety of our planet was no longer an abstraction. At last we perceived the Earth in context, not merely as our own specific location. Traveling to the moon gave us an unprecedented sense of our own uniqueness in space and of the limits of the oasis we inhabit. Even as we continued to reach farther outward beyond our world, much of our imagination was turned inward.

Well into the nineteenth century, the moon, understood as a space of the imagination, had assumed a role similar—metaphorically, at least—to those of earlier unknown landscapes and continents. But conquest of the moon was driven by completely different motives than those of the Portuguese and Spanish seafarers in the fifteenth century.

Was the lunar mission really more than a historical accident? It cannot be simply dismissed as a propaganda coup, as an example of hubris, an American ego trip; those characterizations are not false, but they're no more than half-true. The program was not motivated by a desire for wealth, and innovations it inspired still benefit people who never thrilled to Neil Armstrong's "small step." The fuel cell, based on a controlled hydrogen-oxygen reaction, has various applications and may yet prove to be the most useful "clean" alternative to the internal-combustion engine. It was developed in part to power the life support system in a space capsule. Tiny diodes measuring the astronauts' pulse and blood pressure were forerunners of the medical telemetry equipment used today. Freeze-drying made it possible to preserve and to condense such foods as potatoes, peas, carrots, and minced meat, which could later be restored to their original forms with the help of water and a microwave. In addition to these specific developments, Apollo sparked a general interest in engineering, the results of which later benefited various industries. Some of the gadgets and activities associated with moon flight also found their way into popular culture. Michael Jackson's moonwalk was inspired by the astronauts' mode of moving on the lunar surface, and the Italian designer Giancarlo

Zanatto's moonboots would not have made sense without their real-life predecessors.

But to find the most ubiquitous consequence of Apollo-related research, most of us need look no farther than our desktops. The NASA program drove the miniaturization of information technology crucial to the development of modern computers. The Apollo Guidance Computer developed at MIT—a seventy-pound on-board device with a capacity less than that of a cell phone today—made the safe landing on the moon possible. As David A. Mindell writes in *Digital Apollo,* "Apollo began in a world when hardware and electronics were suspect and might fail anytime. It ended with the realization that as electronics became integrated, computers could become reliable, but that software held promise and peril. It could automate the flying, eliminate black boxes from the cramped cabin, and make the subtlest maneuvers seem simple. Yet it was hideously complex and difficult to manage. If it went wrong at a bad time, it could abort a mission or kill its users."

After 1968, about ninety thousand people registered for Pan American World Airways' First Moon Flights Club. Departure was scheduled for the year 2000; Ronald Reagan was one of the first to reserve a seat. The fare for the trip was projected to be fourteen thousand dollars. In retrospect, it is doubtful that a moon flight could have been provided at such a low price, but Pan Am went out of business in 1991 and never had to

Know All Ye by These Presents that:

Mr. _____

has become a certified member of Pan Am's

"FIRST MOON FLIGHTS" CLUB

Number _____   Vice President, Sales

James S Montgomery

Pan Am's lunar ambitions

fulfill its lunar obligation. A few years later, enthusiasm for space exploration had faded, superseded by more urgent cultural and political issues. America's manned lunar landing project ended in December 1972, when Apollo 17 ended its flight in the Pacific Ocean. In Washington, D.C., NASA officials, astronauts, scientists, and business managers celebrated with what the *Washington Post* called "the last splashdown party."

Depending on political perspectives, the story of the U.S. space program can be portrayed as a feel-good triumph or as a waste of money that should have gone to social programs. But there is a third, less common perspective that has stirred considerable emotion over the decades. According to this revisionist narrative, Apollo moon landings never happened at all, and what we saw was faked at a terrestrial staging area. According to Roger D. Launius, such theories began to be spun early on, "almost from the point of the first spaceflight missions." The explanation offered most commonly for the contrarian view is simple ignorance: for some people this technological accomplishment just didn't fit their worldview, and it was easier for them to imagine a complex hoax than to accept the challenge to their assumptions. For some moon-landing deniers, though, the issue was more complex. They had an almost messianic belief in a conspiracy and often would aggressively resist any discussion of it.

In 1974, at the height of the Watergate scandal, when trust in the American political system was at a low point, Bill Kaysing published *We Never Went to the Moon: America's Thirty Billion Dollar Swindle.* Since then, discussion has raged—often repeating the same arguments ad nauseam. At one point, the former Apollo astronaut Buzz Aldrin was so provoked by Bart Sibrel—who made two films claiming the landings were a fraud and labeled Aldrin "a coward, a liar, and a thief"—that he punched the much younger Sibrel in the face.

Conspiracy theorists disagree about the extent to which the moon landings were faked. A minority believes that the crew actually reached the moon, but that the images were faked to obfuscate the technical details of their journey. Others think that Stanley Kubrick, who directed *2001: A Space Odyssey,* was commissioned by NASA to produce some of the Apollo 11 and 12 footage. According to this scenario, a dummy was launched and allowed to splash down into the ocean. All the other images transmitted into hundreds of millions of homes are said to have been fake footage, produced on a quickly improvised movie set in some remote spot in the Nevada desert. That Kubrick hired former NASA employees for *2001* is, to conspiracy theorists, further evidence. The issue created quite a stir when the Fox television network aired *Conspiracy Theory: Did We Land on the Moon?* in 2001. This hourlong documentary gave a platform to several hoax advocates but offered very little refuting evidence.

Even though a number of independent sources have stressed that the sheer, overwhelming body of physical evidence proves that humans did walk on the moon, the minority view provides admittedly fascinating fodder for publishers, and the nicely packaged story is always guaranteed an audience. If we begin with an open mind, the arguments of the conspiracy theorists may initially seem plausible, but they don't hold up to rigorous scrutiny.

The arguments have been discussed extensively elsewhere, but a taste of the contrarian stance is still in order. Believers in a NASA conspiracy—the Apollo Simulation Project, as some call it—often cite the absence of stars in the jet-black sky photographed by the astronauts. They disregard the explanation that the exposure times were too short to have captured the faint stars. The unusual play of light and shadows in some of the photos, such as the "man on the moon" image of Buzz Aldrin, also draws the ire of conspiracy theorists. Why does it look like a spotlight is directed at him although the sun is clearly behind him or shining from the side? What the conspiracy theorists fail to consider is that the lunar surface has a tendency to reflect light back in the direction of the source. This phenomenon results in a specific glow: a halo or aureole. Often, the deniers are also skeptical about the astronauts' ability to survive exposure to radiation during the trip or the high temperatures on moon's surface. Although radiation and temperature were indeed threats to the astronauts, program scientists charged with astronaut safety clearly resolved the issues. The intensity of denial in the face of all evidence to the contrary reminds us that, for a number of people, the moon landings may have been traumatic. The intrusion of humans onto this numinous orb somehow violated their idea of the natural order. A trip to the moon had been a dream for centuries. Few events are more disturbing than a dream come true.

The issue of the denial also touches upon other, deeper questions. We live in a culture in which the deliberate blurring of boundaries between fact and fiction has achieved a certain artistic legitimacy in popular culture. And while many of us still remember the live images and sounds from the first moon landing, more than half of the world's population is too young to have such an intimate connection with this event. Though not remembering doesn't necessarily translate to denying, they may be less inclined to take its truth for granted.

For those who accept that men have been to the moon, the visits have raised a host of more mundane issues. For example, some people claim to own parts of the moon. But which property laws apply there? And who is authorized to define such laws in the first place? Who—if anyone—should have the right to change the moon's surface? In 1967 the United States, the United Kingdom, and the Soviet Union signed the Outer Space Treaty that declared the moon to be *terra nullius,* a world belonging to no one. About one hundred nations adhere to this agreement today. In 1979 this treaty was supplemented by a more comprehensive one. The Agreement Governing the Activities of States on the Moon and Other Celestial Bodies specifies that the moon should be used for the benefit of all states and peoples and not, for example, as a testing ground for military purposes. The "Moon Treaty" also precludes any state from claiming sovereignty over any territory of a celestial body. However, the treaty has never been ratified by any of the major space-faring powers and remains unsigned by most of them, so it carries no legal weight. Nor is it clear how environmental ethics for the moon can be established and put into practice.

Future projects involving the moon still face major challenges and require piecemeal technological solutions. Companies under contract to NASA are exploring lunar logistics, mining, and spacesuit design. An inflatable test habitat in Antarctica serves as a base for lunar research and a testing station for Mars exploration. Locations such as the Canadian Arctic, the Arizona desert, and the underwater Aquarius habitat serve as key proving grounds. Of course, these extreme terrestrial environments provide only distant approximations of the many challenges that we would confront on future trips to the moon.

The perils of the moon go well beyond the mere terrain. Even though the moon doesn't have the aggressive gases of Venus or radiation contamination of Jupiter, exposure to cosmic rays and solar wind and flares poses much more risk than on Earth, where the atmosphere and the magnetic field act as protective shields. The fine lunar dust, moreover, is highly abrasive and potentially harmful not only to the lungs of the astronauts and to their spacesuits but also to the joints and bearings of lunar robots—a problem that some propose solving by outfitting the robots with disposable coveralls. In any case, robotics will play a major role in lunar exploration, with lightweight vehicles or robotractors moving about the surface and performing tasks that include sampling the lunar soil and extracting any water or oxygen-rich minerals it might contain. Once the water is broken down into hydrogen and oxygen, it might be used to fuel rocket propellants and to create air for breathing.

Champions of the moon's commercial exploitation, such as the Arizona-based Lunar Research Institute, like to stress that even the bulk of the resources needed to build an industrial complex on the moon would not have to be taken there but could be mined on-site. Larry Clark, senior manager for Lockheed Martin's spacecraft technology development laboratory, calculates that processing just the top two inches of soil from an area half the size of a basketball court

A recent
visualization
for a lunar base,
realized by
architects from
the Technical
University of
Darmstadt,
Germany

could yield enough oxygen to keep four astronauts alive for seventy-five days. Silicon can be used to make cells to harvest solar energy, iron to build structures, aluminum, titanium, and magnesium for the construction of spacecraft, and carbon and nitrogen to help grow food. Given the moon's comparably low gravitational pull, transport back to the Earth would be much cheaper than the other way around.

The moon may be a place of infinite possibilities, and some of these projects are reminiscent of the enthusiasm of a gold rush. At the core of these dreams is helium-3, the light variant of a noble gas that is hardly present on Earth but appears to exist in large quantities on the moon. Although Earth's atmosphere and magnetic field shield it from helium-3 in space, the moon lacks these obstacles. Shipped to Earth, the moon's helium-3 could be processed with deuterium in fusion reactors to create helium-4, the energy source of the sun and stars, which could provide power without producing nuclear waste. But such fusion reactors are far from ready for operation; some experts claim that this technology is decades away from commercial viability.

One undeniable resource of the moon is an abundant supply of solar energy, which could be harnessed to power lunar bases or even as an energy source for Earth, thus freeing future generations from dependence on fossil and nuclear fuels. To make the moon's solar energy available on Earth, scientists would have to find a way to convert it into electricity and then into microwaves that could be beamed across space. It is impossible to gauge the effects of sending such concentrations of microwaves through the Earth's atmosphere. Paul D. Spudis, chief of the *Clementine* team (named for a spacecraft launched in 1994 to study the moon's surface by means of special cameras), considers the "Mountain of Eternal Light" near the moon's south pole to be "the most valuable piece of extraterrestrial real estate in the solar system." "The big advantage of this place on the Moon is that it allows you to survive the 14-day lunar night with solar power. You cannot do this on the equator. Moreover, it is close to deposits of excess hydrogen, allowing us to make water, air and rocket propellant. If the Moon is a desert, the poles are its oases," Spudis writes.

All such futuristic projects resonate uneasily

on an Earth that more than ever seems on the brink of ecological disaster. Working on an off-shore oil platform entails many privations, but who would want to spend a substantial part of his or her life on the moon? Who would choose the moon if he or she could have green landscapes, the ocean's shore, and the mountains? Are we really supposed to live somewhere other than Earth? In addition, living in an environment with less gravity than the Earth has severe side effects. A period of rehabilitation is now standard practice after long stays in space. How can detrimental consequences for the circulatory and musculoskeletal systems be limited and psychological and social well-being be maintained? At this point, there still are a lot of open questions.

With no spaceship in sight, the European Space Agency (ESA) can only dream about a manned space mission. Recognizing the immense cost involved, U.S. President Barack Obama has also put the brakes on moon euphoria. The Russians are not in a hurry, either—no cosmonaut will set foot on the moon before 2025. And China doesn't foresee trying to get there before 2030. But there are parts of the world where the notion of a lunar mission still provokes strong emotions.

A case in point is India. When an unmanned spacecraft called *Chandrayaan-1* (moon craft) went to the moon in October 2008, it was a matter of national pride and received extensive coverage in the Indian media. The main aims of this mission were to prepare a three-dimensional atlas in extremely high resolution of both the near and far sides, and to map the lunar surface for the distribution of various elements. For the first time, imaging radar was flown to the moon, resulting in valuable data on the lunar poles. Currently India is planning a manned mission for 2020.

Just before the launch of *Chandrayaan,* millions of Hindu women fasted until the moment when they could first discerned the moon's reflection in a bowl of oil—a rite that they usually perform to safeguard the welfare of their husbands. For many Indians no real contradiction seems to exist between their ancient beliefs and the contemporary scientific pursuits in space. The Indian Space Research Organisation carried the following verse from the *Rig Veda:*

O Moon!
We should be able to know you through
 our intellect,
You enlighten us through the right path.

Epilogue

Moon Melancholia

From humankind's beginnings, the moon was the mysterious light in the night sky, ascribed sundry magical powers and used to structure time. The telescope changed our view, literally and figuratively. Science asserted the moon's status as a satellite of the Earth, and imagination drifted to speculation about whether anyone—or anything—might live there. Following the Age of Discovery—when indigenous people of the New World were displaced or killed, game was slaughtered, and forests were razed—there was hardly anywhere left on Earth where a utopian society could plausibly have been settled. The moon offered a relatively likely venue for utopia. But it turned out to be lifeless—a blank slate for our dreams, and for our depravity.

In essence, humankind's premodern lunar fantasies were part of the nascent desire to explore outer space, the first toddling steps of the space race. But one of the many consequences of the Copernican revolution was that the moon lost its preeminence in the topography of the sky. No longer considered a planet, it became more remote than ever from its ancient position of honor as one of the seven bodies circling the im-

Lunar contemplation

251

mobile Earth. Displaced, disenchanted, debased, the moon lost its status. Poets probably sensed first the tremendous change this implied and began to imagine the moon as a shadow world intimately related to Earth, but following another, ultimately unintelligible logic. The speaker in *The Nighttime Chant of a Wandering Asian Sheep-Herder* (1830) by Giacomo Leopardi directs questions at the moon: "What do you do Moon in the sky? Tell me, what do you do, / silent Moon?" His questions remain unanswered. This silence suggests that although the moon is still regarded as a companion and an object of desire, its significance is no longer self-explanatory. The moon has lost its "voice." But even though the moon is unable to answer the restless shepherd's pressing questions, a sense of intimate connection remains, reminiscent of the time when the moon was an important symbol of our connection with the cosmos.

Our most clearly visible neighbor in space, the moon eventually made plausible the possibilities of space travel. With scientific inquiry taking ever-new directions, new objectives emerge. The more complicated life on Earth becomes, the greater the temptation to long for a place where earthly problems do not exist. But perhaps the moon, as the target of some of our hopeful projections, is not all it's cracked up to be. Our future quests may take us to the most unexpected of places. Cultural, spiritual, or scientific progress or renewal may not even be defined in spatial terms, as a movement from one place to another.

In some ways, the interior of our own planet is more remote than the moon, and there is no inherent reason why the future of mankind should be in outer space or why the popular imagination should target some far-away destination in space as its preferred goal. Many scientists are working on uncovering Earth's deep secrets, and the ocean depths still hold many mysteries as well. Many of science's most pressing challenges for the advancement of human knowledge—or even our survival—may be found around the corner rather than in space. If so, will our impulse to use the moon as the screen onto which we project our hopes and fears fade as well?

Over time, our view of which historical events really matter changes. The veteran jet propulsion engineer Wernher von Braun once compared the landing on the moon with the distant moment in natural history when the first animals left the water to find their way on land. The former U.S. president Richard Nixon said that the flight of Apollo 11 represented the most significant week in the history of Earth since the creation. What a difference time makes. Only a few years later, these pronouncements already sounded somewhat inflated, and from the perspective of the twenty-first century, they seem almost absurd overstatements. Boys no longer list "astronaut" as their first choice of profession, and it has become something of a cliché to say that a trip to the moon is no longer an "adventure." The late Carl Sagan even dared call the moon "boring."

In a not-too-distant future, humanity may re-

A comparison of moon and Earth at diminished distance

member the twentieth century for the nuclear bomb, industrialized genocide, the rising awareness of global warming, the Internet, the achievement of powered flight, and, yes, the first steps into space—but hardly recall the twelve humans who actually ventured to the moon. In an effort to anchor Apollo in memory, in early 2010 a collection of 106 objects left by the Apollo 11 mission—including the bottom stage of the lunar lander, a seismic monitor, the American flag, and, ironically, even bags of human waste—were placed on California's registry of historic landmarks and resources which makes it the first cultural resource listed not located on Earth.

In our age of moral relativism and multiculturalism it has become difficult to speak of eternal truths and universals, but surely the moon remains an inalienable emblem of the human imagination—even "in the year 2525," to quote the lyrics of the 1969 song by Zager and Evans. There are many moons, but Earth's moon remains something special. But to say it still has meaning for us is not to say that it can have an unchanging meaning. Its significance and roles have always varied across cultures and eras—from heavenly god to symbolic guardian or judge, to the scene or stage of spectacular visions and visits, to being "just" an object of scientific investigation.

Humans have never had to travel to the moon to discover it, or to appreciate its fascination. Why long for the moon if we can see it from the Earth? Does walking on the moon guarantee ultimate access to its secrets? Is it not possible that such physical familiarity may make it impossible to understand those secrets? Technology and water may someday make the moon a habitable place for some humans, but it will—for our lifetimes, at least—remain hostile to life. For the vast majority of us, a trip to the moon will be possible only from an armchair. Maybe it's better that way. Just as we were given a new and unique perspective on Earth from space, perhaps we can grasp the moon only by keeping our distance.

Our future moon of the mind may hold many surprises in store. Unless some unforeseen catastrophe alters its surface or changes the spin or position in space, or our view of it becomes obstructed by some impenetrable layer here on Earth, the moon will continue to *look* the same. But the frameworks from which we see it will inevitably be different; its symbolism is apt to change further as well. Consider, for example, that many scientists consider the moon as a part of the terrestrial system, based on the extensive systemic knowledge we possess about Earth and moon. Thus some observers speak not of the

Earth and its moon but rather of a dual planetary system that revolves around the sun, with the center of mass about three thousand miles from Earth's geographical center. If we accept this view, then the moon becomes essentially another continent.

Perhaps the interpretive frameworks of future generations will be religious, perhaps not, but our relationship with the moon will continue to reflect how we understand our own planet and our place in the universe. Today the moon is not an enigma anymore, but maybe the most surprising fact is that—despite all the knowledge we have about it—the moon is still able to amaze us and create a sense of fascination hardly matched by anything else. *Luna Luna in the Sky. Will You Make Me Laugh or Cry?* read the legend on a T-shirt created by the artist Keith Haring (1958–1990). Haring's question articulates the two faces of the moon in our quest to discover what it means to us. Maybe we should try sometimes to un-think our scientific knowledge of the moon. Just as leaving the city and freeing ourselves from pervasive lighting helps us enjoy the privilege of *seeing* the moon again, putting aside our scientific concepts may help us *perceive* it. Perhaps we can learn from José Arcadio Buendia, the memorable character in Gabriel García Márquez's novel *A Hundred Years of Solitude* who locks himself up with his astronomical instruments for several months to observe the sky, fantasizing about imaginary excursions across unknown oceans. The fictional Buendia deserves a special place among more contemporary stargazers. We may not want to pursue his path with the same degree of tenacity, nor become lost as he did, but it's reassuring to remember that looking up at the moon and stars doesn't have to be purely

"The oldest television"

a matter of astrophysics and mathematics — that a real flight of the imagination may still be possible, even in an age saturated by technology. A final suggestion: the Korean video artist Nam June Paik (1932–2006) called a sculpture he completed in 1967 *Moon Is the Oldest Television."* Why not turn off the box and venture into the night's sky? For while the history of the moon reveals a wealth about our past, maybe even more profound lessons about our present and future await us in its face.

Bibliographic Essay

Titles are mentioned on a selective basis, with focus on references in the English language.

Introduction

For further ideas about life on Earth without the moon see Neil F. Comins, *What if the Moon Didn't Exist? Voyages to Earths That Might Have Been* (New York: HarperCollins, 1993). The "Rare-Earth Hypothesis," positing that the development of life on Earth is the result of improbable circumstances, is also interesting in this context; see Peter Ward and Donald Brownlee, *Rare Earth: Why Complex Life Is Uncommon in the Universe* (Berlin: Springer, 2000).

For a collection on issues related to night in the contemporary world see Paul Bogard, ed., *Let There Be Night: Testimony on Behalf of the Dark* (Reno: University of Nevada Press, 2008).

The International Dark-Sky Association tries to raise awareness of "light pollution." See www.darksky.org.

Chapter One. Gazing at the Moon

A very good resource for many issues concerning the observation of the moon is Patrick Moore, *Patrick Moore on the Moon* (London: Cassell, 2001).

For a more detailed discussion of the question of multiple moons circling Earth see Michael E. Bakich, *The Cambridge Planetary Handbook* (Cambridge: Cambridge University Press, 2000), 145 ff.

The reference for the Shona is taken from William M. Clements, ed., *The Greenwood Encyclopedia of World Folklore and Folklife* (Westport, Conn.: Greenwood, 2005).

For a comprehensive treatment of solar eclipses see J. P. McEvoy, *Eclipse: The Science and History of Earth's Most Spectacular Phenomenon* (London: Fourth Estate, 1999).

The description of the solar eclipse in India is taken from www.krysstal.com/eclipses.html. At the time of writing this Web site also displayed many photos from various solar eclipses.

Camille Flammarion's quotations are taken from *Popular Astronomy: A General Description of the Heavens* (London: Chatto and Windus, 1894). Flammarion is also the source of my quotation of François Arago.

For a general discussion of life outside the Earth in historical perspective see Michael J. Crowe, *The Extraterrestrial Life Debate, 1750–1900: The Idea of a Plurality of Worlds from Kant to Lowell* (Cambridge: Cambridge University Press, 1986).

Richard Holmes, *The Age of Wonder: How the Romantic Generation Discovered the Beauty and Terror of Science* (London: Harper, 2008), has a chapter with biographical information on Sir William Herschel, who started out as an organist in Hannover, Germany, and later became an astronomer who built several hundred telescopes.

Valdemar Axel Firsoff, *Strange World of the Moon: An Inquiry into Its Physical Features and the Possibility of Life* (New York: Basic, 1960), is the most recent book I found in my research that still evaluated the possibility of life on the moon. The perspective of Firsoff, an amateur astronomer, was marginal.

Chapter Two. Moon of the Mind

The role of the moon, the sun, and stars in ancient cultures fills volumes. A comprehensive overview of the historical relationship of humans toward the cosmos can be found in:

John North, *Cosmos: An Illustrated History of Astronomy and Cosmology* (Chicago: University of Chicago Press, 2008); Edwin C. Krupp, *Beyond the Blue Horizon: Myths and Legends of the Sun, Moon, Stars, and Planets* (New York: HarperCollins, 1991); and Anthony Aveni, *People and the Sky: Our Ancestors and the Cosmos* (London: Thames and Hudson, 2008). Aveni's older book *Empires of Time: Calendars, Clocks, and Cultures* (New York: Basic, 1989), is also relevant for some of the issues dealt with in this chapter.

Peter Watson's massive *Ideas: A History of Thought and Invention from Fire to Freud* (London: Weidenfeld and Nicolson, 2005) deals with some of the aspects of this chapter in the broader context of a more general history of human ideas and inventions.

The examples for lunar myths of the Maoris, the Tupí, and the Tartars, as well as the various mythical tropes, are taken from Jules Cashford, *The Moon: Myth and Image* (New York: Four Walls Eight Windows, 2003).

On the cult of the moon in Lusitania see Moisés Espírito Santo, *Cinco mil anos de cultura a oeste. Etno-história da religião popular numa região da Estremadura* (Lisbon: Assírio and Alvim, 2004).

For more scholarly information on the significance of the sun, the moon, and the stars in ancient India refer to Georg Feuerstein, Shubhash Kak, and David Frawley, *In Search of the Cradle of Civilization: New Light on Ancient India* (Wheaton: Quest, 2001).

The reference to Roger Bacon's measurement of the distance between Earth and moon is taken from Albert van Helden, *Measuring the Universe: Cosmic Dimensions from Aristarchus to Halley* (Chicago: University of Chicago Press, 1985).

Martin Nilsson, *Primitive Time-Reckoning: A Study in the Origins and First Development of the Art of Counting Time Among the Primitive and Early Culture Peoples* (Lund: Gleerup, 1920), remains a fascinating study of early timekeeping systems.

The information about the Hopi Indians' astronomical

system is taken from Stephen C. McCluskey, "The Astronomy of the Hopi Indians," *Journal for the History of Astronomy* 8 (1977): 174–195.

Chapter Three. Charting the Moonscape

Comprehensive treatments of the evolution of lunar maps are offered by Ewen Adair Whitaker, *Mapping and Naming the Moon: A History of Lunar Cartography and Nomenclature* (Cambridge: Cambridge University Press, 1999) and Scott L. Montgomery, *The Moon and the Western Imagination* (Tucson: University of Arizona Press, 1999). A major part of my treatment in this chapter is based on these books. A more technical resource with a somewhat different selection of historical maps is Zdeněk Kopal and Robert W. Carder, *Mapping the Moon: Past and Present* (Dordrecht: D. Reidel, 1974).

Galileo Galilei, *Sidereus Nuncius; or, The Sidereal Messenger,* trans. Albert van Helden (Chicago: University of Chicago Press, 1989).

The quotation of Martin Kemp related to Galileo is taken from *Seen/Unseen: Art, Science, and Intuition from Leonardo to the Hubble Telescope* (Oxford: Oxford University Press, 2006).

A modern photographic map of the moon, including the locations of the Apollo landings, can be found at www .google.com/moon/.

Chapter Four. Pale Sun of the Night

On the issue of the moon's whiteness: Sobha Sivaprasad and George M. Saleh, "Why Is the Moon White?"; Robert W. Kentridge, "Constancy, Illumination, and the Whiteness of the Moon"; Robert W. Kentridge and Paola Bressan, "The Dark Shade of the Moon"; and David H. Foster, "Confusing the Moon's Whiteness with its Brightness," all in *Clinical and Experimental Ophthalmology* 33 (2005): 571–575.

The quotation of Leonardo da Vinci is from *Philosophical Diary,* trans. Wade Baskin (New York: Philosophical Library, 1959).

The Alexander von Humboldt quotation is from *Cosmos: A Sketch of a Physical Description of the Universe,* trans. E. C. Otte and B. H. Paul (New York: Harper and Brothers, 1868).

On the historical evolution of the measurement of light intensity see J. B. Hearnshaw, *The Measurement of Starlight: Two Centuries of Astronomical Photometry* (Cambridge: Cambridge University Press, 1996).

The quotation about moonless nights in ancient Rome is taken from Jérôme Carcopino, *Daily Life in Ancient Rome: The People and the City at the Height of the Empire,* trans. E. O. Lorimer (New Haven: Yale University Press, 2003).

The source of the anecdote on Jacqueline Kennedy Onassis and the Taj Mahal is Wayne Koestenbaum, *Jackie Under My Skin: Interpreting an Icon* (New York: Farrar, Straus and Giroux, 1995).

The Johann Wolfgang von Goethe quotation is from *The Autobiography of Goethe: Truth and Poetry from My Own Life,* trans. John Oxenford (London: Bell and Daldy, 1867).

The quotation of Henry Matthews can be found in *The Diary of an Invalid; being the Journal of a Tour in Pursuit of Health; in Portugal, Italy, Switzerland, and France in the years 1817, 1818, and 1819* (London: John Murray, 1820).

The quotation of Doreen Valiente is from *Where Witchcraft Lives* (London: Aquarian, 1962).

The quotations of Marcel Proust are from *In Search of Lost Time (Remembrance of Things Past),* trans. C. K. Scott Moncrieff and Terence Kilmartin (New York: Random House, 1981).

For a vivid account of the members of the Lunar Society see Jenny Uglow, *The Lunar Men: Five Friends Whose Curiosity Changed the World* (New York: Farrar, Straus and Giroux, 2002).

A recent study of the moon in Italian culture is Pietro

Greco, *L'astro narrante. La Luna nella scienza e nella lettera-tura italiana* (Milan: Springer-Verlag Italia, 2009).

For a study of Yoshitoshi's moon illustrations see John Stevenson, *Yoshitoshi's One Hundred Aspects of the Moon* (Leiden: Hotei, 2001).

On the specifics of how to create a moon garden, see the popular guide by Marcella Shaffer, *Planning and Planting a Moon Garden* (North Adams, Mass.: Storey, 2000).

For other cultural aspects of the night see the following works:

Christopher Dewdney, *Acquainted with the Night: Excursions Through the World After Dark.* (London: Bloomsbury, 2004).

Roger A. Ekirch, *At Day's Close: Night in Times Past* (New York: Norton, 2006).

Eluned Summers-Bremner, *Insomnia: A Cultural History* (London: Reaktion, 2008).

Chapter Five. Encounters of a Lunar Kind

Charles Morton (1627–1698), an English naturalist, in a remarkable if strange combination of astrology and ornithology, suggested in all earnestness that birds hibernate on the moon. As Thomas B. Harrison has remarked, this was "the earliest treatise on bird migration on England." Morton seems to have been inspired by Godwin's novel—the rather unusual case of an idea taken from literature making its way into science: Thomas B. Harrison. "Birds in the Moon," *Isis* 45 (1954): 323–330.

The following monograph helped me with identifying some of the English and French examples in this chapter: Stephan Edinger, *Literarische Reisen zu fernen Planeten. Eine ideengeschichtliche Untersuchung zur französischen Literatur des 19. Jahrhunderts* (Marburg: Tectum Verlag, 2005).

Here are the original titles and translations in the order they appear in the chapter. Not all of these works have been translated into English.

Bernard De Fontenelle, *Entretiens sur la pluralité des mondes* (Discourses on the plurality of worlds).

Ludovico Ariosto, *Orlando furioso* (The frenzy of Orlando).

Cyrano De Bergerac, *L'Autre Monde: Histoire comique des états et empires de la lune* (The other world: The comical history of the states and empires of the moon).

Madam la Baronne de V***, *Le char volant* (The flying tank).

Anonymous, *Histoire intéressante d'un nouveau voyage à la Lune* (Interesting account of a new trip to the moon).

Alexandre Cathelineau, *Voyage à la Lune* (Trip to the moon).

Jacques Bujault, *Voyage dans la Lune* (Trip to the moon).

Anonymous, *Voyage tout récent dans la lune* (A very recent trip to the moon).

Louis Desnoyer, *Les Aventures de Robert Robert* (The adventures of Robert Robert).

De Sélènes, Pierre, *Un monde inconnu. Deux ans sur la lune* (An unknown world: Two years on the moon).

The classic survey of literary trips to the moon is Marjorie Hope Nicolson, *Voyages to the Moon* (New York: Macmillan, 1948).

A more recent treatment is Aaron Parrett, *The Translunar Narrative in the Western Tradition* (Aldershot and Burlington: Ashgate, 2004).

Lucian Boia's *L'Exploration imaginaire de l'espace* (Paris: Editions La Découverte, 1987) is an amazing collection of literary travels into space and has also been useful in writing this chapter.

The Russian examples are taken from Stephen Lessing Baehr, *The Paradise Myth in Eighteenth-Century Russia: Utopian Patterns in Early Secular Russian Literature and Culture* (Stanford: Stanford University Press, 1991).

Chapter Six. Lunar Passion in Paris

Jules Verne, *From the Earth to the Moon,* 1865 (available in various editions).

William Butcher, *Jules Verne. The Definitive Biography* (New York: Thunder's Mouth, 2006).

Chapter Seven. Accounts of Genesis

For the theories of Darwin and Urey see Stephen G. Brush, *Fruitful Encounters: The Origin of the Solar System and of the Moon from Chamberlin to Apollo* (Cambridge: Cambridge University Press, 1996).

Dana Mackenzie narrates the quest to explain the genesis of the moon in her formidable *The Big Splat, or How Our Moon Came to Be: A Violent Natural History* (Hoboken: Wiley, 2003).

On the possibility of a metallic core of the moon see Ian Garrick-Bethell, Benjamin P. Weiss, David L. Shuster, and Jennifer Buz, "Early Lunar Magnetism," *Science* 323 (2009): 356–359.

Chapter Eight. A Riddled Surface

James Lawrence Powell's readable *Mysteries of Terra Firma: The Age and Evolution of the Earth* (New York: Free Press, 2001) presents a detailed introduction to the question of the origin of and theories about the lunar craters.

For a more comprehensive treatment of the moon from a geological perspective see Stuart Ross Taylor, "The Moon," in *Encylopedia of the Solar System,* 2nd ed., ed. Lucy-Ann McFadden, Paul R. Weissman, and Torrence V. Johnson (San Diego: Elsevier Science/Academic Press, 2007).

For the discussion of Aepinus see Roderick Home, "The Origin of the Lunar Craters: An Eighteenth-Century View," *Journal for the History of Astronomy* 3 (1972): 1–10.

The quotation of Paul D. Spudis is taken from *The Once*

and Future Moon (Washington, D.C.: Smithsonian Institution Press, 1996).

The quotations of Richard Anthony Proctor are from *The Moon: Her Motions, Aspect, Scenery, and Physical Condition* (New York: Appleton, 1886).

On the latest discoveries regarding water on the moon see, e.g., Kenneth Chang, "Water Found on Moon, Researchers Say," *New York Times,* November 14, 2009.

The exotic theories of Peal, Fountain, Fauth, Davis, and Ocampo related to the makeup of the lunar surface are taken from the chapter "How the Lunar Craters *Weren't* Formed" in Patrick Moore, *The Wandering Astronomer* (Bristol: Institute of Physics Publishing, 2003).

Chapter Nine. Lunar Choreography

On the possible influence of the moon on seismic activity see "Can the Moon Cause Earthquakes?" *National Geographic News,* May 23, 2005.

On the research into the condition of trees and wood at different times of the year see Ernst Zürcher et al., "Looking for Differences in Wood Properties as a Function of the Felling Date: Lunar Phase Correlated Variations in the Drying Behavior of Norway Spruce (*Picea abies* Karst.) and Sweet Chestnut (*Castanea sativa* Mill.)," *Trees: Structure and Function,* Springer, August 26, 2009 (online publication).

The references for studies on seals and skylarks are: P. Watts, "Possible Lunar Influence on Hauling-out Behavior by the Pacific Harbor Seal (*Phoca vitulina Richardsi*)," *Marine Mammal Science* 9 (1993): 68–76; D. James, G. Jarry and C. Érard, "Effet de la lune sur la migration postnuptiale nocturne de l'alouette des champs *Alauda arvensis L.* en France," *Sciences de la vie* 323 (2000): 215–224.

On the mass spawning of corals see William J. Broad, "Sexy Corals Keep 'Eye' on Moon, Scientists Say," *New York Times,* October 19, 2007.

On the reproductive behavior of sea urchins and its synchronization with lunar phases see Bill Kennedy and John S. Pearse, "Lunar Synchronization of the Monthly Reproductive Rhythm in the Sea Urchin *Centrostephanus coronatus* Verrill," *Journal of Experimental Marine Biology and Ecology* 17 (1975): 323–331.

On the dung beetle and the role of the moon as visual landmark see Eric Warrant and Dan-Eric Nilsson, eds., *Invertebrate Vision* (Cambridge: Cambridge University Press, 2006).

Chapter Ten. Esoteric Practices

The reference for the account of the moon doctor is M. Herz, "Die Wallfahrt zum Monddoktor in Berlin," *Berlinische Monatsschrift* (1783), 368–385.

The account of medical astrology is mostly based on Mark Harrison, "From Medical Astrology to Medical Astronomy: Sol-lunar and Planetary Theories of Disease in British Medicine, c. 1700–1850," *British Journal for the History of Science* 33 (2000): 25–48.

The quotations of Erasmus Darwin are from *Zoonomia; or, The Laws of Organic Life,* 2 vols. (London: J. Johnson, 1794–1796).

For a study of pagan ritual from a historical perspective see Ronald Hutton, *The Triumph of the Moon: A History of Modern Pagan Witchcraft* (Oxford: Oxford University Press, 1999).

The details of the description of the Wiccan full-moon ritual are taken from http://paganwiccan.about.com/od/moonphasemagic/ht/SpringFullMoon.htm.

Chapter Eleven. Spurious Correspondences

Folklore related to the moon is ubiquitous, and often the tales are surprisingly similar across cultures. The examples of superstitions in German folklore are taken from Werner Wolf, *Der Mond im deutschen Volksglauben* (Bühl: Druck und Verlag der Konkordia AG, 1929). The source of the examples from the Philippines is Francisco R. Demetrio, *Encyclopedia of Philippine Folk Beliefs and Customs* (Cagayan de Oro City: Xavier University, 1991).

For the described case study of the lycanthropic woman see Harvey Rosenstock and Kenneth R. Vincent, "A Case of Lycanthropy," *American Journal of Psychiatry* 134 (1977): 195–205.

On the werewolf issue see Ian Woodward, *The Werewolf Delusion* (New York: Paddington, 1979).

On the topic of the alleged impact of the moon on human behavior also see the following articles and studies mentioned in the chapter:

Scott O. Lilienfeld and Hal Arkowitz. "Lunacy and the Full Moon: Does a Full Moon Really Trigger Strange Behavior?" *Scientific American Mind,* February 9, 2009.

I. W. Kelly, James Rotton, and Roger Culver, "The Moon Was Full and Nothing Happened: A Review of Studies on the Moon and Human Behavior and Human Belief," in *The Outer Edge,* ed. J. Nickell, B. Karr, and T. Genoni (Amherst, N.Y.: CSICOP, 1996).

J. M. Gutiérrez-García and F. Tusell, "Suicides and the Lunar Cycle," *Psychological Reports* 80 (1997): 243–250.

The Swiss study mentioned is Martin Röösli et al., "Sleepless Night, the Moon Is Bright: Longitudinal Study of Lunar Phase and Sleep," *Journal of Sleep Research* 15 (2006): 149–153.

These are the studies concerned with birth and fertility: E. Periti and R. Biagiotti, "Lunar Phases and Incidence of Spontaneous Deliveries: Our Experience," *Minerva Ginecologica* 46 (1994): 429–433; J. M. Arliss, E. N. Kaplan, and S. L. Galvin, "The Effect of the Lunar Cycle on Frequency of Births and Birth Complications," *American Journal of Obstetrics and Gynecology* 192 (2005), 1462–1464.

On the menstrual patterns among the Dogon see B. I. Strassmann, "The Biology of Menstruation in *Homo sapiens:* Total Lifetime Menses, Fecundity, and Nonsynchrony in a Natural Fertility Population," *Current Anthropology* 38

(1997): 123–129; Sung Ping Law, "The Regulation of Menstrual Cycle and Its Relationship to the Moon," *Acta Obstet Gynecol Scan* 65 (1986): 45–48; T. B. Criss and J. P. Marcum, "A Lunar Effect on Fertility," Social Biology 28 (1981): 75–80.

For a current scientific discussion of human biological processes and calendrical cycles see R. G. Foster and T. Roenneberg, "Human Responses to the Geophysical Daily, Annual, and Lunar Cycles," *Current Biology* 18 (2008): 784–794.

On the historical roots of belief in the power of the moon to cause insanity and epilepsy see Charles A. Raison, Haven M. Klein, and Morgan Steckler, "The Moon and Madness Reconsidered" *Journal of Affective Disorders* 53 (1999): 99–106.

The following books explore the field of chronobiology:

Russel G. Foster and Leon Kreitzman, *Rhythms of Life: The Biological Clocks That Control the Daily Lives of Every Living Thing* (New Haven: Yale University Press, 2005).

Russel G. Foster and Leon Kreitzman, *Seasons of Life: The Biological Rhythms That Living Things Need to Thrive and Survive* (New Haven: Yale University Press, 2009).

Klaus-Peter Endres and Wolfgang Schad, *Moon Rhythms in Nature: How Lunar Rhythms Affect Living Organisms,* trans. Christian von Armin (Edinburgh: Floris, 2002). This book relies on the tradition of anthroposophy, which has traditionally been concerned with the study of "rhythmical processes" in nature.

The Bob Berman quotation is from *Secrets of the Night Sky: The Most Amazing Things in the Universe You Can See with the Naked Eye* (New York: William Morrow, 1995).

Chapter Twelve. Visions of the Moon

For a detailed analysis of James Nasmyth's photographs see Frances Robertson, "Science and Fiction: James Nasmyth's Photographic Images of the Moon," *Victorian Studies* 48 (2006): 595–623.

The description of A Trip to the Moon in Coney Island's Luna Park is based on a text by Jeffrey Stanton at www .westland.net/coneyisland/articles/lunapark.htm.

The original German title of Karl-August Laffert's novel is *Der Untergang der Luna* (Berlin: Stilke, 1921).

For a more general discussion of technology in American culture, including the role of world's fairs to promote space travel, see David E. Nye, *American Technological Sublime,* (Cambridge: MIT Press, 1994).

For a thorough discussion of science in movies see Sidney Perkowitz, *Hollywood Science: Movies, Science, and the End of the World* (New York: Columbia University Press, 2007).

The full reference for the text about the vile race of lunarians on the far side of the moon: Kenneth Heuer, *Men of Other Planets* (New York: Pellegrini and Cudahy, 1951). My thanks go to the staff of the Strand Bookstore for pointing this out to me.

The information for the segment about the plan for a nuclear bombing of the moon is from Antony Barnett, "U.S. Planned One Big Nuclear Blast for Mankind," *Observer,* May 14, 2000.

The description of the moon ride in Disneyland is based on "Feeling of Space Journey Accompanies Trip to Moon," *Disneyland News,* January 1957.

At the time of writing, the movies by Georges Méliès, Fritz Lang, and Arthur Hilton, as well as the Walt Disney program *Man and the Moon,* were available on YouTube. com. In one of the sections of the latter program Wernher von Braun explains the principles of rocket propulsion.

For an interesting introduction to space-age euphoria by one of its major protagonists see Arthur C. Clarke, *The Promise of Space* (New York: Harper and Row, 1968). For ideas on lunar stations see also Fred Warshofsky, *The 21st Century: The New Age of Exploration* (New York: Viking, 1969).

On Fritz Zwicky and his "experimental astronomy": Richard Panek, "The Father of Dark Matter Still Gets No

Respect," *Discover,* January 2009, http://discovermagazine
.com/2009/jan/30-the-father-of-dark-matter-still-gets-
no-respect.

Chapter Thirteen. Before and After Apollo

On the effect of railway travel on the human imagination
see Wolfgang Schivelbusch, *Railway Journey: The Industrial-
ization of Time and Space in the 19th Century* (Berkeley: University of California Press, 1986).

The quotation of Walter A. McDougall is from *The
Heavens and the Earth: A Political History of the Space Age*
(New York: Basic, 1985).

The quotation of Michael J. Neufeld about Wernher
von Braun is from *Von Braun: Dreamer of Space, Engineer of
War* (New York: Knopf, 2007).

Space travel, including rocketry development, is the
subject of: Wernher von Braun and Frederick I. Ordway
III, *History of Rocketry and Space Travel* (New York: Crowell,
1975).

Frederick I. Ordway III, *Blueprint for Space: Science Fiction to Fact* (Washington, D.C.: Smithsonian Institution
Press, 1992).

Howard E. McCurdy, *Space and the American Imagination* (Washington, D.C.: Smithsonian Institution Press,
1997).

For more scholarly treatments of space travel and the
Apollo program see Roger D. Launius, *Frontiers of Space
Exploration* (Westport, Conn.: Greenwood, 2004), and
Everett C. Dolman, *Astropolitik: Classical Geopolitics in the
Space Age* (London: Frank Cass, 2001). And, by the same
author, another essay dealing with the various historical
spaceflight narratives and with the issue of conspiracy in
the context of the moon landings: "American Spaceflight
History's Master Narrative and the Meaning of Memory,"
in *Remembering the Space Age: Proceedings of the 50th Anniversary Conference,* ed. Steven J. Dick (Washington, D.C.:
NASA, 2008), www.asiaing.com/remembering-the-space-

age-proceedings-of-the-50th-anniversary-conference.html (a free e-book).

For a scholarly survey of Russian space history see Brian Harvey, *Russian Planetary Exploration: History, Development, Legacy, Prospects* (Berlin: Springer, 2007).

Books about the Apollo program and spaceflight abound. Here is a small selection:

Andrew Chaikin, *A Man on the Moon: The Voyages of the Apollo Astronauts* (New York: Viking, 1994).

Bob Berman, *Shooting for the Moon: The Strange History of Human Spaceflight* (Guilford, Conn.: Lyons, 2007).

Craig Nelson, *The Epic Story of the First Men on the Moon* (New York: Viking, 2009).

Gerard J. DeGroot, *Dark Side of the Moon: The Magnificent Madness of the American Lunar Quest* (New York: New York University Press, 2006).

On the integration of man and computer systems in the context of Apollo, and on various other, more technical issues and questions, such as the need for human presence in future space exploration, see David A. Mindell, *Digital Apollo: Human and Machine in Spaceflight* (Boston: MIT Press, 2008).

This book deals in more detail with the idea that the lunar landings never happened: Philip C. Plait, *Bad Astronomy: Misconceptions and Misuses Revealed, from Astrology to the Moon Landing "Hoax"* (New York: Wiley, 2002).

Several sites on the Internet claim that humans never landed on the moon, for example: www.moonmovie.com.

Roger D. Launius has written an interesting article about the ways in which the history Apollo program has been told, depending on political inclinations: "American Spaceflight History's Master Narrative and the Meaning of Memory."

Information about future lunar stations can be found in Alan Boyle, "Wanted: Home-builders for the Moon. NASA's Post-2020 Plan Involves the Usual (and Unusual) Space Suspects," MSNBC, February 1, 2007.

On the physical consequences of space travel see

Jerome Groopman, "Medicine on Mars: How Sick Can You Get During Three Years in Deep Space?" *New Yorker,* February 14, 2000, 36–41.

On the topic of water on the moon see Marc Chaussidon, "Planetary Science: The Early Moon Was Rich in Water," *Nature* 454 (2008): 170–172.

References to Lewis Mumford are from Donald L. Miller, *Lewis Mumford: A Life* (New York: Grove, 2002).

The astronautical engineer Robert Zubrin's *Entering Space: Creating a Spacefaring Civilization* (New York: Jeremy P. Tarcher/Putnam, 1999) is a passionate plea for the colonization of space, financed by private enterprise. It includes ideas for colonizing Mars, for mining the solar system, and for building fusion-powered spaceships. A helium-3 mining facility on the moon is one of the first steps.

On the Indian space mission see Tunku Varadajan, "Fly Me to the Deity," *New York Times,* October 29, 2008.

For a thorough discussion of how the view of our planet from space changed our view of the Earth see Robert Poole, *Earthrise: How Man First Saw the Earth* (New Haven: Yale University Press, 2008).

Epilogue

Richard Nixon made his statement on July 20, 1969.

While I was completing this manuscript, Barack Obama was calling on NASA to end the program that was designed to bring humans back to the moon and focus instead on radically new space technologies.

In addition to the above references, the following books have been helpful with a number of aspects:

Marta Erba, Gianluca Ranzini, and Daniele Venturoli, *Dalla Luna alla Terra. Mitologia e realtà degli influssi lunari* (Turin: Bollati Boringhieri, 2010).

Daniel Grinsted, *Die Reise zum Mond. Zur Faszinationsgeschichte eines medienkulturellen Phänomens zwischen Realität und Fiktion* (Berlin: Logos, 2009).

Acknowledgments

I would like to thank above all my editor Jean E. Thomson Black for her continuing enthusiasm and support, Dan Heaton for his intensive and thoughtful work on the final manuscript, Sonia Shannon for designing this book, Jaya Chatterjee for helping me with the organization of the illustrations, and the staff of Yale University Press in the New Haven and London offices. Particular thanks are due to David Luljak for his index. Many thanks also to my parents, Helgard and Siegfried Brunner, and to the following friends and associates: Rathnayake M. Abeyrathne (University of Peradeniya, Kandy, Sri Lanka), Ines Cabral, Alexandre Cabrita, José João Dias Carvalho, Francesco Paolo de Ceglia (University of Bari Aldo Moro, Italy), Dan Delany, Detlef Feussner, Jacinto José Gomes, Joseph P. Grubb, Michael Hely, Klaas Jarchow, Brendan Kenney, Lori Lantz, Ulrich Meyer, Olaf Oberschmidt, Scott W. Perkins, Eva Schoening, Julien Sialelli, Benjamin A. Smith, Keijiro Suga (Meiji University, Tokyo, Japan), and a very dear friend who prefers to remain unnamed here.

My gratitude also extends to the many often unknown artists whose fine illustrations are featured in this book and — for a range of reasons — to Margaret Adamic (Disney Publishing Worldwide, Inc.), Paul Brown (National Park Service), Jules Cashford, Nina Cummings (The Field Museum, Chicago), Volker Dehs, Ian Garrick-Bethell (MIT), Pietro Greco, Jeannine Green (Bruce Peel Special Collections Library, University of Alberta), Daniel Grinsted, Ove Hoegh-Guldberg (Global Change Institute, University of

Queensland), Eric van den Ing, Edwin C. Krupp (Griffiths Observatory, Los Angeles), Jacques Laskar (Observatoire de Paris), Benjamin Lazier (Reed College, Portland), Vera Martinez (Technical University Darmstadt), Nancy O'Shea (The Field Museum, Chicago), Jeff Papineau (Bruce Peel Special Collections Library, University of Alberta), Bernd A. Pflumm, Marco Scola, Paul D. Spudis (Lunar and Planetary Institute, Houston), Stuart Ross Taylor (Australian National University, Canberra), Ana Tipa, Joe Tucciarone, Mark Wieczorek (Institut de Physique du Globe de Paris), the librarians of the Berlin State Library, the New York Public Library, and the National Library of Portugal in Lisbon, as well as several sellers of antique books, especially Christoph Janik. I would also like to thank the readers who evaluated the manuscript and gave constructive criticism.

A number of books have been influential in my writing; the most important of these are included in the bibliographic essay.

Any problems that remain are, of course, my own responsibility.

If you want to drop me a line, please do so by E-mail: bbrunner@gmx.net.

Illustration Credits

4, 57, 84, 87, NASA

16, The British Library Board.1023910.271

18, Courtesy of E. C. Krupp

21, University of Glasgow, Department of Special Collections

31, The British Library Board. 1023910.271

33, Owner: Wallraf-Richartz-Museum and Fondation Corboud; photo credit: Rheinisches Bildarchiv, Cologne

34, University of Glasgow, Department of Special Collections

43, Courtesy of E. C. Krupp

49, Joe Tucciarone

53, Ludek Pesek

68, *Fate* magazine, P.O. Box 460, Lakeville, Minn., 55044, www.fatemag.com

70, Bruce Peel Special Collections Library, University of Alberta

78, Actionpress, Hamburg, Germany

74, Klaus Bürgle

83, Landov Media

87, Owner: Wallraf-Richartz-Museum and Fondation Corboud; photo credit: Rheinisches Bildarchiv, Cologne

89, Technical University Darmstadt, Germany

124, Joe Tucciarone

130, Von Del Chamberlain

136, Ludek Pesek

153, 234, 236, NASA

193, *Fate* magazine, P.O. Box 460, Lakeville, MN 55044, www.fatemag.com

200, Landov Media

206, Klaus Bürgle

215, Actionpress, Hamburg, Germany

246, Technical University Darmstadt, Germany

Every reasonable effort has been made to determine copyright holders of illustrations and permissions secured as needed. If any copyrighted materials have been inadvertently used in this work without proper credit being given, please notify Bernd Brunner at bbrunner@gmx.net so that future printings of this work may be corrected accordingly.

Index

Note: Illustrations are indicated by page numbers in *italic* type.